Geological Structures and Moving Plates

R.G. PARK, BSc, PhD
Reader in Geology
University of Keele

BLACKIE ACADEMIC & PROFESSIONAL
An Imprint of Chapman & Hall

London · Glasgow · New York · Tokyo · Melbourne · Madras

Published by Blackie Academic & Professional, an imprint of
Chapman & Hall, Wester Cleddens Road, Bishopbriggs, Glasgow
G64 2NZ, UK

Chapman & Hall, 2-6 Boundary Row, London SE1 8HN, UK

Blackie Academic & Professional, Wester Cleddens Road,
Bishopbriggs, Glasgow G64 2NZ, UK

Chapman & Hall, 29 West 35th Street, New York NY10001, USA

Chapman & Hall Japan, Thomson Publishing Japan, Hirakawacho
Nemoto Building, 6F, 1-7-11 Hirakawa-cho, Chiyoda-ku, Tokyo 102,
Japan

DA Book (Aust.) Pty Ltd, 648 Whitehorse Road, Mitcham 3132,
Victoria, Australia

Chapman & Hall India, R. Seshadri, 32 Second Main Road, CIT East,
Madras 600 035, India

First edition 1988
Reprinted 1993

© 1988 Blackie Academic & Professional

Typeset by Best-set Typesetter Ltd
Printed in Great Britain by Thomson Litho Ltd, East Kilbride, Scotland

ISBN 0 7514 0143 9

A catalogue record for this book is available from the British Library
Library of Congress Cataloging-in-Publication Data available

Preface

The great classical tectonicians, such as Suess, Argand and Wegener, attempted to understand, without the benefit of the plate tectonic theory, the workings of the Earth engine as a whole, and the part that deformation played in that whole. In my student days, I derived great pleasure and benefit from De Sitter's textbook on structural geology where the study of geological structures and major Earth structure received more or less equal treatment. Since then, until relatively recently, there has been a tendency for structural geology to become more parochial and inward-looking, despite the enormous advances in understanding that the plate tectonic revolution has brought about. I have long felt the need, therefore, for a book that would give students a tectonic overview in which geological structures and deformation could be seen in their context as byproducts of the plate tectonic system.

This book attempts to integrate structural geology and plate tectonics (often taught quite separately) by dealing with the theoretical background knowledge necessary to understand plate movements and plate interactions. Thus the mechanical properties of plates, sources of stress and stress distribution in the lithosphere, and the causes of plate motion, are examined first, followed by a discussion of the kinematic aspects of relative plate movements and interactions.

The second part of the book deals with some modern case studies — examples where present-day structures can be related with some degree of confidence to plate movements, such as the Central Asian collision zone, the Lesser Antilles subduction zone in the Caribbean, and the Rhine–Ruhr rift system. These are discussed in terms of four main types of plate tectonic regime: divergent, convergent, strike-slip, and intraplate.

In the third section of the book, examples of classical orogenic belts, of both Phanerozoic and Precambrian age, are discussed and interpreted in the light of the principles established in the earlier chapters. Thus the Alps are discussed in terms of African–European plate interactions, and the Cordilleran orogenic belt in terms of Mesozoic subduction and subsequent strike-slip collage tectonics. A more speculative approach is necessary in the Precambrian examples, where the differing tectonic styles of, for example, the mid-Proterozoic Grenville Province and the Archaean greenstone belt terrains may reflect genuine differences in lithosphere behaviour.

The book is aimed at readers who are already familiar with the basic principles and nomenclature of geotectonics and structural geology, who understand plate tectonic theory and its supporting evidence, and who are familiar with its central role in modern geology.

In order to keep the book to a reasonable length, I have deliberately concentrated on the role of geological structure in plate tectonics and resisted the temptation to include more than passing reference to other relevant topics, such as petrogenesis.

Finally, I am indebted, firstly, to two anonymous reviewers who read the first draft of the manuscript and made a number of helpful suggestions for its improvement; secondly, to the many authors who have allowed me to reproduce diagrams from their published work (these are individually acknowledged in the figure captions); and thirdly, to my wife, friends and colleagues who have borne with me through the trauma of the writing of this work.

RGP

To those structural and tectonic geologists who, through their stimulating lectures and papers, have nourished my interest in geotectonics over the years, and provided the motivation to write this book

Contents

1 Introduction

The theme of this book is the relationship between geological structures and plate tectonic theory.

The plate-tectonic 'revolution'

It is now universally acknowledged that the plate tectonic theory has brought about a revolution in our perception of geology in almost all its branches. In the case of structural geology, the applications of the plate tectonic model are particularly obvious, and have affected in a fundamental way our interpretation both of geological structures in the narrow sense, and of orogenic belts. The two key disciplines of *structural geology* and *plate tectonics* (or *geotectonics*) are usually taught separately, and are often dealt with in different textbooks. Yet the great expansion in research publications dealing with the applications of plate tectonic theory to orogenic belts, and to the interpretation of geological structures, demands an integrated approach which it is the aim of this book to provide.

Classical geological theorists were fascinated by orogenic belts and other major earth structures, and speculated widely on their origin. Şengôr presents an interesting overview of the evolution of classical ideas on orogenesis in the opening chapter of '*Orogeny*' by Miyashiro *et al.* (1982). This work is a very readable account of the way in which the concept of orogeny has been transformed by plate tectonic theory. The weakness in pre-1970 theory lay in the absence of a generally accepted tectonic model that could satisfactorily explain not only geological structures, but also the distribution and variation both of igneous and metamorphic activity, and of sedimentary facies. It is in the successful linkage between previously unrelatable phenomena that the strength and success of the plate tectonic model lie.

The concept of *plates* arose from the observation that large areas of the crust have apparently suffered very little lateral distortion although they have travelled several thousand kilometres, if the evidence of continental drift is accepted. The detailed and accurate 'jigsaw' fit of the opposing coastlines of America and Africa for example, after 200 Ma and 4000 km of drift, testifies to this lack of distortion. In the oceans also, are found regular linear magnetic stripes and faults that have maintained their shape for tens of millions of years. This evidence reinforces the conclusions reached by studying the distribution of tectonic movements that there are large stable areas (continental cratons and deep ocean basins) that suffer little internal deformation and exhibit only slow vertical movements, while at the same time moving laterally as coherent units at rates 10 to 100 times faster.

The recognition of the lithosphere plate as the fundamental kinematic unit now underlies the study of all surface tectonic processes. Tectonic activity is now divisible into *plate boundary* effects and *intraplate* (or within-plate) effects, and this subdivision has replaced the old dichotomy between *orogenic* and *anorogenic* regions or activity.

The role of the lithosphere

The vertical extent of a plate is defined by the base of the *lithosphere*, the strong outer layer of the Earth that rests on the underlying weaker *asthenosphere*. The lithosphere includes both the crust and part of the upper mantle, and has an average thickness of around 100 km (see Figure 2.1). The concept of the plate is therefore bound up with the properties and behaviour of the lithosphere and asthenosphere, discussed in Chapter 2.

The lithosphere is usually defined on seismological criteria that reflect relatively gradual changes in various physical properties (see 2.2). The most useful distinction between the lithosphere and the underlying asthenosphere is in terms of the higher viscosity of the former. Since the viscosity is controlled by the temperature, the position of the base of the lithosphere, and the variation of viscosity (and hence strength) within the lithosphere, are controlled by the geotherm. Thus ocean ridges are characterized by warm, thin lithosphere, and subduction zones by thick, cool lithosphere. The lithosphere may be regarded as the cool surface layer of the Earth's convective system. It is therefore necessary to discuss both the thermal structure of the lithosphere (2.3) and mantle convection (2.4) in order to understand the properties and behaviour of the lithosphere itself.

Geological structures are created by forces acting on rock. To investigate the nature and origin of these forces, we must enquire, firstly, what forces are available to act on the lithosphere; in other words, what are the sources of *stress* within the lithosphere, and how large are they (see 2.5). The study of the nature and influence of forces or stresses on a body is known as *dynamics*. It turns out that the largest stresses are associated with subduction zones and with isostatically compensated loads (e.g. plateau uplifts and ocean ridges). A number of methods are available for the determination of stress within the lithosphere, particularly in the upper part of the crust, where stress can be directly measured (see 2.6). These methods yield a bewildering variety of estimates, both of the actual magnitude of stress carried by the lithosphere, and of the strength of the lithosphere. The orientation of stress on the other hand, usually bears a simple relationship to the more obvious local stress sources. The distribution of measured stress can only be understood by considering the long-term strength of the lithosphere, that is, its strength over periods of the order of tens of Ma (see 2.7). The ductile behaviour of the lower part of the lithosphere over such long periods of time produces a concentration of applied stress in the upper part of the lithosphere (the principle of *stress amplification*). This explains why the magnitudes of the available stress sources are so much smaller than actual measured values near the surface. Knowing the temperature structure of a given piece of lithosphere, and

assuming a likely value for the magnitude of the available stress sources (forces) on it, it is possible using numerical modelling techniques to predict the strength of that piece of lithosphere and the time that will elapse before failure. In this way, we can ultimately estimate the stress conditions accompanying the formation of geological structures, both at plate boundaries and also within plates, at the time when new plate boundaries are initiated.

Plate motion

Geological structures are controlled both by stress (or force) and by relative motion. The *dynamic* study of the conditions under which deformation takes place must be complemented by the study of the relative motions of the various plates or blocks concerned. This type of approach is termed *kinematic*. For all practical purposes, we can regard relative plate motion as taking place at constant velocity. The accelerations and decelerations that occur take place over such long periods of time that the forces generated are much too small to be significant. Plate movements at constant velocity do not of themselves create forces, or produce deformation. The forces associated with relative motion are created by resistances across planar boundaries of relative motion or viscous drag between two opposed moving blocks. In this way the kinematic and dynamic approaches to deformation are linked.

If it is accepted that plates can be regarded as strong rigid shells, their relative motion across the surface of the Earth can be described in terms of the simple rules of motion on a sphere. Any relative motion between two plates on the surface of a sphere becomes an angular rotation about an axis through the centre of the Earth, which intersects the Earth's surface at two points called the *poles of rotation* for that movement.

The direction of movement at the surface of the Earth is parallel to a set of small circles about the axis of rotation. As realized originally by Tuzo Wilson (1965) in his classic paper on transform faults, any such faults displacing the boundary between two plates must be parallel to these small circles; that is, parallel to the direction of relative motion between the two plates.

This principle was used by McKenzie and Parker (1967) in their key paper on the movement of the Pacific plate. They analysed the direction of motion from focal-plane-mechanism solutions of earthquakes along trenches, transform faults and ridges bordering the Pacific plate, and showed that they gave a common movement direction (see 3.1). This direction is parallel to the small-circle arcs obtained from the orientation of the San Andreas fault. The position of the pole of rotation was located by finding the intersection of the radii drawn through a number of small-circle arcs in different positions.

Relative plate velocities can be found by analysing the magnetic stratigraphy, so that for any plate pair sharing a spreading ridge, the movement vector can be found (see Figure 3.2). If the movement vector between these two plates and a third plate is known, the relative motion of the third plate relative to either of the other two plates can be calculated using a vector triangle. In this way the relative motions of all six major plates were determined by Vine and Hess (1970) — see Figure 3.1. The principles governing plate kinematic behaviour are discussed in 3.1.

The importance of plate boundaries

The obvious link between seismicity and present-day tectonic activity suggested that the seismic zones must represent the boundaries of the stable blocks, and that each block or plate could be defined by a continuous belt of seismic activity (Isacks *et al.*, 1968). Since the seismic activity represents fault movements with high strain rates, each plate must be in a state of relative motion with respect to each of its neighbours. Three types of plate boundary are recognized: (i) *constructive boundaries* where adjoining plates are moving apart and new plate is being created at ocean ridges; (ii) *destructive boundaries* where adjoining plates are moving together and plate material is being destroyed by subduction at ocean trenches; and (iii) *conservative boundaries* where adjoining plates are moving laterally past each other with a horizontal strike-slip sense of displacement along transform faults. The sense of displacement at these boundaries can be deduced from first-motion studies of individual earthquakes that in general confirm the relative movements inferred from other evidence such as palaeomagnetism and magnetic stratigraphy.

Following these principles, a network of boundaries may be drawn dividing the present Earth's surface into six major plates (see Figure 3.1): the Eurasian, American, African, Indo-Australian, Antarctic and Pacific plates, together with a number of smaller plates associated mainly with destructive boundaries, especially around the margins of the Pacific Ocean. Continental margins may or may not correspond with plate boundaries. Those that do, such as the western margin of the American continents, are termed *active margins*; those that lie within plates are termed *passive margins*.

The position of the plate boundaries determines the location of the more significant tectonic activity, and the type of boundary controls the nature of the tectonic processes operating there. The threefold division into constructive, destructive, and conservative plate boundaries is reflected in the organization of this book. Structures of currently active plate boundaries are considered in terms of three fundamental types of tectonic regime: *divergent regimes* relating to constructive boundaries (Chapter 4); *convergent regimes* relating to destructive boundaries (Chapter 5); and *strike-slip and oblique regimes* relating to conservative boundaries or to those with a component of strike-parallel motion (Chapter 6). Intraplate regimes are discussed in Chapter 7.

Geological structures

It is assumed that the reader has an adequate working knowledge of structural geology. A familiarity with the various types of geological structure and their origin is implicit in the description of structures in Chapters 4–9. In considering the relationship between geological structures and plate movements, certain aspects of structural geology are obviously more relevant than others. Orientation of major structures (folds and faults), intensity of deformation, magnitude and orientation of the bulk strain axes, and structural symmetry all convey important information about the way in which the crust responds to relative plate movements at plate boundaries.

In assessing the regional tectonic significance of geological structure therefore, the orientation of folds and faults is critical, and also particularly the orientation of the transport direction, since this will be related to the kinematic convergence (or divergence) direction. In high-strain zones, this direction will be close to the maximum principal strain axis.

In terms of strain, we are concerned essentially with bulk properties and bulk geometry, and with how these relate to the large-scale kinematic pattern. For this purpose, a geometrical overview or over-simplification of the large-scale structural pattern is more useful than a consideration of the structural detail, at outcrop scale for example. The same principle applies to strain rates — a very important control on deformation. Bulk strain rates relating to mountain belts or large zones are cited, and can be compared with theoretical strain rates derived from the mechanical behaviour of the lithosphere, but strain rates on smaller scales are generally not discussed.

Symmetry is another important aspect of the large-scale structure. The vergence direction, or facing direction, of thrusts, overfolds and other asymmetric structures is of fundamental significance in understanding the way in which a piece of crust has been deformed. Some structural geologists would argue that movements on low-angle faults and shear zones are the dominant mechanism in the bulk deformation of the crust. Such structures impart an obvious asymmetry to the structural pattern. Tectonic processes driven by simple shear are favoured in convergent regimes because of the basic asymmetry inherent in the subduction process. A similar basic asymmetry characterizes all strike-slip regimes. Recent studies of extensional regimes indicate that quasi-symmetrical fault-block arrangements at high crustal levels, by detachment at lower levels on low-angle décollement planes, are part of an asymmetrical system overall.

Orogenic or mobile belts in the geological past

Tectonic effects of great interest to the structural geologist are produced in active convergent regimes (see Chapter 5). It is those that provide working models that can be used to interpret the orogenic belts, the convergent regimes of the geological past. Chapter 8 examines a selection of such belts of Phanerozoic age, commencing with Mesozoic-Cenozoic examples where the plate-tectonic setting is reasonably well constrained. By showing how various workers have interpreted structure in terms of plate tectonic processes, the principles underlying the relationship between plate motion and structure may be illustrated.

The problem of the Precambrian

Examples of Precambrian mobile belts are discussed in Chapter 9. Despite the fact that the Precambrian occupies approximately eight-ninths of geological time, remarkably little is known about how the plate tectonic process operated during that period, or indeed whether it operated at all in the earlier part of Earth history. The examples discussed show sufficient similarity to Phanerozoic systems from the Early Proterozoic onwards to suggest that the plate tectonic model is probably applicable, albeit in modified form, for the last 2500 Ma. There is as yet no general consensus concerning the type of plate regime in the Archaean, and even those indications that we have are biased towards the later part of the Archaean. Little or nothing is known about large-scale structure prior to about 3000 Ma BP.

2 The lithosphere: some important properties

2.1 Lithosphere, asthenosphere and mesosphere

In his model of continental drift, Wegener (1929) originally visualized pieces of continental crust moving across a plastic oceanic crust. However this idea was abandoned many years ago when it was realized that oceanic rocks could not behave in a sufficiently ductile manner near the surface. When the plate tectonic model was being developed in the 1960s, it was realized that the moving plates included oceanic as well as continental crust. The oceanic crust is only about 7 km thick and could not remain undistorted when subjected to the horizontal stresses associated with plate tectonic processes. Plates must be therefore be considerably thicker than the crust and include part of the upper mantle as well.

The *lithosphere* is defined on seismological criteria as the strong outer layer of the Earth; it can equally be regarded as the cool surface layer of the Earth's convective system. This convective system is discussed further below (see 2.4) but it is important at the outset to point out that the lithosphere cannot be considered in isolation; the boundary between it and the underlying *asthenosphere* is transitional, and is continually changing; and there is a constant interaction between the lithosphere and the other parts of the convective system (Figure 2.1).

The base of the lithosphere is usually defined on the basis of a relatively rapid change in seismic wave velocity (of both P and S waves), and specifically of a fall in the rate of increase of V_p and V_s, which takes place at depths of around 100–150 km within the upper mantle. This change in velocity is related to changes in density and rheology, which in turn are related to the *geotherm* — the profile of temperature variation downwards through the crust and upper mantle. Although there may be petrographic differences between the lithospheric and asthenospheric mantle, these are not the main factor in differentiating the two layers, which are distinguished most conveniently in terms of their *viscosity*. The viscosity of the asthenosphere is usually estimated to be around 10^{21}–10^{22} poise in contrast with that in the lower part of the lithosphere which probably varies from 10^{23} poise upwards (see McKenzie, 1967). It is this rapid decrease in viscosity that enables the solid material of the asthenosphere to flow at a geologically significant rate, carrying the more viscous and therefore stronger lithosphere above it.

The relative plasticity of the asthenosphere is due mainly to the effect of elevated temperature on the rheology of the material, which is governed by a flow law in which the strain rate is both temperature- and stress-dependent (see 2.2). A significant fraction of asthenosphere material is believed to be composed of melt

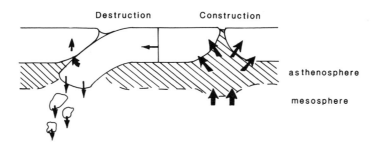

Figure 2.1 Asthenosphere–lithosphere–mesosphere interaction. The lithosphere gains material from the asthenosphere at constructive plate boundaries; during plate destruction, material is lost by the lithosphere to the mesosphere, and to the asthenosphere. Material lines in the lithosphere migrate laterally relative to the sites of construction and destruction.

(perhaps as much as 10%) although this may decrease away from sites of upwelling convection currents.

The asthenosphere is continuously fed by uprising material from the underlying *mesosphere*. Mesosphere is transformed to asthenosphere simply by a change in rheology — the same piece of mantle material may commence in the mesosphere, be changed into asthenosphere at a site of convective upwelling, with a decrease in viscosity and increase in flow rate. It may subsequently be transformed again into lithosphere by a reversal of this process — by cooling and a consequent rise in viscosity. In addition, of course, there are material changes: partial melting causes magmas to leave the system and move up into the lithosphere, volatiles may enter and leave the system, etc., so that the composition of the asthenosphere presumably varies significantly at the sites of upwelling, but is probably much more uniform beneath stable plate interiors.

The lithosphere grows at the expense of the asthenosphere mainly at ocean ridges. Here new lithosphere is generated by the cooling of asthenospheric material as it is carried laterally away from the hotter ridge axes. Again the system is very complex in detail, but can be simplified in terms of a model of continuous accretion of lithosphere along an essentially thermal boundary in the flanking regions of an ocean ridge (Figure 2.1). As material is carried across this boundary, it changes from asthenosphere to lithosphere. Thus although new material is being continuously emplaced below the ridge, and is thereafter being transported laterally away from the ridge axis, the actual boundary to the lithosphere is stationary with respect to the upwelling thermal source, as long as the thermal conditions are unchanged. Material points within the lithosphere move although the lithosphere boundaries themselves may remain fixed.

Away from the ridge, the cooling of oceanic lithosphere proceeds more slowly, although differences can be detected in normal ocean basin lithosphere thickness that are related to

its age (see 2.3). The older a piece of oceanic lithosphere, the cooler it will be, and the deeper will be the thermal boundary which defines its base. It has been estimated that ocean-basin lithosphere varies in thickness from around 50 km at the ridge crest to about 150 km in the oldest parts furthest from the crest. There is thus a simple relationship between thickness and age that can be determined fairly accurately for most parts of the ocean using the methods of magnetic stratigraphy.

The oldest and coolest parts of the oceanic lithosphere are either attached to continents and continue to migrate laterally with them, or form subducting slabs which descend through the asthenosphere to merge indistinguishably with the mesosphere. These slabs then form the cool downward-flowing limbs of convective 'cells'. They sink because they are cooler and therefore more dense than the surrounding material. An oceanic lithosphere plate may thus extend from a ridge crest to a trench and therefrom down to the base of the asthenosphere (Figure 2.1). As it descends, it becomes warmer and will lose material to the asthenosphere by partial melting. At its base, it may disappear, or break into sections, but the material of which it is made will generally descend, carried down by the cool return limb of the convective circulation, but indistinguishable seismically from the adjacent mesospheric material. It thus appears that the lithosphere plates, because of their strength and continuity, display a pattern of movement which is part of a more fundamental mantle circulation to be discussed in 2.4.

2.2 Some short-term mechanical properties of the lithosphere

The formation of geological structures is ultimately dependent on the mechanical properties of the material in which they are formed. It is essential therefore to discuss the mechanical properties of the lithosphere in order to discover how, and under what conditions, geological deformation is produced. The important mechanical properties which control deformation are elasticity, viscosity, fracture strength and yield strength. These properties vary with rock composition, depth and temperature. The information provided by laboratory experiments on rock materials can provide estimates for these mechanical parameters under a limited range of conditions. However, a more useful method of studying the mechanical behaviour of the lithosphere as a whole is to make simplifying assumptions about bulk properties, and to use mathematical models to determine how these properties interact and vary in changing physical conditions (see 2.7).

The main source of information about the short-term mechanical properties of lithosphere plates comes from indirect geophysical methods, particularly the study of *seismic waves*. The velocity of seismic waves provides information about the elastic properties of a plate (elasticity and rigidity) and also, of course, its effective thickness, as explained above. The values of elasticity or rigidity derived from seismic wave velocities define the elastic properties of the lithosphere over very short time periods (0.1 s to 1 h). However the lithosphere strength calculated in this way (often called the 'instantaneous strength') is very much greater than its strength when subjected to forces for periods of tens of Ma.

Useful information may also be gained from the study of lateral variations in both P- and S-wave velocity. Figure 2.2*A* is a map of the North American continent showing the variation in mean P_n-wave velocity (P waves propagated in the uppermost mantle). Figure 2.2*B* shows in addition the variation in mean crustal velocity together with contours of crustal thickness. There is clearly a crude relationship between these; most of the P-wave velocity variation in the crust appears to be due to the crustal thickness variations. Lateral differences in the lithosphere, however, are more evident when the velocity of the P_n waves is studied (Figure 2.2*B*). The map shows a decrease in P_n velocity from about 8.2 km/s on the North American craton to below 7.8 km/s

in regions of current or recent volcanism along the central Cordilleran orogenic belt (for example in the Basin-and-Range province and the Cascades volcanic arc). These are regions of warmer lithosphere with higher surface heat flow and steeper geothermal gradients (see 2.7) which act to reduce P-wave velocities in the mantle part of the lithosphere. Very similar effects are seen in other currently or recently active regions. For example the velocity of P_n waves beneath the Japanese arc is 7.5–7.7 km/s compared with 8.0–8.1 km/s for those below the adjoining Pacific ocean. In continental rift zones and also at ocean ridges, there is a corresponding reduction of P_n-wave velocity compared with nearby stable regions. Thus the study of the regional variation of P_n-wave velocity reveals zones of anomalously weak, warm lithosphere which must persist for long periods of time and are thus relevant to the longer-term strength properties of the lithosphere as well as to its instantaneous strength.

S_n waves are S waves propagated in the uppermost mantle. Like P_n waves, their velocity is affected by change in elastic properties. Unlike P_n waves though, the S_n waves do not penetrate the low-velocity zone due to their short wavelength. The efficiency of transmission of these waves is shown on a world-wide scale in Figure 2.3. It is clear that stable plate interiors are zones of efficient transmission, whereas active tectonic zones (island arcs, ocean ridges etc.) correspond to zones of inefficient transmission.

Another geophysical method of studying lateral variations in lithosphere strength is through the study of the *dispersion of surface waves* (Rayleigh and Love waves). Dispersion is a measure of the spread in amplitude and wavelength within a wave train, and occurs because of the variation in elastic modulus and density with distance from the source. The depth of penetration of a surface wave is directly dependent on its wavelength. Thus if the velocities are calculated as a function of wavelength, the rigidity of the material can be

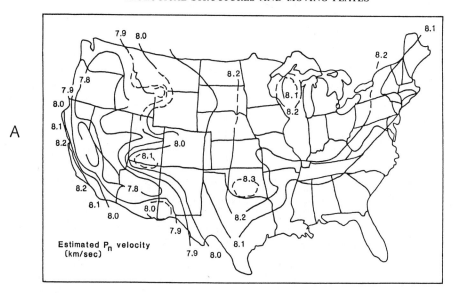

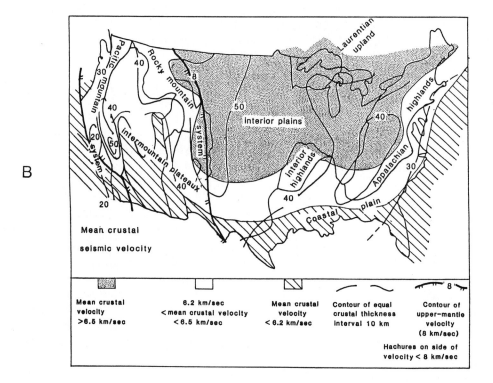

Figure 2.2 (*A*) Estimated P$_n$ seismic wave velocity for upper mantle in the USA. After Wyllie (1971), from Herrin (1969). (*B*) Variations in crustal thickness, mean crustal velocity, and upper mantle velocity (see *A*) in the USA. After Wyllie (1971), from Pakiser and Zietz (1965).

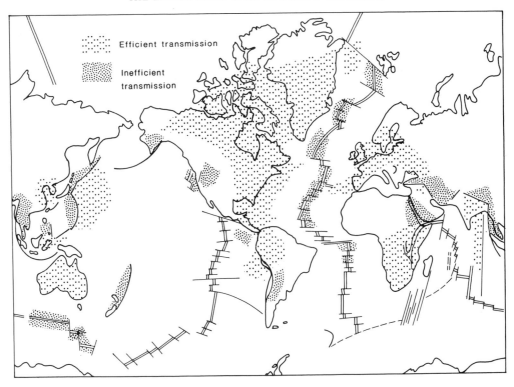

Figure 2.3 Regions of efficient and inefficient propagation of S_n seismic waves in the upper mantle (see text). After Molnar and Oliver (1969).

estimated as a function of depth. This method depends on the simultaneous study of body waves (e.g. P waves) and surface waves, and reveals zones of anomalously weak mantle below currently active volcanic regions corresponding to those revealed by P_n wave analysis.

Seismic wave attenuation is the reduction of amplitude with distance and time due to energy loss, and is measured by the quantity Q. The amount of attenuation Q is related to the strength of the material through which the wave is propagated. The 'low-velocity zone' (asthenosphere) is characterized by low values of Q.

Further evidence as to the presence of anomalous zones comes from the study of *electrical conductivity* in the Earth. Rocks are weak conductors of electric currents. It is possible by studying rapid variations in the Earth's magnetic field to isolate effects caused by electric currents in the Earth's upper atmo-

sphere which are due to changes in the flow of radiation and charged particles from the Sun. The resulting magnetic field induces electric currents within the Earth, which in turn cause rapidly changing modifications to the magnetic field. By isolating the secondary variation, the strength of the electric currents, and hence the electrical conductivity of the Earth, can be estimated.

Anomalies in electrical conductivity were found to be associated with island arcs and continental rift zones (see e.g. Hermance, 1982). These anomalies are zones of poor conductivity that correlate with regions of unusually warm lithosphere.

Thus several independent geophysical methods indicate that active tectonic zones (volcanic arcs, ocean ridges and continental rifts) display anomalous physical properties that correspond to those exhibited by the asthenosphere. We may conclude that such zones possess abnormally low instantaneous strength and can be regarded as zones of anomalously thin and weak lithosphere.

Summary

For time periods in the range seconds to hours, the lithosphere generally behaves as a strong elastic and rigid body with effectively infinite viscosity. Active tectonic zones exhibit anomalous physical properties indicated by low P_n wave velocity, greater surface-wave dispersion, low Q (high degree of attenuation), inefficient propagation of S_n waves, and low electrical conductivity. The strength of the lithospheric mantle in these anomalous zones is in many cases (e.g. Japan) comparable with that in the asthenosphere, implying that the 'instantaneous' thickness of the lithosphere is considerably reduced. In simple terms, plates are thinner and weaker in active tectonic zones. As we shall now see, this is related to their thermal structure.

2.3 Thermal structure of the lithosphere

The average total rate of heat loss through the Earth's surface is about 2.4×10^{20} cal/year. This represents an enormous loss of energy, several orders of magnitude greater than the total loss associated with earthquake or volcanic activity.

Mean values of heat flow per unit surface area for the different continents and oceans are shown in Table 2.1. It is clear that the average continental and oceanic heat flow is essentially the same. It appears that more than 99% of the Earth's surface has a 'normal' heat flow of around 1.5 HFU (= 60 mW m^{-2}), with anomalous zones of very much higher heat flow. Note that 1 HFU = 1μcal cm^{-2} s^{-1} = 40 mW m^{-2}. These localized zones can only be explained by the transfer of material bringing thermal energy from deeper sources and correspond mostly to areas of current or recent volcanic activity, such as island arcs, major continental rift zones and ocean ridges.

The measurement of heat flow on land is

Table 2.1 Mean heat flow for continental and oceanic regions.

Region	N	$\bar{q}$	S.D.
Von Herzen and Lee, 1969			
All continents	255	1.49	0.54
All oceans	2329	1.65	1.14
Atlantic	406	1.43	1.07
Indian	331	1.44	1.09
Pacific	1232	1.71	1.24
Arctic	29	1.23	0.33
Mediterranean seas	71	1.33	0.89
Marginal seas	260	2.13	0.63
Girdler, 1967			
Africa	13	1.20	0.21
Japan	38	2.21	2.73
Australia	20	1.76	0.62
Europe	31	1.91	1.70
N. America	44	1.26	0.57

N = number of observations; $\bar{q}$ = arithmetic mean of heat flow in μcal cm^{-2} s^{-1} (HFU); S.D. = standard deviation in HFU (heat-flow units).

much more difficult than at sea because of various surface effects, which are minimized by the blanketing effect of the overlying water in the oceans.

It is necessary to drill to depths of about 300 m or more on land to produce accurate results. In general, estimates of heat flow in the oceans are accurate to within 10%, but continental estimates are very variable in their accuracy. Regional estimates of mean heat flow based on more than 10 observations are believed to be accurate to within 0.2 HFU. Local departures from these mean values of more than this amount are considered to be significant.

Table 2.2 shows selected heat flow values for the major types of tectonic province, from which it may be seen that low mean heat flow values characterize the Precambrian shields and that high mean heat flows are associated with Cenozoic volcanic areas. High heat flows are also associated with ocean ridges and island arcs. Profiles of heat flow values across the mid-Atlantic ridge, and the Kurile and Japan island arcs (Figure 2.4) show very marked but relatively narrow anomalies with a rather rapid

Table 2.2 Continental heat flow data divided into shield, intermediate (Younger Proterozoic to Phanerozoic crust) and thermally active regions. After Kusznir and Park (1984).

Region	Heat flow mW m^{-2}	HFU
A. Shield		
Superior Province[2]	34 ± 8	0.85
West Australia[2]	39 ± 8	0.98
West Africa (Niger)[2]	20 ± 8	0.50
South India[2]	49 ± 8	1.23
Mean Archaean + older Proterozoic[5]	41 ± 10	1.03
B. Intermediate		
Eastern USA[2]	57 ± 17	1.43
England and Wales[2]	59 ± 23	1.48
Central Europe (Bohemian massif)[2]	73 ± 18	1.83
Northern China[1]	75 ± 15	1.89
Mean Younger Proterozoic[5]	50 ± 5	1.25
Mean Palaeozoic[5]	62 ± 20	1.55
C. Thermally active		
Rhine graben[3]	107 ± 35	2.68
(flanks-Rhenish massif)[3]	(73 ± 20)	(1.83)
Baikal rift[3]	97 ± 22	2.43
(SE flank — Lower Palaeozoic)[3]	(55 ± 10)	(1.38)
East African rift[3]	105 ± 51	2.63
(flanks)[3]	(52 ± 17)	(1.30)
Basin-and-Range Province[2]	92 ± 33	2.30
(E. flank-Colorado plateau)[4]	(60)	(1.50)

Heat flow data from Pollack and Chapman (1977)[1], Vitorello and Pollack (1980)[2], Morgan (1982[3], 1983[4], in press[5]).

change within a few 100 km to normal regional values.

Sclater and Francheteau (1970) compared the profile of average measured heat flow values across the East Pacific ridge with a theoretical profile expected from a 75 km-thick cooling lithosphere model (Figure 2.5*A*). The close correspondence was held to support the ocean-floor spreading model of the growth of lithosphere at the ridge by addition of new hot mantle material. As the new ocean floor spreads away from the ridge axis, it cools from the surface downwards and gradually thickens (see Figure 2.5*B*). The decrease of heat flow with distance from the ridge axis is therefore a direct consequence of a cooling and thickening lithosphere plate. This relationship may be checked by plotting the variation of heat flow

with age of the ocean floor (Figure 2.5*C*). The heat flow decreases steeply until an age of about 50 Ma, after which it becomes more or less constant.

The profiles across the volcanic arcs show anomalously low values at the trenches as well as high values over the arcs. The distribution of these zones of high and low heat flow are shown for the north and west Pacific region in Figure 2.6*A*. The relationship with trenches and volcanic island arcs is very clear. Figure 2.6*B* illustrates in profile the depression of isotherms associated with the subducting slab.

It can be seen from Table 2.2 that there is a tendency for the heat flow to increase with decrease in age of orogeny for Phanerozoic and Precambrian orogenic belts, but the differences are small in relation to the error. This decrease of continental heat flow with age was investigated in more detail by Vitorello and Pollack (1980). They showed (Figure 2.7) that the decrease with tectonic age (i.e. age since the last major tectonothermal event) could be approximately fitted to an exponential curve representing the gradual decay of a transient thermal perturbation together with a component of heat loss due to gradual erosion of radiogenic crust.

Summary

There is no significant difference between mean continental and mean oceanic heat flow. Large heat flow anomalies occur over very narrow zones making up less than 1% of the Earth's surface area. Zones of high heat flow comprise the active tectonic regions where magma has been introduced into high levels in the lithosphere — the active volcanic arcs, continental rifts and ocean ridges. Low heat flow zones are associated with the trenches. Both oceanic and continental crust show a progressive decrease in heat flow with increase in thermal age. These variations in heat flow reflect variations in geothermal gradient that have important implications for lithosphere strength, as we shall see in 2.7.

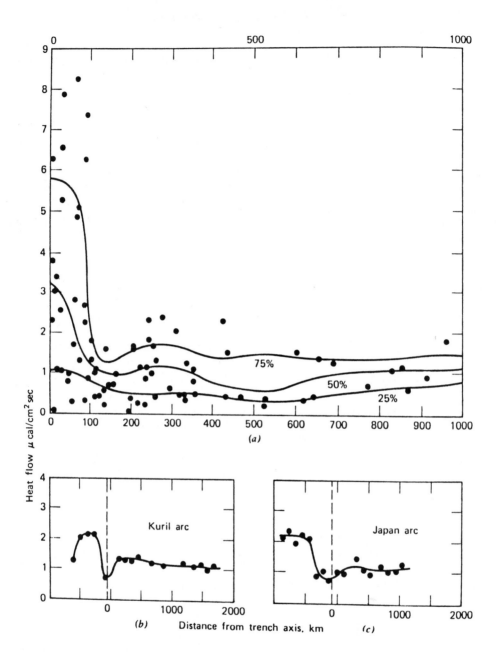

Figure 2.4 (*a*) Heat flow values *v*. distance from the crest of the mid-Atlantic ridge. Percentile lines enclose 75%, 50% and 25% of the data respectively. From Wyllie (1971), after Lee and Uyeda (1965). (*b,c*) Heat flow profiles across the Kurile and Japan arcs, averaged in 100 km intervals. From Wyllie (1971), after Vacquier *et al.* (1966)

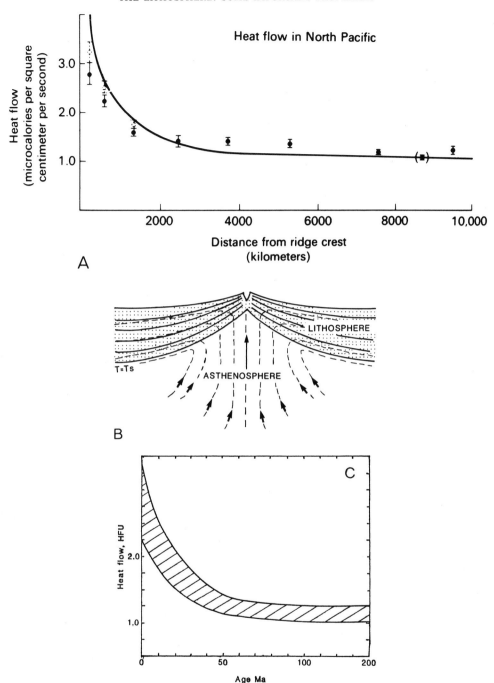

Figure 2.5 (*A*) Heat flow profile from the East Pacific ridge crest across the North Pacific. Mean observed values (closed circles) are compared with a theoretical profile for a cooling lithosphere 75 km thick. Open circles represent mean values increased by 15% to allow for a possible bias near the ridge. From Uyeda (1978). (*B*) Schematic diagram showing material flow lines (dashed) and isotherms (solid) at an ocean ridge. The solidus temperature T_s is the isotherm representing the boundary between the lithosphere and the asthenosphere. After Uyeda (1978). (*C*) Variation of heat flow with age of ocean crust, showing cooling with increasing age. After Sclater (1972).

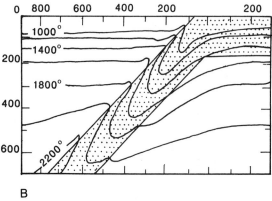

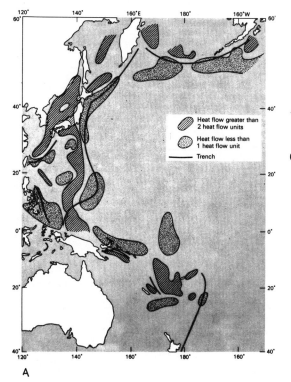

Figure 2.6 (*A*) Simplified map of heat flow distribution in the West Pacific showing decreased heat flow at trenches and increased heat flow at volcanic arcs and back-arc regions. From Uyeda (1978). (*B*) Schematic profile across a subduction zone showing depressed isotherms along the subducting slab. After Wyllie (1971).

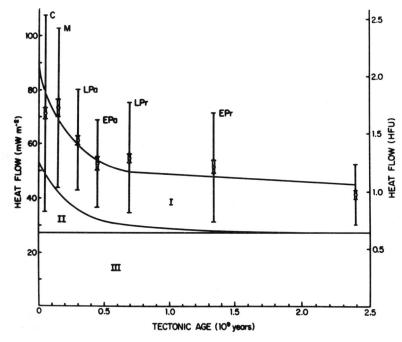

Figure 2.7 Decrease of continental heat flow with tectonic age of crust. The upper curve (visually fitted) is interpreted as made up of three components: I is radiogenic heat from the crust, II is heat from a transient thermal perturbation associated with a tectonothermal event, and III is background heat flow from a deeper source. Double bars represent the standard error, single bars the standard deviation, of the means. From Vitorello and Pollack (1980).

2.4 Mantle convection: the role of the lithosphere

The Earth is a heat engine, and the ultimate source of almost all the energy required in geological processes is thermal in origin, arising out of the transfer of heat (partly original, partly radiogenic) from the interior of the Earth to its exterior. Arthur Holmes (1929) proposed thermal convection currents as the driving mechanism for continental drift, but the suggestion was only seriously considered by most geologists thirty years later, when enough evidence had accumulated to convince them that continental drift and ocean-floor spreading were viable hypotheses and that convection was the only adequate explanation for these processes.

It is now generally accepted that the solid material of the Earth is capable of flow given a long enough time scale. The rate of flow is governed by the *effective viscosity* of the material, which is strongly temperature-dependent. An estimate of the mean viscosity of the outer shell of the Earth can be determined from the isostatic response of the lithosphere. For example, the upward movement of Scandinavia in response to the removal of the Pleistocene ice-sheet can be accurately measured by observations on raised beaches, which show a maximum elevation of $c.300$ m in 1000 years. This is extremely rapid in geological terms and indicates a mean viscosity of about 10^{21} poise. This mean value represents a range in viscosity that must decrease downwards with increase in temperature.

Flow in the upper mantle has, until recently, been assumed to be governed by ductile creep in either dry or wet olivine. Flow laws based on dislocation creep have been provided by Kohlstedt and Goetze (1974) and Goetze (1978) for dry olivine, and by Post (1977) for wet dunite, based on laboratory data. These flow laws are of the form

$$\dot{e} = A \, \exp B/T\sigma^3 \, \mathrm{s}^{-1}$$

where $\dot{e}$ is strain rate, T temperature and σ stress. A and B are constants and are strongly temperature-dependent.

It was originally thought that olivine was the dominant mineral throughout the upper mantle, and was replaced by a mineral with similar composition but with a spinel structure (i.e. a denser phase) in the lower mantle. However, petrologists have more recently suggested a more complex compositional structure. Anderson and Bass (1986) summarize the evidence in favour of a heterogeneous upper mantle dominated in the lower part by clinopyroxene and garnet, the base of which may be composed largely of subducted material, and suggest that olivine is dominant only in the upper part of the upper mantle. Olivine is now considered to be unstable below about 400 km. Below 650 km there is a major change in physical properties which is explained as the result of a change to dense perovskite -like mineral. The flow laws for these materials are not known. Although it has been suggested that the lower mantle may not be capable of convective flow, the balance of opinion now is in favour of convection in both upper and lower mantle but probably not directly linked in a large mantle-wide cell.

The ground rules for the plate tectonic theory having been laid down in 1968, the key role of convection in this new theory was analysed in a classic paper by McKenzie (1969) entitled 'Speculations on the consequences and causes of plate motions'. McKenzie pointed out that large-scale convection throughout at least the upper mantle, as originally envisaged by Holmes, provided the only adequate mechanism for plate movements. Viscous forces were required to couple the plates to the moving mantle below, and vertical movements of both hot and cold material must also be involved to complete the convective circulation. Plate movements are thus seen as part of a large-scale convective process that is in a sense both a consequence and a cause of the motion.

This concept was at variance with a suggestion by Elsasser (1967) that the motion of the plates was governed by the pull of the cold sinking slabs in subduction zones. Although the 'slab-pull' force is an important element in lithosphere dynamics, it can easily be shown that it is only one of several forces of comparable magnitude acting on the lithosphere, and that much of the oceanic lithosphere is under compression rather than tension as required by the Elsasser model (see 2.5, 2.6). The implication of McKenzie's argument is that such forces are secondary, that no single force provides the driving mechanism for plate tectonics, and that all such forces arise from the convective flow system itself.

The early models of Holmes (1929) and Runcorn (1962) (Figure 2.8) implied a link between ocean ridges (or more generally, constructive plate boundaries) and rising currents on the one hand, and between subduction zones (or destructive boundaries) and falling currents on the other. It became apparent, however, that such a link could only be transitory, since distances between ridges and trenches are continuously changing on a geological time-scale, and that all plate boundaries migrate with respect to a fixed mantle reference frame.

It was considered by Runcorn and others that the positions of major convective upflows and downflows in the mantle should be reflected in distortions of the *geoid*. The geoid (Figure 2.9) is the sea-level equipotential surface of the gravity field of the Earth, and this surface defines the departures from radial symmetry of the distribution of mass within the Earth. It is measured by extremely accurate satellite observations. The geoid surface reveals departures from a simple spheroid of flattening which are of several orders of magnitude. The largest type of anomaly has a half-wavelength of about 4000 km, and is comparable in dimensions with the major plates. It produces elevations and depressions of the order of 50–100 m on the geoid surface. Rather smaller-scale distortions with half-wavelengths in the range 1000–1500 km correspond to the major oceanic 'swells' like those of Hawaii and Bermuda. These are accompanied by positive geoid anomalies of 6–8 m. A third scale of anomaly can be detected with a half-wavelength of 100–250 km.

The long-wavelength components of the Earth's gravity field are considered to result mainly from density contrasts associated with mantle convection. Thus a mass excess is associated with a cold dense downward current (giving a geoid swell), and a mass deficiency with a warm, less dense upward current, giving a geoid depression. However, these effects are largely offset by the dynamically maintained deformation of the surface topography. (e.g. ridges and trenches) that produces effects on the geoid that are opposite in sign but comparable in magnitude. In other words, the existence of general isostatic compensation in these major topographic structures reduces the net geoid anomaly to around zero, particularly in the case of the ridges.

The long-wavelength anomalies probably relate to swells and depressions of perhaps

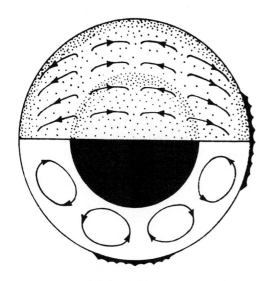

Figure 2.8 The Runcorn (1962) model of mantle convection. From Runcorn (1962).

Figure 2.9 The shape of the geoid. Different views (*a–f*) of the geoid shape (see text). Contours at 20 m intervals show departures of the Smithsonian Standard Earth II geoid from a spheroid of flattening 1/298.25. Positive elevation is shaded. From Gough (1977).

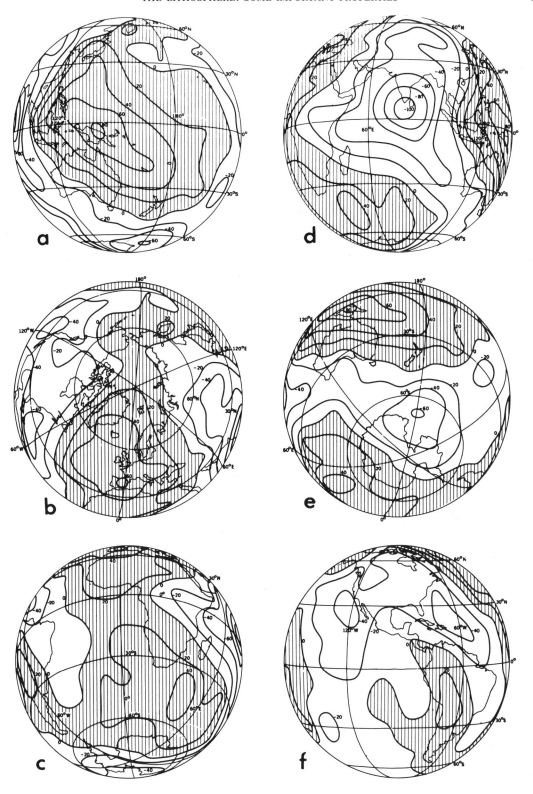

several km magnitude on the core–mantle interface, that would be associated with rising and falling convective columns. Anomalies with a deep source should primarily reflect the density imbalance; thus upflows in the lower mantle produce negative and downflows positive anomalies. The opposite appears to be true of the shallow disturbances rooted in the asthenosphere, where the anomalies reflect the deformed shape of the lithosphere; for example the Hawaii swell is associated with a *positive* anomaly.

Figure 2.10 compares the large-wavelength anomalies with the present plate boundary network. Although there is a poor correlation in general, and a total lack of correlation between ridges and negative anomalies, the major subduction zones are situated in regions of positive anomaly. McKenzie (1969) presumed that ridges must exist over both falling and rising currents. This was explained because the creation of lithosphere at a ridge requires a large volume of hot mantle material that,

because it rises more or less adiabatically, convects little heat except within the lithosphere, and so produces little distortion of the isotherms within the sub-lithospheric mantle. In subduction zones however, the opposite is the case. No change in temperature structure within the lithosphere occurs until it commences its descent into the mantle. This motion distorts the isotherms by several hundred km throughout the upper mantle (see Figure 2.6*B*). Thus the sub-lithospheric mantle loses material beneath ridges but loses heat beneath island arcs. The opposite is true for the lithosphere, as explained in 2.1.

McKenzie concluded that the cold descending slabs produce large horizontal and vertical temperature gradients and should therefore govern the position of the descending limb of any convection cell. If the subduction zone moves, the sinking current should move with it.

Cool oceanic lithosphere is able to sink because it is denser than asthenospheric man-

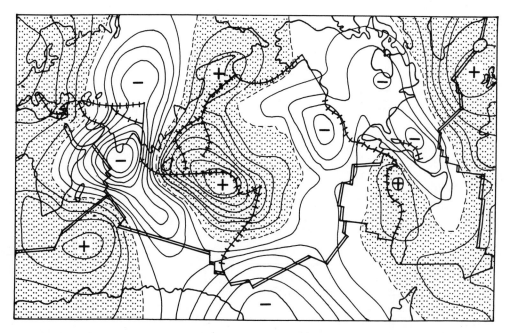

Figure 2.10   Contours at 10 m intervals of the height of the non-hydrostatic geoid compared with the plate boundary network. Trenches, hatched lines; ridges, double lines; faults, single lines. Major trenches all occur in positive areas. After McKenzie (1969).

tle, despite the negative buoyancy of the oceanic crust, which is probably overcome by the transformation of gabbro to eclogite as it sinks. Continental lithosphere, on the other hand, has a much lower mean density than oceanic lithosphere, since about 20–30% of it is composed of crustal material, compared with only about 5% in the case of oceanic lithosphere. Continental lithosphere will therefore resist subduction, and continents, once formed, will be very difficult to destroy.

Dewey and Bird (1970) showed how continent–island arc and continent–continent collisions are the inevitable consequence of continued subduction, and how subduction must cease when collision takes place. Thereafter, for continued convergence to occur, the site of subduction must move away from the collision zone in order that further subduction of oceanic material may take place. The pattern of mantle convection must therefore depend to a large extent on the position of the continents and must move with the continents, since the latter control the position of the subduction zones. However, convective changes may lag behind the continent movements because of the time taken for the old cold slab below the collision zone to warm up.

The last major reorganization of the convective pattern was probably associated with the continental amalgamation that formed Pangaea. The supercontinent appears to have been almost surrounded by subduction zones. The Pacific Ocean of that time formed about half the surface area of the Earth (Figure 3.5), and probably possessed a simple symmetrical ridge system that may reflect the contemporary rising convective limbs. Since that time, the Pacific ridge system has moved northwards and been overridden by continental plate, particularly on its eastern side. Geometric consequences of plate movements have thus tended to move ridges away from their longer-lived mantle roots. However, the link between falling currents and subduction zones may have been longer-lasting, and may partly explain the relationship between geoid anomalies and trenches seen on Figure 2.10.

Gough (1977) has compared the shape of the long-wavelength anomaly (Figure 2.9) to the two spiral strips covering a tennis ball. The high, or positive anomaly, has one lobe covering the western Pacific with its trench system, and including most of Australia, and is joined across the Arctic and northern Atlantic region to a second lobe covering the southern Atlantic and southern Indian oceans, extending westwards over the Andes. The low, or negative anomaly, has one lobe centred south of India and covering much of Asia and the Indian Ocean, which is joined across the Antarctic to a second lobe covering the eastern Pacific, North America, the Brazilian shield and the western Atlantic. Thus the low strips surround the high lobes and vice versa. Gough argues that this pattern may indicate a global single-cell convective system where the upcurrents probably underlie the low geoid strip (although the opposite could be the case).

Whole-mantle convection cells were envisaged by earlier workers (e.g. Holmes, 1929; Runcorn, 1962) and are still advocated by some (e.g. Kenyon and Turcotte, 1983). However, many investigators now believe that the change in chemical and physical properties at 650 km depth, which produces the seismic discontinuity separating upper from lower mantle, represents a barrier to complete circulation, and that separate convective patterns may be independent or only partially coupled (see e.g. Le Pichon and Huchon, 1984). It is possible that the rising currents are controlled by the temperature distribution at the base of the mantle, but are uncoupled from the lithosphere at the surface, whereas the falling currents are more strongly influenced by lithosphere movements. Figure 2.11 shows diagrammatically how a possible mantle convective system might link with plate movements. The position of hot columns in the mantle is governed largely by the sites of upwelling liquid outer core material.

An important source of information on the Earth's convective system is provided by the so-called 'hot spots' or *mantle plumes* (Wilson, 1965; Morgan, 1972). These are long-lived

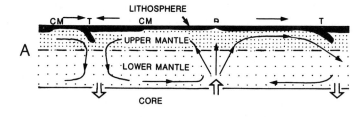

A

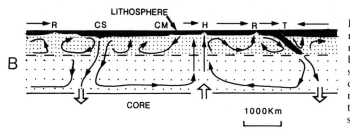

B

Figure 2.11 Cartoon representing possible mantle convective flow patterns. A simple whole-mantle pattern (*A*) degenerates into partly-linked upper and lower mantle systems after some plate movement, in which the plate boundaries have migrated away from their original mantle roots (*B*). *CM*, continent margin; *T*, trench; *R*, ridge; *CS*, collision suture; *H*, hot spot.

sources of volcanic activity which are 'stationary', or at least move much more slowly with respect to a fixed mantle reference frame, than the plates. Consequently they leave a linear 'track' in the form of a chain of volcanoes on the surface of the moving oceanic plates. A good example is the Hawaii–Emperor chain in the Pacific Ocean (Figure 3.10). Iceland is another example, actually situated on a ridge, so that there is a symmetrical arrangement of linear chains running towards the present position from each side. Morgan used the hot-spot frame of reference to determine the 'absolute' motions of the plates (Figure 3.12).

The oceanic hot spots are associated both with topographic swells (with amplitudes of about 1 km) and also with positive geoid anomalies (with amplitudes of around 10 m) — see Parsons *et al.* (1983). The topographic swells are caused by warmer and less dense lithosphere similar to that of the ocean ridges. Both swells and anomalies are elongated in the direction of plate movement. It is believed that the size and spacing of the hot spots indicates the presence of a convective circulation pattern which is much smaller in scale (i.e. with a wavelength of 2000–3000 km) than the major circulation responsible for the plate movements, and comparable in size to the thickness of the upper mantle.

An even smaller-scale convection is suggested by the pattern of short-wavelength (200–500 km) undulations of the geoid observed in the central Pacific ocean (Figure 2.12; Haxby and Weissel, 1983). These features are elongated parallel with the 'absolute' motion of the plates, like the hot-spot chains, but are volcanically quiet. This pattern is interpreted as the result of small-scale instabilities in cooling of the plate as it moves away from the ridge, and is thought to be associated with flow patterns within the asthenosphere.

Modelling convection

There have been many attempts is simulate mantle convection by means of models of both experimental and mathematical type. Both kinds of model are highly artificial and give only a rough guide to the way convection might operate. A stimulating account of the physics of the behaviour of the Earth is provided by Elder (1976) in his book, *The Bowels of the Earth*. This account serves as a useful illustration of both the possibilities and limitations of modelling the Earth's internal behaviour.

Elder shows that thermal conduction is totally inadequate to explain the heat flow and temperature distribution in the uppermost parts of the Earth, and that convective flow is essential to transport the heat and energy

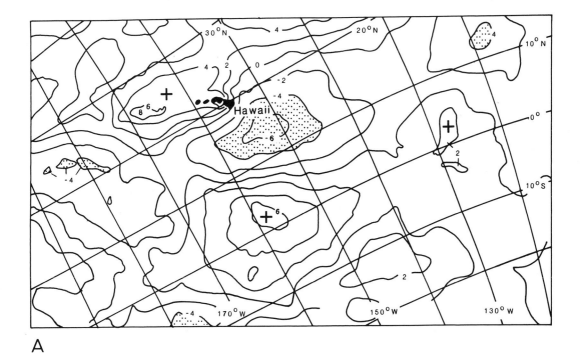

A

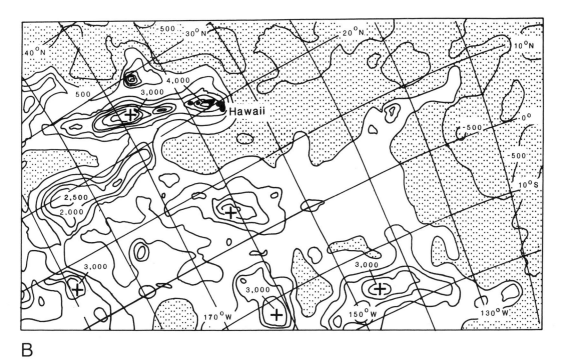

B

Figure 2.12 (A) Gravity anomaly pattern over part of the Pacific plate, observed as contours on the sea surface, measured by satellite. (B) Bathymetric contours of the same region as (A), plotted as residual depth (the difference between the depth attributable to cooling oceanic plate and the observed depth). Both maps are smoothed to eliminate fluctuations with wavelengths shorter than 500 km. Lows on both maps are dotted. Slightly elongate positive geoid anomalies correlate with topographic highs. The maps are oriented such that the motion of the plate is towards the left, in the direction of elongation of the elliptical anomalies. After McKenzie (1983).

required for surface processes. Furthermore, the remarkable similarity over geological time of suites of basic igneous rocks indicates a well-mixed ('well-stirred') source, which again indicates convection.

Elder uses the principles of fluid dynamics to treat the Earth as an example of thermal turbulence in a medium with variable viscosity. The fundamental driving mechanism for convection is a horizontal temperature gradient which provides changes in density. These in turn generate buoyancy forces which initiate and maintain motion. In an unstable system above a critical *Rayleigh number*, any small perturbation in the temperature field will become amplified and generate motion. The Rayleigh number (Ra) is a measure of the effectiveness of the buoyancy forces acting against the combined resistance to motion of viscosity and the diffusion of temperature variations by thermal conductivity.

For a body of given shape and boundary temperature distribution, the Rayleigh number is given by:

$$Ra = \gamma g \triangle Th^3 kv$$

where γ is the coefficient of cubical expansion, g is the local acceleration due to gravity, $\triangle T$ the temperature difference, h the width of the body (radius of the Earth), k the thermal diffusivity and v the kinematic viscosity. Elder assumes the following values for these parameters: $\gamma = 10\,K$, $g = 10\,m\,s^{-2}$, $\triangle T = 2500\,K$, $h = 6370\,km$, $k = 10^{-6}\,m^2\,s^{-1}$, and $v = 6.46 \times 10^{16}\,m^2\,s^{-1}\,cm^2\,s^{-1}$, giving a Rayleigh number of 10^9. Considerable uncertainly attaches to the value of v, which may lie in the range 10^{15}–$10^{18}\,m^2\,s^{-1}$ for the mantle. However, the relevant values of the kinematic viscosity lie in the upper parts of the mantle and are known more accurately. Values for γ, $\triangle T$ and k are probably accurate within a factor of 2. However, it is clear that the Rayleigh number must be large. For bodies with $Ra \geqslant 10^5$, vigorous non-steady convection is indicated, dominated by chaotic eddying motion with eddies of a wide range of size. This model is quite different from the regular cell model of Runcorn (1959)

but similar to the kind of model inferred by McKenzie and others from the relationship between plate boundaries and geoid anomalies.

Elder's model predicts a thermal sub-layer (corresponding to the lithosphere) which is a cooled and highly viscous buffer between the well-mixed interior and the constant-temperature surface. The thickness of this layer is a function of the Rayleigh number. Vigorous interior flow tends to thin the sub-layer, which must periodically overturn for convection to be maintained. The behaviour of the sub-layer is simulated using a pot of hot oil, broader than its depth, insulated on its base and sides, and cooled from above. The cooled sub-layer grows to a maximum thickness determined by the value of Ra, then overturns. Before it does so, it forms a series of eddies of the same dimensions as the depth of the sub-layer (Figure 2.13A). These eddies produce large distortions of the initially vertically stratified temperature field. Each eddy entrains cool fluid from the surface and brings hot material closer to the surface. Eventually, blobs of cool fluid fall out of the sub-layer into the interior and initiate disruption of the sub-layer. This process, which is effectively random, simulates the behaviour of the lithosphere as it is attacked by ridges and plumes, and eventually by subduction.

A further useful analogue of Earth behaviour is provided by experiments simulating the effects of a continental slab on the convective system (Figure 2.13B). Elder finds that an asymmetrical eddy is generated by a continental edge; this provides a lateral force that could act to propel the continent (cf. the *subduction suction force*, 2.5). There is a strong upwelling immediately behind the leading edge that simulates volcanic arc production. Any thermal perturbation below the continental sheet would tend to propel the sheet laterally, unless it were centrally situated.

Summary

The study of the geoid anomaly pattern suggests that the mantle contains at least three scales of convective circulation. The first, or large-scale circulation is comparable with the dimensions of the plates and returns material from trenches to ridges. The second, of intermediate scale, is comparable to the depth of the upper mantle in scale, and controls the hot spots. The third has a small scale comparable to the width of the asthenosphere.

In terms of plate kinematics, mantle convection does not appear to provide a direct link between asthenosphere flow and plate movement. Any large plate would be expected to have different directions of asthenosphere flow beneath it. This observation is of critical importance in evaluating the mechanisms of plate movement and the forces acting on the plates, as we shall now see (2.5).

The conclusions that can be drawn from Elder's model bear an interesting resemblance to those of McKenzie, and can be summarized as follows.

(i) Descending cool currents (subduction zones) exert a major control on the convective system
(ii) Continents generate their own system of small-scale eddies which generate lateral forces and volcanic activity
(iii) Continental plates do not ride passively on the back of a horizontal mantle flow
(iv) There is no direct link between asthenosphere flow and plate movement (plates are not driven by 'mantle drag')

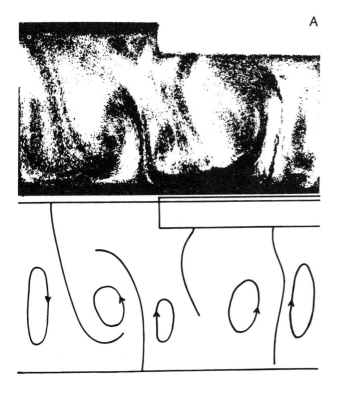

A

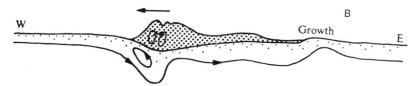

B

Figure 2.13 (*A*) Laboratory simulation (using hot oil cooled from above) of convection in the upper mantle near a stationary continental margin. The sketch shows the direction of fluid motion. From Elder (1976). (*B*) Sketch of convective flow pattern for an isolated migrating (continental) sheet with attached, trailing oceanic crust, based on a numerical simulation. From Elder (1976).

(v) Plate movement is not directly controlled by upwelling currents (i.e. plates are not primarily driven by 'ridge push').

2.5 Sources of stress in the lithosphere

Because of its comparative strength, the lithosphere can support substantial deviatoric stresses. Estimates of maximum stress differences produced by surface topography yield values of 200–300 MPa (= 2–3 kilobars; 1 MPa = 10 bars) in the upper crust (Birch, 1964; Jeffreys, 1970). However, such large stress differences cannot be maintained in the lower part of the lithosphere or in the asthenosphere because of viscous creep. Kusznir and Park (1984) estimate that average stress levels within the present continental plates probably lie within the range −25 to +25 MPa distributed over the whole thickness of the lithosphere.

Intraplate extensional tectonic effects in the form of graben, rifts and sedimentary basins are widespread in the major continental plates, but compressional intraplate effects appear in contrast to be relatively uncommon. The deformation of oceanic plates seems to be neither pervasive nor extensive, being generally confined to brittle deformation in the upper crust associated with insignificant strains, or to lithosphere flexure associated with sediment or seamount loading (see Bodine et al., 1981).

The magnitude of the stresses involved within lithosphere plates must clearly be large enough to promote frequent though localized extensional failure in continental lithosphere, but too small for compressional failure except under exceptionally favourable circumstances.

The sources of stress in the lithosphere have been investigated by Forsyth and Uyeda (1975), Turcotte and Oxburgh (1976), Richardson et al. (1976) and Bott and Kusznir (1984) (see Table 2.3). Stress systems affecting the lithosphere can be conveniently divided into renewable and non-renewable types (Bott, 1982). Renewable stresses are those that persist as a result of the continued presence or reapplication of the causative forces, even although the strain energy is progressively dissipated. The two main examples are stresses arising from plate boundary forces and from isostatically compensated surface loads. Non-renewable stresses are those that are dissipated by release of the strain energy initially present. Bending stresses, membrane stresses and thermal stresses are examples of this type. The value of stress at any point in the lithosphere results from the superimposition of several different stresses and is affected by local variations in mechanical properties.

It is convenient to treat the lithosphere as a single unit of varying ductility in which the applied forces are averaged over the whole lithosphere thickness. This simplification may only be applied to intraplate lithosphere. At plate boundaries, or where major internal deformation is taking place, it will no longer be applicable.

Plate boundary forces

As inertial forces are negligible, the driving and resistive forces acting on a moving plate must be in dynamic equilibrium. The stress distribution within plates depends critically upon whether the plates are driven by forces on their edges or by the underlying mantle drag. The various types of driving and resistive forces that may act on a plate have been summarized by Forsyth and Uyeda (1975) and Bott and Kusznir (1984). These are as follows (Table 2.3). (i) The slab-pull force (Figure 2.14A) acts on a subducting plate and results from the negative buoyancy of the cooler, denser lithosphere of the sinking slab. This force is potentially the largest of the plate boundary forces, but is partly counteracted by resistances produced by sinking, downbending and collision. Bott and Kusznir estimate a magnitude of 0–50 MPa for this force. (ii) The subduction suction force, originally recognized by Elsasser (1971) and named the 'trench suction force', is caused by the effect of subduction on the overlying plate, and is estimated to be around 20 MPa in magnitude. Both the slab-pull and subduction-suction forces will produce tensional stresses in adjacent litho-

Table 2.3 Summary of the principal mechanisms of stress generation in the lithosphere, with their more important properties and estimated magnitudes. From Bott and Kusznir (1984).

Mechanism	Renewable or non-renewable stresses	Compression or tension	Approximate level of stress difference	Stresses subject to amplification effects	Stresses significant in tectonics
Subduction slab pull (plate boundary force)	Renewable	Tension (normally)	0–50 MPa (?)*	Yes	Yes
Subduction trench suction (plate boundary force)	Renewable	Tension (normally)	0–30 MPa (?)*	Yes	Yes
Ridge push (plate boundary force)	Renewable	Compression	20–30 MPa	Yes	Yes
Mantle convection or asthenosphere drag	Renewable	Both	1–50 MPa	Yes	Probably not
Lithosphere loading (uncompensated)	Renewable	Both (mainly tension)	35 MPa **	No	Locally perhaps
Lithosphere loading (compensated)	Renewable	both (mainly tension)	50 MPa **	Yes	Locally yes
Lithosphere bending (due to loads)	Non-renewable	Both	Up to 500 MPa	No	No (?)
Lithosphere bending (due to subduction)	Non-renewable but continually generated	Both	Up to 1000 MPa	No	No (?)
Thermal effects (cooling and subducting lithosphere)	Non-renewable	Both	Up to 500 MPa	No	No (?)
Membrane effects	Non-renewable	Both	Up to 100 MPa	No	No (?)

*Highly variable in space and time because of variation in resistances, subduction velocity, length of slab, etc.
**For 2 km elevation.

sphere, provided that the resistance forces are sufficiently low. These resistances are highly dependent on the length and velocity of the subducting slab. (iii) The *ridge-push force* (Figure 2.14*B*) acts at ocean ridges, helping to force the plates apart and causing lateral compression within the adjacent ocean plates. This is a buoyancy force arising from the mass of hotter, less dense material making up the ridge, and is calculated to be 20–30 MPa in magnitude. (iv) The *mantle drag force* is the force acting on the base of a moving plate. Because of the relatively low viscosity of the asthenosphere, this force is considered to be small compared to the plate boundary forces. According to Schubert *et al.* (1978), the shear

stresses produced by a major convective cell of the same dimensions as the plate would produce a maximum stress of about 40 MPa in the lithosphere. However, the evidence discussed in 2.4 suggests that it is much more likely that smaller convective cells underlie the large lithosphere plates, and that these will exert shear forces in varying directions, whose effects will probably largely cancel out. It would appear that plates are driven more by edge forces than by mantle drag. (v) *Resistance forces* are produced at ocean ridges and at transform faults. These appear to be small compared with the resistance at convergent plate boundaries.

It is possible to calculate theoretical stress

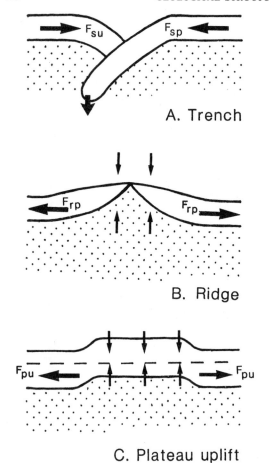

A. Trench

B. Ridge

C. Plateau uplift

Figure 2.14 Origin of the main plate driving forces acting on the lithosphere, derived from density imbalances and gravitational loading. *B* and *C* are examples of topographic loads isostatically compensated by a lower-density asthenospheric root. Thin arrows show gravitational and isostatic forces, and thick arrows show derived forces acting parallel to the plates. Lithosphere, blank; asthenosphere and mesosphere, dotted; F_{su}, subduction-suction force; F_{sp}, slab-pull force; F_{rp}, ridge-push force; F_{pu}, plateau uplift force (see Table 2.3).

distributions for certain simplified plate scenarios. Bott and Kusznir consider three (Figure 2.15). (*a*) A plate with ocean ridges on both sides should be in compression throughout. An example is the present African plate with ridges on three sides. (*b*) A plate with an ocean ridge on one side and a subduction zone on the other might be expected to show a gradation from compression near the ridge to tension near the subduction zone, although the effect

of resistance might offset the tensional force. An example is the Pacific plate, with a ridge on its southeast side and trenches on its western and northern sides; another is the South American plate with a ridge on its east side and a subduction zone on its west. (*c*) A plate with subduction zones on both sides might be expected to be in tension throughout. Although there is no present-day example of this, the early Mesozoic Pangaea is presumed to have formed a large continental plate of this kind, with subduction zones around the margins. Tensional stresses generated by these subduction zones, in conjunction with continental hot-spot activity, may have caused the break-up of Pangaea.

It follows that widespread extensional effects may be confined to circumstances like those of case (*c*), and would be replaced as soon as splitting and rifting commenced by gradational stress systems of type (*b*).

Local modifications of the stress system within plates may be caused by irregularities of plate boundary shape, by variations in physical properties, by lines of weakness, and by geometrical incompatibilities of shape as plates evolve, all of which may cause stress concentrations that tend to localize deformation if it occurs.

Loading stresses

Local stress fields are produced when the lithosphere is loaded by surface topography (e.g. mountain ranges) or by lateral density variations. Small topographic loads, less than about 50 km wide, produce insignificant stresses. Loads greater than 50 km that are not isostatically compensated will produce bending stresses (see below). However, significant stresses are produced where a large topographic load is isostatically compensated by lower-density material at depth, for example the thickened continental crust below a mountain range, or an ocean ridge, which is compensated by less dense asthenospheric mantle. The combined effects of the surface load and the upthrust of the low-density compensating material pro-

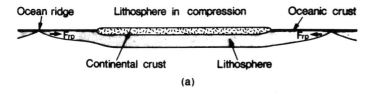

Ocean ridge   Lithosphere in compression   Oceanic crust

Continental crust   Lithosphere

(a)

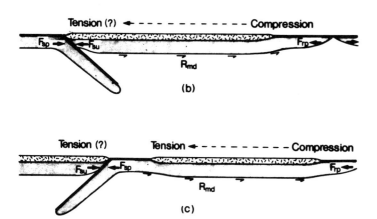

Tension (?) ← − − − − − − − − − − − Compression

F_{sp} F_{su} F_{rp}

R_{md}

(b)

Tension (?)   Tension ← − − − − − − − − − Compression

F_{su} F_{sp} F_{rp}

R_{md}

(c)

Continental lithosphere in tension

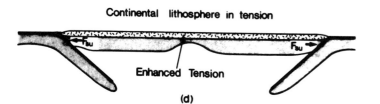

F_{su} F_{su}

Enhanced Tension

(d)

Figure 2.15 Examples of simple stress systems in lithospheric plates generated by plate boundary forces. (*a*) Ridge-push force developed at ridges on opposite sides of a continental plate, causing the whole plate to be in compression; an example is the present African plate. (*b*) Ridge-push force on one side of a plate and subduction suction force on the other, causing a stress system grading from compression at the ridge to tension, possibly, at the trench; an example is the present South American plate. (*c*) Ridge-push force on one side of a plate and slab-pull on the other, causing a gradational stress system as in (*b*); a possible example is Carboniferous basin formation in Britain. (*d*) Subduction suction on both sides of a continental plate, producing tension throughout; a possible example is Pangaea immediately prior to its break-up. F_{rp}, ridge-push; F_{sp}, slab-pull; F_{su}, subduction suction; R_{md}, mantle drag. From Bott and Kusznir (1984).

duce horizontal deviatoric tension in the region between the load and the compensatory mass (Figure 2.14*C*). Such a stress system will cause a piece of continental crust in regions of mountain ranges or plateau uplift to be in tension relative to adjoining regions, with stress differences of the order of 50 MPa. Crust in the adjoining regions would suffer corresponding compression. The same effect will occur at passive continental margins.

Non-renewable stress sources

Several different sources of stress exist that theoretically can produce large stress differences in the crust, giving initial strains of 1% or more. However, because they are non-

renewable, they will be substantially relieved by creep and brittle failure over geologically rather short periods of time. Such sources will never produce large stresses over the whole thickness of the lithosphere, since they will be partly relieved before they are able to reach their theoretical value. Examples of this kind of stress are bending stresses, membrane stresses and thermal stresses.

Bending stresses arise from flexure of the lithosphere resulting from uncompensated loads and downbending at subduction zones. Horizontal compression occurs on the concave side of the bend, and tension on the convex side. Because the lithosphere shows visco-elastic behaviour over geological time periods, the initial stresses calculated using an elastic model (cf. Walcott, 1970) of around 500 MPa for a load of about 500 km width, would rapidly decay to about half that value in the upper part of the lithosphere and totally disappear in the lower part, if the visco-elastic behaviour of the lithosphere is taken into account (Kusznir and Karner, 1985). This problem is considered at greater length in 2.7. Because the stress system is non-renewable, it is not comparable with the plate boundary forces in its long-term effects.

Membrane stresses are caused by changes in the radius of curvature of a plate as it migrates from the pole to the equator or vice versa (Turcotte, 1974a). However, in a very similar way to bending stresses, membrane stresses will never reach their theoretical maximum (calculated at around 100 MPa by Turcotte) because of continuous viscous relaxation in the long time period taken for the plate to move to its new position.

A third common source of non-renewable stress is caused by temperature changes within the lithosphere that give rise to *thermal stresses*. These are most important in the oceanic lithosphere as it cools after formation at a ridge, and again as it heats up after subduction. Tensile stresses parallel to the ridge axis of up to 400 MPa are calculated by Turcotte (1974b) and both compressive and tensile stresses of the order of 600 MPa in subducting lithosphere are suggested by House and Jacob (1982). Like the other non-renewable stresses, however, it appears unlikely that they play an important role in tectonic deformation, although they may have a significant local weakening effect on the lithosphere.

Other sources of stress include changes in volume associated with phase transitions in the mantle, erosion (surface unloading), which is locally important in joint formation, tidal deformation, which yields negligible stresses, and lateral variations in strength caused by major compositional boundaries (e.g. margins of magma chambers or of sedimentary basins). None of these is a possible source of major tectonic deformation.

Amplification of stress

The stresses generated by the major plate boundary and loading forces have been calculated as applying to the whole thickness of the lithosphere. However, because the lithosphere behaves as a visco-elastic body, stress will decay in the lower, warmer, and less viscous part, and become concentrated in the upper part where elastic/brittle behaviour is more important. In the case of renewable stresses, since the force is constantly applied, and retains the same value, the effective stress can be doubled if the effective thickness of the lithosphere is halved. Kusznir and Bott (1977) and Kusznir and Park (1984) have developed a model which investigates mathematically how a visco-elastic lithosphere with downward variation in viscosity responds to a constant applied force. They show that, depending on the temperature structure of the lithosphere and on the size of the applied force, there is an amplification of the stress in the upper part of the lithosphere of about ×2 for cool continental shield regions after 1 Ma, and about ×8 for very warm lithosphere, like that of the Basin-and-Range province of the USA, after only 1000 years. With larger initial stresses or a longer time of application, the lithosphere fails completely, when the strength of the strongest part is overcome, and large strains can develop.

This principle of *stress amplification* is critically important in explaining how the relatively small forces available from renewable plate boundary sources (giving stresses in the range 10–30 MPa when applied over the whole thickness of the continental lithosphere) are nevertheless able to overcome the known strength of rocks in the middle and upper crust, which is in the range 100–400 MPa.

Summary

The sources of stress in the lithosphere that are likely to be of tectonic significance are renewable buoyancy forces arising from density contrasts acting at plate boundaries. These buoyancy forces provide the principal dynamic operating mechanism of the plate system. They are opposed by various resistance forces due to subduction and collision, and along transform faults. Mantle drag forces may assist or oppose them. Buoyancy forces also arise from isostatically compensated loads on the lithosphere. Continents, plateau uplifts and mountain ranges all generate their own internal tensional stresses and produce corresponding compressional stresses in the adjoining lithosphere.

These forces combine to produce average stresses in the lithosphere that are probably in the range -25 to $+25\,\mathrm{MPa}$. However, because of the visco-elastic properties of the lithosphere, these rather small stresses become amplified over periods of 10^6-10^8 years to levels that are capable of locally overcoming the strength of the lithosphere and causing geologically significant strains. The extent to which this happens depends on the local strength of the lithosphere, which is in turn controlled by the geothermal gradient. This subject is explored in more detail in 2.7.

Many other sources of stress exist that are not considered to be of tectonic significance. Examples of these are non-renewable forces, which give theoretically large values, such as thermal, bending and membrane stresses. However, the stresses produced by these effects will be, in geological terms, rapidly dissipated by viscous creep and, being non-renewable, will not produce important long-term forces.

2.6 The determination of stress in the lithosphere

There are number of ways of obtaining estimates of the present state of stress in the lithosphere. Some methods provide measurements of both the magnitude and the orientation of the principal stresses; others measure orientation or magnitude only.

Direct measurement of the magnitude and orientation of the principal stress axes (σ_1, σ_2, σ_3) in the uppermost crust (*in-situ stress*) is carried out by several different methods. Some employ a *strain gauge*, which records small elastic strains produced within rock across a cavity. Measurements are made either at the surface or at depth, in mines or boreholes. *Overcoring* with a large-diameter drill bit is employed to relieve the stress around the strain gauge. *Flatjack* measurements are also widely used. The flatjack is a thin hydraulic-pressure cell in which the pressure is increased until it cancels the strain displacements created by cutting a hole or slot in the rock. Both overcoring and flatjack techniques employ the principle of *stress relief*. An alternative type of method is the *hydrofracture* technique in which hydraulic fracturing is induced artificially in the rock, in a borehole section.

Stress magnitude

Near-surface measurements of in-situ stress give widely varying magnitudes (Table 2.4). The use of the hydrofracture technique appears to have given more consistent results than the stress-relief methods. Available hydrofracture data suggest stresses of several hundred bars (tens of MPa) to depths of several km (Haimson, 1977; McGarr and Gay, 1978). McGarr (1980) shows, on the basis of determinations from North America, southern Africa and Australia, that, on average, maximum shear stress increases linearly with depth at a rate of $38\,\mathrm{MPa/km}$ in soft rocks and $6.6\,\mathrm{MPa/km}$ in hard rocks.

Brace and Kohlstedt (1980) point out that the limits of upper lithosphere stress can be determined on the assumptions (i) that rocks are fractured, and that friction on the fractures controls the stress at shallow depths, and (ii) that upper crustal strength is based on the strength of quartz. At a given depth, depending on the temperature gradient, strength becomes dependent on the creep properties of minerals, particularly quartz, olivine and feldspar (see 2.7). For dry rocks, the maximum strength in

Table 2.4 Some strain-relief in-situ stress measurements in North America. After Sbar and Sykes (1973). Data sources: 1, Hooker and Johnson (1969); 2, Hooker and Johnson (1967); 3, Sellars (1969); 4, Obert (1962); 5, Moruzi (1968); 6, Eisbacher and Bielenstein (1971); 7, Agarwal (unpubl.).

Location	σ_1 (bars)	σ_3 (bars)	Depth, if > 50 m	σ_1/σ_3	Trend of σ_1	Reference	Rock type
Barre, Vermont	118	54		2.2	N. 14° E.	1	Granite
Proctor, Vermont	90	35		2.6	N. 4° W.	1	Dolomite
Tewksbury, Massachusetts	81	45		1.8	N. 2° W.	1	Paragneiss
W. Chelmsford, Massachusetts	145	76		1.9	N. 56° E.	1	Granite
Nyack, New York	12	5		2.4	N. 2° E.	2	Diabase
St. Peters, Pennsylvania	56	23		2.4	N. 14° E.	1	Norite
Rapidan, Virginia	114	94		1.2	N. 6° E.	1	Diabase
Mt. Airy, North Carolina	168	81		2.1	N. 87° E.	1	Granite
Lithonia, Georgia	102	68		1.5	N. 8° E.	1, 7	Granite
Lithonia, Georgia	111	64		1.7	N. 49° E.	1, 7	Gneiss
Douglasville, Georgia	35	19		1.8	N. 64° W.	1, 7	Gneiss
Carthage, Missouri	217	95		2.3	N. 67° E.	1	Limestone
Graniteville, Missouri	73	53		1.4	N. 2° E.	1	Granite
Troy, Oklahoma	73	35		2.1	N. 84° W.	1	Granite
Marble Falls, Texas	151	101		1.5	N. 33° W.	1	Granite
Niagara Falls, New York	68	−0.7*			N. 55° E.	3	Dolomite
Barberton, Ohio	440	230	850	1.9	N. 90° W.	4	Limestone
Sudbury, Ontario	510	−440*			ENE.	5	
Elliot Lake, Ontario	210	180	300–400	1.2	E.	6	Sandstone
	370	200	300	1.8	NE.	6	Sandstone
	370	230	700	1.6	NE.	6	Sandstone
Morgantown, Pennsylvania	510	40	700	13.0	N. 27° E.	7	Diabase

the crust is given by a linear friction law, Byerlee's law (Byerlee, 1968) which gives:

$$\tau = 0.85\,\bar{\sigma}_n, \text{ for } 3 < \bar{\sigma}_n < 200\,\text{MPa},$$
and $\tau = 60 \pm 10 + 0.6\,\bar{\sigma}_n$ for $\bar{\sigma}_n > 200\,\text{MPa}$

where τ and $\bar{\sigma}_n$ are the shearing and normal stresses respectively at which friction is overcome on a fracture. Brace and Kohlstedt find, using mainly hydrofracture stress measurements, that Byerlee's law appears to predict horizontal stresses satisfactorily down to about 4 km. In dry rocks the maximum strength is considered to be 850 MPa in compression and 300 MPa in extension. The influence of pore-fluid pressure is important, and the values of stress obtained assuming that the pore-fluid pressure is hydrostatic, are reduced to 600 MPa for compression and 200 MPa for extension. These results apply down to a level of about 25 km, below which a temperature-dependent flow law will become increasingly important. The calculation of strength distribution with depth, and the influence of the thermal gradient on it, are dealt with in 2.7.

Another approach in estimating stress magnitudes in the upper lithosphere is through laboratory measurements of common rock strengths (see Table 2.5). These can give us minimum estimates of the strength of fractured upper crust. Many regions exhibit upper-crustal tensile fracturing but no evidence of compressional failure, in which case the avail-

Table 2.5 Uniaxial compressive strength, C_0, for a number of rocks. Multiply by 100 to convert to MPa.

Rock	C_0 (kbars)	Reference
Granite, Westerly	2.29	Brace (1964b)
Quartzite, Cheshire	4.60	Walsh (1965c)
Diabase, Frederick	4.87	
Marble, Tennessee	1.52	Cook (1965)
Granite, Charcoal	1.73	
Shale, Witwatersrand	1.72	Cook et al. (1966)
Granite Aplite (Chert)	5.87	
Quartzite, Witwatersrand	2.00	Wiebols et al. (1968)
Dolerite, Karroo	3.31	
Marble, Wombeyan	0.77	
Sandstone, Gosford	0.37	
Limestone, Solenhofen	2.24	

able stress levels are bracketed by the values of compressive and tensile strength of the appropriate materials.

These figures may be contrasted with the strength estimates calculated from various topographic loads (mountains, sea mounts, etc.) supported by both continental and oceanic crust, which yield strength estimates of around 100 MPa (Heard, 1976; McNutt, 1980). It is clear that the values of stress magnitude obtained in these various ways give us no indication of the real long-term strength of the lithosphere as a whole, as explained in 2.4.

Stress magnitude in the mantle

Estimates of stress magnitude at various depths in the lithospheric mantle and below have been made by studying sub-grain size and dislocation density in samples of mantle peridotite brought to the surface as xenoliths in lavas, kimberlites, etc. Recrystallization at high temperatures and stresses takes place by grain-boundary migration, and by sub-grain rotation at low temperatures, or low stresses, or both (Sellars, 1978; Guillope and Poirier, 1979). Since both processes are highly stress-dependent, analysis of olivine grain size may be used to estimate stress magnitude. Plots of grain size against depth of origin show a prominent grain-size discontinuity at a critical depth that can be interpreted as the level of change-over from the sub-grain rotation mechanism to grain-boundary migration. The system can be calibrated in terms of stress using the *piezometer* (stress meter) developed by Ross *et al.* (1980), consisting of an experimentally derived grain-size/stress relationship. The results of this method are applied to

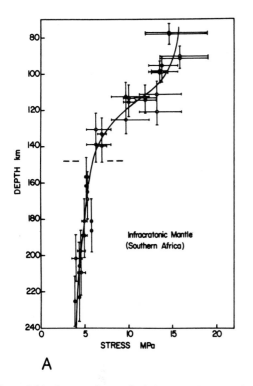

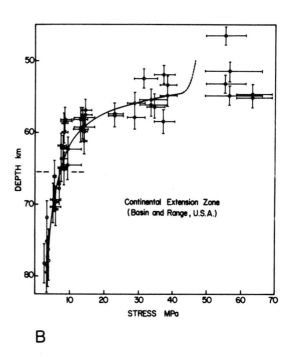

A   B

Figure 2.16 Stress estimates for infra-cratonic mantle (southern Africa) (*A*), and for a continental extension zone (Basin-and-Range Province, USA) (*B*), based on the olivine grain-size piezometer. Estimates for shallow samples (above the dashed line) are based on the SGR piezometer (Mercier, 1980) and those for the deep samples on the piezometer of Ross *et al.* (1980). From Mercier (1980).

various tectonic provinces by Mercier (1980) in the form of stress–depth curves (Figure 2.16). These may be compared with the theoretical stress–depth curves given by Kusznir and Park (see 2.7). There are significant differences between the stress–depth profiles for the relatively cool continental shield lithosphere of southern Africa (Figure 2.16A) and the warmer Basin-and-Range lithosphere (Figure 2.16B) — compare Figure 2.26.

The stress estimates for the infra-cratonic mantle below southern Africa vary from 15 MPa at 80 km depth to 5 MPa at 140 km, then decrease very slowly with depth to 4 MPa at 240 km, indicating a lithosphere base at around 140 km. The low stress in the asthenosphere is a useful confirmation of the minor role of the mantle drag force in plate dynamics. The stress estimates for the mantle below the Basin-and-Range province reveal a much thinner lithosphere with a base around 56 km, a comparable value of asthenosphere stress, but a much higher value of stress in the lithosphere — up to 70 MPa.

Summary: stress magnitude

Estimates for the uppermost crust vary widely but are of little significance for bulk lithosphere strength. The strongest part of the lithosphere for average geothermal gradients lies at mid-crustal levels (around 20–25 km) and depends on the shear strength of quartz. For rocks with hydrostatic fluid pressures, the maximum strength would be around 600 MPa in compression and 300 MPa in extension. Intraplate stress levels must commonly be bracketed by these values, but will rarely exceed the higher.

Below the zone of maximum strength, the strength depends on temperature-dependent flow laws for minerals such as quartz, feldspar and olivine, and stress levels decrease with depth to levels of 4–5 MPa in the sub-lithospheric mantle.

The long-term strength of the lithosphere is clearly much lower than these stress estimates suggest, as indicated by estimates of 100 MPa from the effects of topographic loading. A realistic model of lithosphere strength must take account of stress, depth, temperature gradient and time, and a model of this type developed by Kusznir and Park is discussed in 2.7.

Stress orientation

Measurements of stress orientation are more directly relevant to tectonic analysis than are magnitude estimates since they can be more easily related both to visible structures and to the kinematic pattern. The most widely used method for obtaining stress orientation data is by analysing *earthquake focal mechanisms*. An individual fault-plane solution gives a choice of two possible vectors, one of which lies in the fault plane, and the other perpendicular to it. Where the orientation of the fault plane is known, therefore, a unique determination of the slip vector is possible. Some areas show remarkable consistency of slip vectors over a large area, and the orientation of slip vectors around the various plate boundaries was one of the important criteria used by McKenzie and Parker (1967) and by Isacks *et al.* (1968) to demonstrate the rigid behaviour of plates (see 3.1).

The calculation of the orientation of the principal stress tensors requires more information than the slip vectors. If the shear plane responsible for the earthquake was initiated by the stress system being investigated, and not by some previous event, the stress orientation can be calculated approximately by assuming a value for the angle between the stress orientation and the shear plane. Over large areas, the method may give satisfactory and consistent results. For example, the compressive stress distribution in the region of the Japanese arc gives results which are generally consistent with a model where the horizontal component of principal stress acts parallel to the direction of relative plate motion (Figure 2.17). Where fault planes with varying orientation exist, however, the method gives a regionally varying stress orientation, which seems unlikely.

Gephart and Forsyth (1985) have discussed

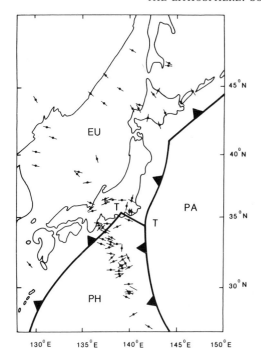

Figure 2.17 Directions of maximum horizontal compressional stress for intermediate and deep earthquakes in the Japanese are system. Toothed lines, trenches; *T–T*, transform fault; *PA*, Pacific plate; *PH*, Philippine plate; *EU*, Eurasian plate. After Uyeda (1978).

this problem in relation to the focal mechanism pattern of recent earthquakes in New England. They show that, if the earthquakes are assumed to result from movements on a set of variably oriented planes of weakness, a mean orientation for the principal stress axes can be found. They use a method developed by Angelier (1979) for determining principal stress directions from slickenside orientations on a set of fault surfaces of widely varying orientation. Gephart and Forsyth show that their data are consistent with a single regional stress orientation with ENE–WSW to NE–SW maximum compressive stress, and near-vertical minimum stress, which is consistent with estimates from other sources.

A method of determining in-situ stress orientations from borehole wall fractures (*borehole breakouts*) is described by Gough and Bell (1982). Breakouts were first identified in oil wells in western Canada by Cox (1970) as borehole intervals with elongate cross-sec-

tions, whose long axes were aligned regionally in a NW–SE direction. Gough and Bell explain the breakouts as zones of spalling along shear fractures, developed because of the increased stress difference caused by the borehole. The breakouts are arranged in pairs at 180° to each other and define an azimuth parallel to the smaller horizontal principal stress. They generally lengthen the diameter by 8–10%. The authors discuss results from three separate areas: the Rangely oilfield, Colorado, the east Texas basin, and the Norman Wells area in Northwest Territories, Canada. At the Rangely oilfield, breakout stress orientations show good agreement with direct measurements made by induced hydraulic fracturing, and with earthquake focal mechanism solutions. Dula (1981) compares the in-situ stress determinations in the Rangely basin with stress orientations determined from fabric elements (e.g. quartz deformation lamellae and microfracture orientation) of Laramide age and concludes that the present WNW–ESE orientation of σ_1 may have been constant since late Laramide times.

A large number of breakout data in the east Texas basin collated by Brown *et al.* (1980) show a consistent pattern (Figure 2.18) interpreted by Gough and Bell as the result of a NE–SW maximum horizontal stress. At Norman Wells, breakout data from two wells show good agreement with fracture orientation and indicate a NE–SW maximum horizontal compressive stress also. This direction seems to be uniform over a large part of the North American craton (see Figure 2.20).

Stress measurements using this technique have also been made in the oceanic crust at holes drilled in the Deep-Sea Drilling Program. Newmark *et al.* (1984) describe two examples in the East Pacific. The first is on the Nazca plate west of the Peru–Chile trench (Figure 2.19*A*) and indicates a maximum horizontal principal stress oriented NE–SW. The minimum horizontal stress is nearly parallel to the predicted tensional stress caused by the slab-pull force at the trench. The second example is from the Pacific plate to the west of the last example and 1800 km west of the East Pacific

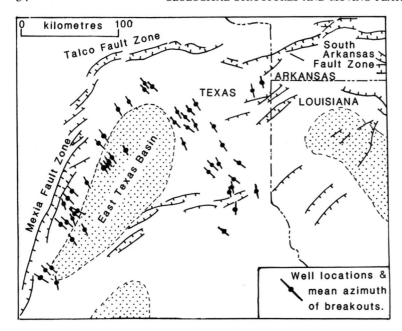

Figure 2.18 Mean orientations of borehole break-outs, in sandstones of the Schuler formation of E. Texas, in relation to the local extensional fault pattern. The break-out orientations indicate a uniform NE–SW maximum horizontal compressive stress. After Gough and Bell (1982).

ridge (Figure 2.19*B*). Here the maximum horizontal compressive stress indicated by the breakout data has an azimuth of 120° (NW–SE) which agrees well with the focal-plane solutions of local transform faults. These are oriented WNW–ESE with a strike-slip sense, parallel to the relative plate movement vector. This result is interpreted as confirmation that the stress system in an oceanic plate near the ridge is dominated by the compressive ridge-push force.

Regional pattern of stress orientation

The most complete regional data coverage is of North America. Zoback and Zoback (1980) present a generalized stress map (Figure 2.20) based on over 200 measurements. Several distinct stress provinces can be distinguished which can readily be correlated with the well-known tectonic provinces. The most important of these are as follows.

(i)   Mid-continent or stable continental interior: NE–SW compression.
(ii) Atlantic coastal province (Appalachian

orogenic belt): NW–SE to WNW–ESE compression.
(iii) Gulf coast province: extension perpendicular to continental margin.
(iv) Cordilleran orogenic province: this is a complex zone containing several subprovinces, but the most consistent stress pattern is the E–W extension seen in the active tectonic areas of the Basin-and-Range, Rio Grande rift and northern Rocky Mountains.

The transition between the stress provinces in the west is relatively sharp (less than about 75 km typically) but more gradual in the east, particularly between the mid-continental and Atlantic coastal provinces. Within each of these provinces, the stress orientation is uniform to within about ±15°. The correspondence between the data derived from the relatively shallow borehole measurements and those from earthquake focal mechanism solutions (from depths in the range 5–15 km) suggests that this uniform stress field is representative of the whole of the strong upper crust.

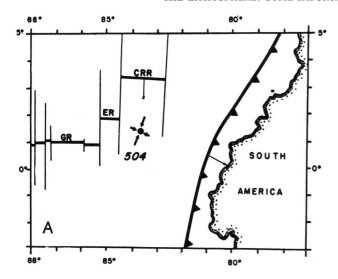

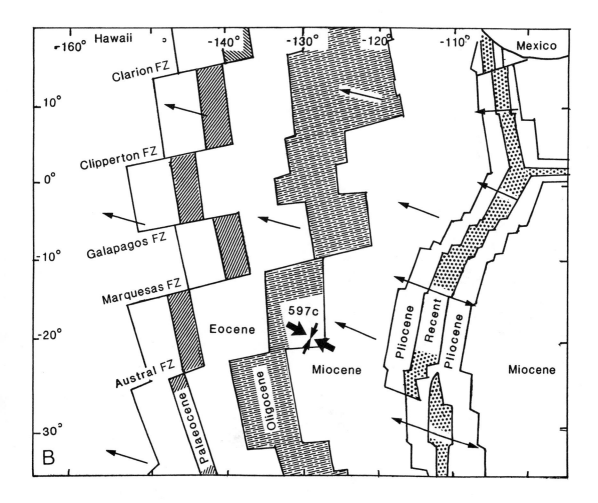

Figure 2.19 (*A*) Stress orientation determined from borehole breakouts near site 504, between the Cocos ridge and the Peru–Chile trench. The longer arrows represent the maximum, and the shorter the minimum, principal horizontal stress. From Newmark *et al.* (1984). (*B*) Stress orientation determined from breakout directions at hole 597c, west of the East Pacific ridge. Heavy arrows indicate maximum, and light short arrows minimum, principal horizontal stress directions. Light long arrows indicate the direction of relative Pacific – Nazca plate motion. After Newmark *et al.* (1984).

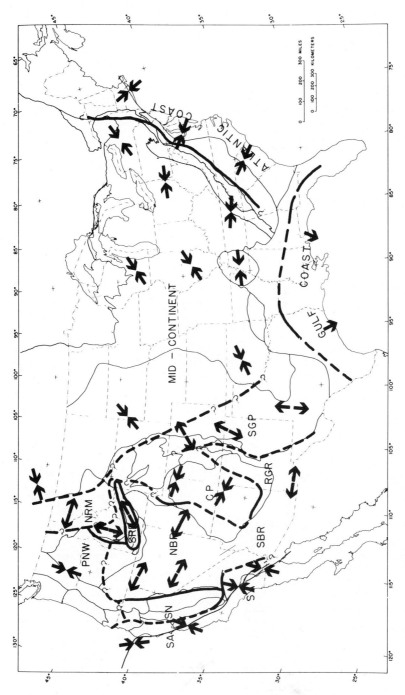

Figure 2.20 Generalized stress map of the USA. The least principal stress is always horizontal, but may be either extensional (outward pointing) or compressional (inward-pointing). Stress provinces, bounded by heavy lines, are as follows: PNW, Pacific NW; NRM, N. Rocky Mountains; SA, San Andreas fault zone; SN, Sierra Nevada; NBR, N. Basin-and-Range; SBR, S. Basin-and-Range; CP, Colorado Plateau; RGR, Rio Grande Rift; SGP, S. Great Plains. From Zoback and Zoback (1980).

The stress-field changes at some boundaries merely involve a swapping of stress axes. For example, in the change from mid-continental through Cordilleran to San Andreas provinces, σ_3 changes only slightly in orientation although it varies from compressional to extensional.

This complex stress field can be broken down into three simple elements: (i) a western or Pacific superprovince related to Pacific plate-boundary processes; (ii) a central stable intraplate field controlled by both Pacific and Atlantic plate boundary forces; and (iii) the marginal fields of the Atlantic and Gulf coast provinces, which probably represent local modifications of the intraplate field. The extensional Gulf Coast field is attributed to the effect of the passive continental margin (see 2.5). The compressional Atlantic province field may possibly reflect an original NW–SE ridge-push force during the earlier phases of Atlantic opening.

It is convenient to consider intraplate stress fields and plate boundary stress fields separately.

Intraplate stress

The mid-continent province discussed above, and its extension into Canada, may be regarded as a typical continental intraplate region. This was shown originally by Sbar and Sykes (1973) who demonstrated the consistent NE–SW to E–W maximum stress orientation pattern using data derived from overcoring, hydrofracture and focal-mechanism methods. The stress magnitudes obtained from the in-situ measurements show a wide range, from 10 to 600 bars (1–60 MPa), but the region overall is one of high horizontal compressive stress.

A similar remarkable regularity of σ_1 orientation is demonstrated by Froidevaux *et al.* (1980) in a study of eight sites in France using the flatjack method. The NW–SE σ_1 orientation obtained by them compares closely with the results for the Rhine–Ruhr rift system (see Figure 4.16B).

Richardson *et al.* (1976) and Solomon *et al.* (1980) present a compilation of world-wide intraplate stress orientation data obtained primarily from earthquake focal-mechanism analysis (Figure 2.21A). The most reliable results come from Europe and North America, as we have seen. In these areas the focal-plane solutions can be checked by in-situ measurements. The South American continent shows a similar orientation to North America. Note also the N–S to NNE–SSW compression in India and the E–W compression in southern Africa. Most oceanic lithosphere older than about 20 Ma also appears to be in horizontal compression.

The authors compare this stress distribution with a number of theoretical models making different assumptions about the various driving forces available (see 2.5, Table 2.3). The most successful fit is obtained using a model (Figure 2.21B) which employs equal ridge-push and slab-pull driving forces, ignores the subduction-suction force, and has a very low value for the drag force resisting plate motion. Although the measurements of intraplate stress are not nearly good enough for a convincing analysis, the evidence so far suggests that intraplate stress can be explained to a first approximation by the simple interaction of these two main driving forces, slab-pull and ridge-push.

Stress at plate boundaries

Oliver *et al.* (1973) investigated the stress orientation in sinking slabs for fourteen different *subduction zones*, using earthquake focal mechanism data. They found that in the great majority of cases there was a major component of either down-dip compression or down-dip extension, parallel to the inclination of the slab (Figure 2.23). Moreover the extensional solutions were confined to high-level earthquakes in slabs where either there were no deep earthquakes or there was a gap between the upper and lower earthquake zones. Compressional solutions were associated either with the lower parts of the slab, or with the whole slab, where seismicity was continuous. This distribution was explained by

MIDPLATE EARTHQUAKES

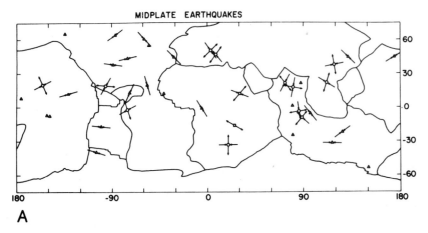

A

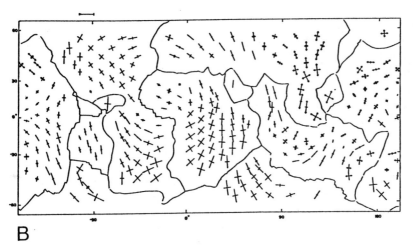

B

Figure 2.21 (*A*) World distribution of principal horizontal deviatoric stresses inferred from mid-plate earthquake mechanisms. Triangles represent thrust fault mechanisms, squares normal fault, and circles strike-slip. Arrows denote the horizontal projection of the inferred stress axes. From Richardson *et al.* (1976), with permission. (*B*) Principal horizontal deviatoric stresses in the lithosphere for a model of plate driving forces (see text). Axes without arrows indicate compression and those with arrows, tension. Relative magnitude indicated by length of axis. From Solomon *et al.* (1980), with permission.

Oliver *et al.* on the basis of the relative strengths of the slab-pull force, which exerts a down-dip tension on the slab, and an opposing resistance force. The resistance produces a compressional stress which becomes larger as the slab sinks deeper into the mesospheric mantle with its increased strength (Figure 2.23*B*). This work was very influential in persuading geologists that plate motion could not be simply explained either by slab-pull or by ridge-push, but was controlled by the interaction of several primary forces (see 2.5).

The orientation of the maximum horizontal stress across subduction zones generally shows a consistent parallelism with the convergence direction. This regular pattern is seen for example in both the Aleutian arc and the Japanese arc (Figure 2.17) (Nakamura and Uyeda, 1980). However the compressive stress field is often confined to the volcanic arc itself, and is replaced by an extensional stress field in the back-arc region. Nakamura and Uyeda suggest that where there is no active opening or spreading of the back-arc region (e.g. in the Peru–Chile, Japan and Kurile zones) the compressive stress associated with the subduction zone may be transmitted directly to the interior of the overriding plate. Where back-arc extension is occurring, however, either the σ_1 stress weakens and swaps over to become σ_3 without

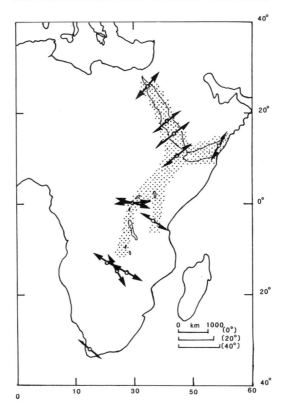

Figure 2.22 Extensional and strike-slip focal mechanism solutions for earthquakes in the African rift system. After Fairhead and Girdler (1971).

any change in direction, or there may be a gradual swing in direction of σ_1, as seen in the Aleutian arc.

The orientation of the maximum horizontal stress at collisional boundaries is generally perpendicular to the boundary or in the direction of relative plate motion. The focal-plane solutions to shallow earthquakes both in the northern part of the Indian plate and in Central Asia north of the India–Asia collision suture show a consistent NNE–SSW orientation (Figure 2.21A).

The stress orientation at *constructive boundaries* (i.e. ocean ridges and major continental rift zones) is well established from earthquake focal-mechanism studies. These show predominantly normal fault movement with a sub-horizontal extensional stress parallel to the direction of relative plate motion, as demon-

strated by Sykes (1967) from more than 50 focal-plane solutions for the world rift system. Fairhead and Girdler (1971) show both extensional and strike-slip solutions for the African rift system (Figure 2.22). The extensional stresses are approximately parallel to the relative plate movement vectors: NE–SW in the Red Sea and NW–SE in the African rifts. The lower part of the Red Sea shows predominantly strike-slip solutions.

Conservative boundaries. Transform faults are associated with earthquakes with a strike-slip sense of movement, opposite to the direction of apparent displacement (Tuzo Wilson, 1963). Sykes (1967) shows a good example from the Gulf of California (Figure 6.10). The orientation of the principal stress axes cannot be simply derived from the slip vectors, how-

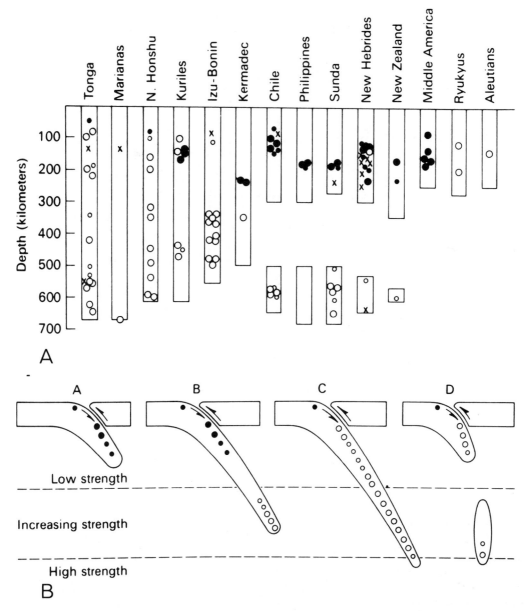

Figure 2.23 (*A*) Down-dip stress plotted as a function of depth for fourteen subduction zones. Filled circles, down-dip extension; open circles, compression. Crosses represent other orientations. Small symbols represent less accurate determinations. The enclosed rectangular areas represent the approximate distribution of earthquakes in the slabs. (*B*) Diagram showing possible explanation of the distribution of stress types in (*A*). Slabs which just penetrate the asthenosphere may be characterized only by down-dip extension (1); as the slabs penetrate further into the higher-strength mesosphere, they are subjected to compression in their lower part (2) and eventually throughout (3). Break-up of the slab is envisaged in (4). From Uyeda (1978), after Isacks and Molnar (1969).

ever, since a unique solution cannot be provided from a set of parallel faults. Examination of the stress orientation on intraplate sea floor (see e.g. Figure 2.19*A*, *B*) suggests that σ_1 is

generally oblique to active transform fracture zones in accordance with classical fault theory. A similar arrangement can be demonstrated in continental strike-slip zones which form plate boundaries. For instance, in the San Andreas fault zone, the arrangement of faults and of individual slip vectors is very complex in detail, but the overall stress field appears to be more uniform, with a NE–SW σ_1 orientation, oblique to the trend of the San Andreas fault, as shown in Figure 2.20 (Zoback and Zoback, 1980).

Summary

The rather sparse data available for intraplate stress fields show considerable regularity. Stress orientations can be explained to a first approximation by the combined effects of neighbouring plate boundary forces. The important forces appear to be a symmetrical ridge-push force and a symmetrical 'trench-pull' force combining the effects of slab pull, subduction suction, and the various opposing resistances.

Stress data for the plate boundaries themselves are abundant and show a consistent pattern of extensional stress parallel to divergent motion, compressional stress parallel to convergent motion, and oblique stress fields across transform faults. Extensional stress fields at divergent boundaries are restricted to the region of the rift zone, and are rapidly replaced by compressive intraplate stresses on either side. Compressive stress fields at convergent boundaries are replaced by extensional stress fields in many back-arc regions on the upper plate, but normally continue into the typical compressive intraplate stress field on the oceanic subducting plate.

Stress distributions in subducting slabs confirm the picture obtained by world-wide stress modelling, of a combined slab-pull/ridge-push driving mechanism for plate motion, with extensional stresses confined to the upper parts of slabs only in the early stages of subduction. The effect of the mantle drag force appears to be minimal.

2.7 The long-term strength of the lithosphere

The strength of the lithosphere controls both the initiation and subsequent evolution of major zones of deformation. The response of a piece of lithosphere to an applied tectonic force is dependent on the vertical distribution of both ductile and brittle strength, which in turn is controlled by the varying rheology with depth shown by lithosphere material. Whereas brittle strength is controlled primarily by lithostatic pressure and increases with depth, ductile strength is controlled by temperature and decreases with depth because of the geothermal gradient.

In tectonically stable lithosphere subjected to an applied force, an upper region of brittle deformation and a lower region of ductile deformation will be separated by a strong competent elastic region (Figure 2.24A). If we can assume that the various layers of the lithosphere are welded together, the strength of this elastic region controls the bulk strength of the whole lithosphere and only very small strains can occur initially. If the applied force is increased, or merely with the passage of time under a constant applied force, the region of brittle deformation will extend downwards, and the region of ductile deformation will extend upwards, eventually reducing the competent elastic core to zero. When this happens, a rapid increase in strain across the whole thickness of the lithosphere can take place, producing geologically significant levels of deformation (Figure 2.24B). This process has been termed *whole-lithosphere failure* (Kusznir, 1982). This process will be accelerated if the geothermal gradient becomes steeper, because of the temperature control over ductile strength.

The vertical distribution of stress which results from an applied force is controlled by the variation of both brittle and ductile strength with depth. This variation in strength is therefore critically important in defining the value of the force which must be applied to the lithosphere in order to produce significant deformation. Evidence for the variation of strength with depth is provided by the stress-depth curves of Mercier (1980) discussed earlier (see Figure 2.16).

Kusznir and Park (1982, 1984a,b) use a mathematical model to calculate this stress distribution assuming Maxwell visco-elastic properties for the lithosphere. Details of the model are given in Figure 2.24. Important assumptions are that the total horizontal force arising from the initial applied force is conserved, and that the lithosphere undergoes a uniform strain with depth.

Because of the siliceous nature of most upper-crustal rocks, and the relatively low strength of quartz, it is assumed that deformation in the upper crust is controlled by the behaviour of this mineral (see White, 1976). The rheology of the lower crust is uncertain, but it appears likely that, in view of the probable importance of basic material, defor-

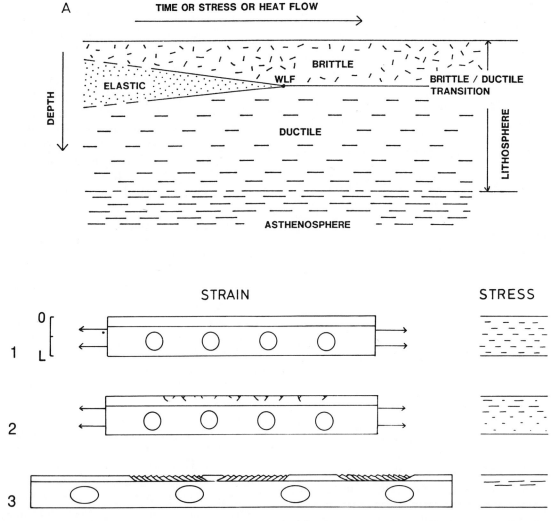

Figure 2.24 (*A*) The regions of brittle, ductile and elastic behaviour in the lithosphere shown diagrammatically. Given a large enough applied stress, the elastic core will reduce to zero with time as the brittle and ductile deformation spreads downwards and upwards respectively. WLF (whole-lithosphere failure) occurs when the elastic region disappears. Increasing heat flow has the same effect as increasing stress. (*B*) Schematic representation of lithosphere response to an applied horizontal stress. (1) Initial elastic response causes uniform distribution of strain and stress with depth. (2) Ductile creep in the lower lithosphere causes stress decay there, and results in stress amplification in the upper lithosphere, sufficient to cause fracture in the uppermost parts. (3) Further stress amplification results in stress levels in the strong upper part of the lithosphere sufficient to cause complete failure and consequential large strains. From Kusznir and Park (1984).

Mathematical model for lithosphere deformation

Lithosphere of initial thickness L is subjected to an initial applied horizontal stress, σ_0, in the x direction. Plane strain ($\sigma_y = 0$) is assumed in the perpendicular horizontal direction and the resulting vertical stress σ_z is assumed to be zero. Conservation of horizontal force and the assumptions that the various layers of the lithosphere are welded to one another provide the additional equations:

$$\int_0^L \dot{\sigma}_x dz = 0 \text{ and } \frac{d\varepsilon_x}{dz} = 0$$

where σ_x is horizontal stress, ε_x is total horizontal strain and z is depth.

These equations, together with the constitutive equations for a viscoelastic material, allow the following integral equation to be formulated, giving σ_x as a function of time and depth:

$$\sigma_x = \int_0^t \left(\frac{1}{L} \int_0^L k\dot{\varepsilon}_v dz - k\dot{\varepsilon}_v \right) dt' - \frac{1}{L} \int_0^L \sigma_x^0 \, dz + \sigma_x^0$$

where

$$k = \frac{E}{1 - v^2} \text{ and } \dot{\varepsilon}_v = \frac{\sigma_x(2 - v) - \sigma_y(1 - 2v)}{6\eta}$$

where E is Young's modulus, v Poisson's ratio and η apparent viscosity.

A similar equation exists for σ_y. Fracture has been predicted using Griffith failure theory and release can be incorporated in the model using the initial stress terms, σ^0. The detailed formulation of these equations and the stress- and temperature-dependent rheology are described by Kusznir (1982). The following parameter values have been used: $E = 10^5$ MPa; $v = 0.25$; tensile strength = 20 MPa and μ (coefficient of friction) = 0.5.

The following relationships are used to derive the creep strain rate. The equations for the creep rates for olivine are taken from Bodine *et al.* (1981) and represent a compromise between the dry olivine creep rates of Goetze (1978) and the wet dunite creep rates of Post (1977).

Dislocation creep:

$$\dot{\varepsilon} = 7 \times 10^{10} \exp\left(\frac{-53030}{T}\right)(\sigma_1 - \sigma_3)^3 \text{ s}^{-1}$$

for $(\sigma_1 - \sigma_3) < 2$ kbar

Dorn Law:

$$\dot{\varepsilon} = 5.7 \times 10^{11} \exp\left(\frac{-55556}{T}\left(1 - \frac{(\sigma_1 - \sigma_3)}{85}\right)^2\right) \text{ s}^{-1}$$

for $(\sigma_1 - \sigma_3) > 2$ kbar
where $(\sigma_1 - \sigma_3)$ is in kbar (1 kbar = 100 MPa).

The ductile deformation of quartz within the crust is assumed to correspond to that of dislocation creep. Creep rates in quartz are controlled strongly by the amount of water. The continental upper crust is assumed to deform according to a wet quartz rheology with 50% quartz. The lower crust is assumed to deform according to a dry quartz rheology with 10% quartz, or a plagioclase rheology with 40% or 50% plagioclase. Creep rates for wet and dry quartz are based on the experimental work of Koch *et al.* (1980).

Wet quartz:

$$\dot{\varepsilon} = 4.36 \exp\left(\frac{-19332}{T}\right)(\sigma_1 - \sigma_3)^{2.44} \text{ s}^{-1}$$

Dry quartz:

$$\dot{\varepsilon} = 0.126 \exp\left(\frac{-18245}{T}\right)(\sigma_1 - \sigma_3)^{2.86} \text{ s}^{-1}$$

Plagioclase:

$$\dot{\varepsilon} = 8.2 \times 10^2 \exp\left(\frac{-28788}{T}\right)(\sigma_1 - \sigma_3)^{3.2} \text{ s}^{-1}$$

where $(\sigma_1 - \sigma_3)$ is in kbar.

mation is controlled by the ductile behaviour of plagioclase, which deforms more readily than pyroxene or olivine. Flow curves plotting strain-rate against temperature for various appropriate minerals and rocks are compared on Figure 2.25. The role of water is critical; wet quartz is much weaker than dry, and whereas wet quartz may be assumed to control upper-crustal rheology, we might expect a mid-crustal region of essentially granitoid composition to be controlled by dry quartz deformation. Kusznir and Park investigate two crustal strength models, one with a 50% wet quartz rheology overlying a 50% plagioclase rheology, and the other a three-layer model comprising a 50% wet quartz layer overlying a 50% dry quartz layer overlying a 40% plagioclase layer. The rheology of the upper mantle is assumed to be controlled by the behaviour of dry olivine.

The critical temperatures required to generate significant strain rates for these different rheologies are reached at depths that are dependent on the geothermal gradient. Different geothermal gradients may be characterized by the surface heat flow, q. Figure 2.26A shows calculated stress–depth (or strength–depth) profiles at various times after the initial application of a tensile force of 10^{12} N/m (i.e. equivalent to a force of 1.5×10^{17} N applied over the whole thickness of the lithosphere). This force is applied to continental lithosphere with a surface heat flow $q = 60\,\text{mW}\,\text{m}^{-2}$ (corresponding to the continental average — see Table 2.1). As time progresses, ductile creep in the lower lithosphere results in dissipation of stress there and its transfer upwards to the cooler non-ductile upper lithosphere, where it becomes amplified to a level sufficient to generate brittle failure in the topmost levels of

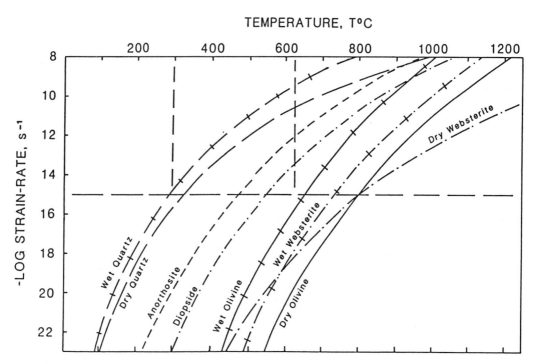

Figure 2.25 Variation in log strain rate with temperature for a number of minerals and rocks important in ductile lithosphere deformation. The curves are derived from experimental data from the following sources: quartz (Koch *et al.*, 1980); anorthosite (Shelton and Tullis, 1981); diopside and websterite (Avé Lallament, 1978). From Kusznir and Park (1987)

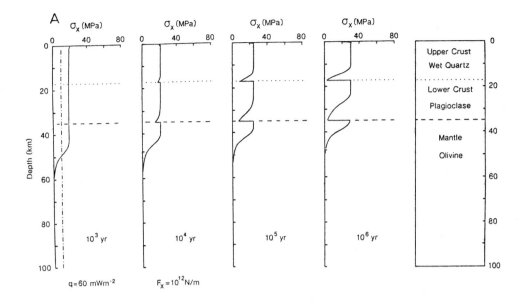

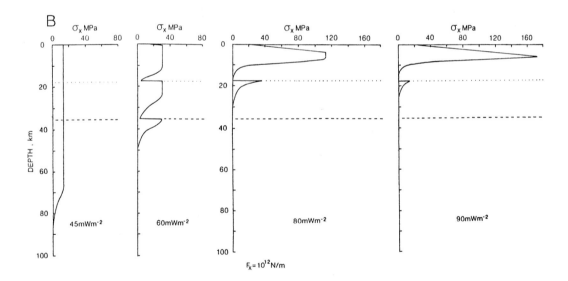

Figure 2.26 (*A*) Stress plotted against depth at various times after the application of a tensile tectonic force of 10^{12} N/m to continental lithosphere with a surface heat flow of 60 mW m^{-2}. Note the development of low-stress (low-strength) regions above compositional (and therefore rheological) boundaries. (*B*) Stress plotted against depth for a range of geothermal gradients, corresponding to surface heat flows of 45 to 90 mW m^{-2}, at 1 Ma after the application of the same tensile force as in (*A*) Note that in this, and in subsequent figures, applied force is given in newtons per unit length of a section of lithosphere plate (N m^{-1}) rather than as a stress applied to unit area of lithosphere. In lithosphere 100 km thick, a stress of 10 MPa = 1 N m^{-1}. Data from numerical model; from Kusznir and Park (1987).

the lithosphere. Two major stress–depth discontinuities are apparent at the changes in rheology between upper and lower crust, and between crust and mantle. These are important weak zones which would assume critical tectonic importance if large strains were to occur.

The effect of different temperature gradients on the stress–depth relationship is illustrated in Figure 2.26B, for the same initial tensile force applied for a period of 1 Ma. Very little upper-lithosphere stress increase is evident in the coolest lithosphere model (corresponding to average continental shield with $q = 45\,\text{mW}$ m^{-2}). However, as the geothermal gradient steepens, the region of ductile deformation extends progressively upwards, concentrating the stress in the upper lithosphere. For heat-flow levels of 80 and $90\,\text{mW}\,\text{m}^{-2}$, corresponding to the hottest regions of the continental lithosphere (e.g. the Basin-and-Range province of the western USA) the stress is entirely confined to the crust, and the strength discontinuity at the mid-crustal rheology change is still evident. At $q = 90\,\text{mW}\,\text{m}^{-2}$, whole-lithosphere failure is about to take place as the

elastic core has been reduced to zero (see Figure 2.24A).

These results may be conveniently summarized by employing the concept of critical stress, defined as that level of stress required to produce whole-lithosphere failure within 1 Ma, and which would therefore be expected to produce geologically significant deformation. Figure 2.27 shows critical stress plotted against heat flow for extensional and compressive stress compared with the theoretical stress levels calculated for various stress sources (see Table 2.3). Figure 2.27A shows the critical stress/heat flow curve for extension, together with theoretical maximum stress levels produced by plateau uplift (50 MPa) and subduction suction (30 MPa), the two most important sources of lithosphere extensional stress (see 2.5). A theoretical maximum net extensional stress of 80 MPa could be produced by combining the two effects. However, as discussed earlier, a more realistic maximum net stress is considered to be about ± 25 MPa. Such a stress intersects the critical stress curve at a heat flow value of about $70\,\text{mW}\,\text{m}^{-2}$, defining a region of potential deformation.

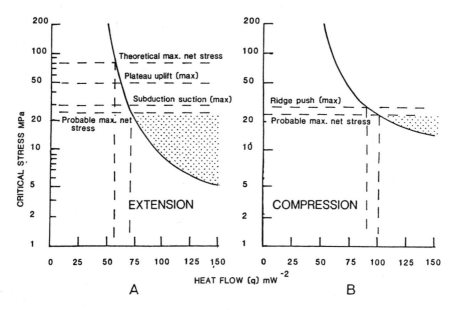

Figure 2.27 Curves of critical stress v. lithosphere heat flow for tensional (A) and compressional (B) applied stress, compared with various possible stress sources and with probable maximum net stress levels. The region of potential deformation is stippled. Numerical model; after Kusznir and Park (1987).

These results demonstrate that, for reasonable estimates of the likely available extensional force, significant extensional deformation may be expected in lithosphere with heat-flow levels found for example in typical regions of Palaeozoic orogenic crust (see Table 2.2). This conforms with the observed occurrence of intraplate extensional zones in such regions.

Compressional strength

Because rocks are much stronger under compression than under tension, the compressive strength of the lithosphere as a whole is considerably greater than its extensional strength. Figure 2.27B shows that for a heat flow q of $80\,mW\,m^{-2}$, sufficient to promote significant extension with a stress of around $20\,MPa$, a stress of over $40\,MPa$ would be required to produce significant compressional deformation — a value that is outside the range of normal compressive stress sources. The probable maximum available net stress of $25\,MPa$ requires a heat flow of over $90\,mW\,m^{-2}$ to produce significant deformation. Such a high heat flow is currently found only in the hottest regions, which are currently undergoing extension, rather than compression. The model therefore predicts that major compressional deformation should not normally occur within plates and we might expect it to be restricted to convergent plate boundaries where large collision resistance forces can occur. That intraplate compressional deformation is uncommon is widely acknowledged. The few cases where it seems to occur need to be examined closely. Perhaps the lithosphere is unusually weak in such zones.

Oceanic lithosphere

Although oceanic lithosphere is much thinner than continental, its composition renders it significantly stronger, since its rheology is controlled almost entirely by the deformation of olivine. The critical stress curve for typical ocean-basin lithosphere therefore corresponds to that for an all-olivine lithosphere. With typical ocean-basin heat flow values in the range $40-50\,mW\,m^{-2}$ (similar to the continental shield values), it can be seen from Figure 2.27A that an unrealistically high value of critical stress would be required to produce deformation. No likely source of stress exists, therefore, that will produce significant compressional intraplate deformation — indeed there is no evidence of such deformation in the ocean basins at present.

Evolution of strength during lithosphere extension and compression: crustal thickness

Because of the significant differences in rheology between the quartzo-feldspathic material of the crust and the olivine-rich material of the mantle, the bulk strength of the lithosphere depends very much on the relative proportions of these two different materials, and thus on crustal thickness. Figure 2.28A shows stress–depth profiles, calculated from the Kusznir–Park lithosphere deformation model, for lithosphere with crustal thicknesses of 20, 35 and 60 km. These show very significant differences. The thin-crust model is stronger overall and shows no strength discontinuities. The thick-crust model shows more than twice the stress level, but the stress is concentrated in the upper crust. These results are for an average geothermal gradient and a force of $10\,N/m$ after 1 Ma. For higher heat flows, the thin-crust model is 3–4 times stronger than average crust.

The above relationship assumes that the geothermal gradient is constant during the application of the force. If, however, the crustal thickness changes as a result of deformation, for example by extensional thinning, the process itself changes the thermal structure of the lithosphere. This situation has been investigated in a modified version of the lithosphere model described above (Kusznir and Park, 1987) which considers the evolution of lithosphere strength during extension, taking account of the changing temperature structure.

The process of lithosphere extension results

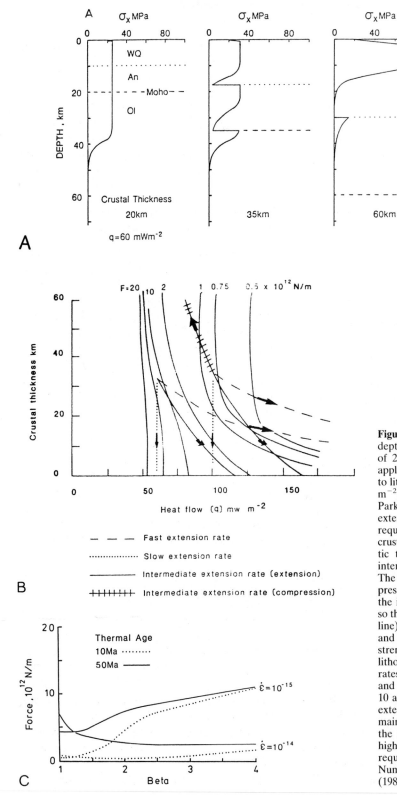

Figure 2.28 (*A*) Stress plotted against depth for models with crustal thicknesses of 20, 35 and 60 km at 1 Ma after the application of a tensile force of 10^{12} N/m to lithosphere with heat flow, $q = 60$ mW m^{-2}. Numerical model; from Kusznir and Park (1987). (*B*) Contours of lithosphere extensional strength (the critical force required to generate WLF in 1 Ma) in a crustal thickness/heat flow plot. Schematic trajectories indicate fast, slow and intermediate lithosphere extension rates. The evolution of crustal strength in compression may be visualized by extending the intermediate extension rate upwards so that the crust is thickened (see hatched line). Numerical model, after Kusznir and Park (1987). (*C*) Plot of lithosphere strength (shown as applied force) against lithosphere extension factor, β, for strain rates of 10^{-14} and 10^{-15}, and for warmer and cooler lithosphere (thermal ages of 10 and 50 Ma respectively). Note that as extension proceeds, the strength depends mainly on the strain rate rather than on the initial thermal structure, and that a high strain rate in cool lithosphere would require an unrealistically large force. Numerical model; from Kusznir and Park (1987).

48

in two opposing effects: (i) a steepening of the geotherm brought about by bringing the hotter asthenosphere nearer to the surface, which weakens the lithosphere; and (ii) a thinning of the crust which, as we have seen, will act to strengthen the lithosphere. Thus the lithosphere may show either a net weakening or a net strengthening during extension depending on which effect dominates.

If the extension takes place slowly, the geotherm may have time to re-equilibrate, that is, the base of the lithosphere will move downwards to compensate for the thinning effect as the extra heat is lost. Slow rates of extension will therefore result in a net strengthening ('strain hardening') of the lithosphere because the crustal thinning effect is dominant. Rapid extension on the other hand will lead to net weakening, since the temperature rise will more than balance the effect of crustal thinning. Figure 2.28B shows contours of lithosphere strength in a crustal thickness/heat flow plot. Trajectories of changing crustal thickness at constant heat flow clearly lead to an increase in strength. Trajectories which show a large change in heat flow (corresponding to a fast extension rate) produce a decrease in strength. An intermediate rate of extension would cause no or very little net change.

Figure 2.28C shows the quantitative results of the lithosphere strength model modified to take account of changing temperature structure. The results are shown in the form of a plot of lithosphere strength (critical force) against beta value (β) for extensional strain rates of 10^{-14} and $10^{-15}\,s^{-1}$, and 'thermal ages' of 10 and 50 Ma. Note that β, the lithosphere stretching factor (McKenzie, 1978a) is equivalent to the *strength* = $(1 + e)$, where e is the *extension*, in the terminology used in structural geology. Thus a value of $\beta = 2$ corresponds to a doubling of the original width and a halving of the original thickness of the lithosphere segment in question. The thermal age is defined as the time since the last major tectonothermal event (orogeny).

The evolution of extensional strength is strongly dependent on the initial thermal state. The faster strain rate in the warmer lithosphere produces approximately constant strength, whereas the slower rate causes rapid strain hardening after $\beta = 1.5$. Fast strain rates cannot be initiated in the cooler lithosphere model (thermal age of 50 Ma) because of the unrealistically high initial strength required.

The model therefore predicts, firstly, that fast extension rates ($\geqslant 10^{-14}\,s^{-1}$) are only possible for hot, thermally young lithosphere that will produce locally intense extensional deformation, with strain softening, leading to large β values and ultimately, if the force persists, to the complete rifting of the continental crust and the formation of an ocean. Secondly, slower extension rates ($\leq 10^{-15}\,s^{-1}$) will produce strain hardening and generate a finite β value of around 1.5. As each section of lithosphere hardens, the locus of intense deformation would be expected to spread laterally to involve a much wider region of extensional deformation (see Figure 2.29).

This critical β value of 1.5 is in remarkable agreement with the estimated β values from a wide range of intra-continental extensional basins (Table 2.6) which show an average β value of 1.4–1.5.

Evolution of strength in compressive deformation

The progressive increase in crustal thickness which results from compressive deformation theoretically produces the reverse situation to

Table 2.6 Estimated values of extension in various continental basins. From Kusznir and Park (1987) (data from G.D. Karner).

	β
North Sea Basin	
Central Graben	1.55–1.9
Flanks	1.2–1.3
Rhine Graben	1.1–1.3
Pannonian Basin	1.8–2.7
Aegean	1.4
Vienna Basin	1.6–1.8
Paris Basin	1.3
Wessex Basin	1.1–1.25
Worcester Basin	1.2
Bass Basin	1.25–1.5
Gippsland Basin	1.8

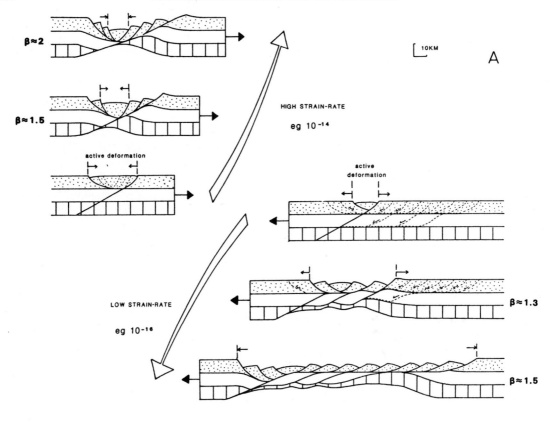

β≈2

β≈1.5

active deformation

HIGH STRAIN-RATE

eg 10^{-14}

10KM

A

active
deformation

β≈1.3

LOW STRAIN-RATE

eg 10^{-16}

β≈1.5

B

Fast Strain Rate

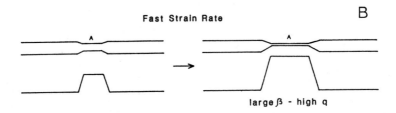

large β – high q

Slow Strain Rate

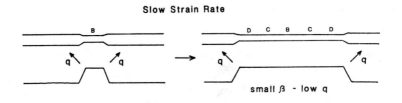

small β – low q

that just described for extensional deformation. That is, fast strain rates produce strain softening, whereas slow strain rates produce strain hardening. However, fast strain rates are prohibited by the high initial force required (see Figure 2.28) and realistic strain rates will produce strain hardening after the crust has thickened beyond about 50 km, that is, after a finite amount of shortening (perhaps 50%) has taken place.

Compressive deformation of the lithosphere is therefore a self-limiting process, in contrast to extensional deformation.

Rheological control of detachment horizons

We have seen that the strength distribution with depth depends in detail on the position of major rheological changes in the lithosphere, and particularly in the crust. Kusznir and Park consider two crustal models, both of which have apparent counterparts in nature. The first (see Figure 2.26) is a two-layer crust with a granodiorite rheology, controlled by wet quartz deformation, overlying a gabbroic rheology controlled by plagioclase deformation. This type of crust may correspond to many parts of the stable continental lithosphere with a well-developed Conrad seismic discontinuity at $c.17$ km depth. The second is a three-layer crust with a middle-crustal layer of broadly granitic composition controlled by dry-quartz deformation (Figure 2.30). This type of crust may correspond to that of the northern Scotland Caledonian terrain as shown on the LISPB seismic refraction line (Bamford, 1979). Other types of crust undoubtedly exist with a much more complex layered structure and several significant rheological changes.

Major rheological changes of this type form zones of low ductile strength, as shown in Figure 2.30, which would be expected to provide detachment horizons during major lithosphere deformation. The importance of the various zones of potential low strength depends very much on the temperature gradient. Figure 2.30A shows that for a three-layer crust in extension, all three zones are well developed only for intermediate heat-flow regimes ($q \sim 70\,\mathrm{mW\,m^{-2}}$). At higher heat flows, only the upper zones are significant. Thus we might expect that a major detachment horizon would be most likely to develop at mid- to upper-crustal levels for warm lithosphere like that of the Basin-and-Range province.

Blundell *et al.* (1985) show that in the BIRPS 'MOIST' deep-reflection line across northern Scotland (Brewer *et al.*, 1984) the major Mesozoic extensional faults flatten out around the 20 km-deep horizon corresponding to the base of the middle-crustal layer identified by Bamford (1979) from the LISPB profile (Figure 2.30B). In this case therefore, we might conclude that the geothermal gradient was not unusually steep and that heat-flow levels perhaps corresponded to those of present-day Hercynian orogenic regions in Europe, with a thermal age of $c.200\,\mathrm{Ma}$. It is interesting that the extensional faults penetrate much deeper in the Precambrian shield region west of the Caledonian front where the lithosphere would have been much cooler during the early Mesozoic extension.

Summary

The rheology and strength of the lithosphere can only be understood in a precise quantitative way by means of mathematical modelling.

Figure 2.29 Cartoon crustal models showing the different styles of extensional deformation expected with fast and slow rates of lithosphere extension. At fast rates (e.g. 10^{-14}) strain softening might be expected to localize deformation near the original site of WLF, causing progressive narrowing and intensification of the active deformation, leading to high β values and eventually to crustal separation. At slow strain rates (e.g. 10^{-16}) local strain hardening might be expected to transfer deformation laterally to previously undeformed areas, thus progressively widening the zone of active deformation, but with a limiting β value of around 1.5. Note the use of detachment horizons in the slow-strain model to transfer the deformation. (B) Summary model of the effect of slow and fast strain rates on the whole lithosphere, and the relationship of β value and heat flow, q. From Kusznir and Park (1987).

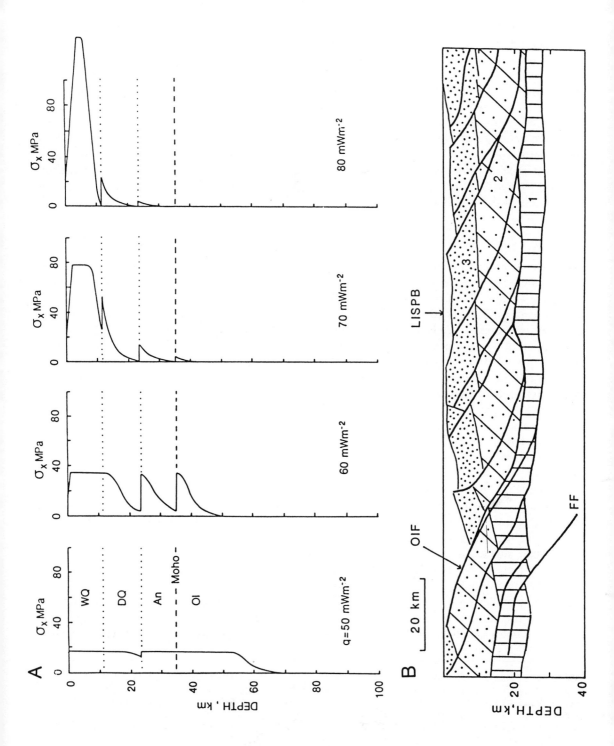

There exists a critical and complex interplay of applied force, thermal state, strength and strain-rate that determines how deformation is initiated and how it proceeds. The results of the particular modelling exercise just described give precise estimates of the strength of the lithosphere under different thermal regimes. These estimates are in close agreement with observations of tectonic behaviour and heat flow at the present time, and with theoretical estimates (see 2.5) of the likely force levels available to deform the lithosphere. A modification of the model allows the evolution of lithosphere strength during extension and compression to be examined. The model predicts that fast initial strain rates $\dot{e} \geqslant 10^{-14} s^{-1}$ and warm lithosphere are necessary to produce large extensions, and eventually oceans; that strain rates of $c.10^{-15} s^{-1}$ will produce broad zones of extensional deformation with a finite limiting β value of about 1.5 (corresponding closely with observations); and finally that initial strain rates of $10^{-16} s^{-1}$ or slower will never succeed in achieving any significant level of extension.

In the case of compressive deformation, much higher levels of applied force are required to initiate deformation than would normally be expected away from plate boundaries, and realistic strain rates are predicted to cause strain hardening after a finite amount of shortening, possibly 30–50%).

During extensional or compressional deformation, major rheological boundaries in the lithosphere (especially at mid-crustal levels) form important zones of low ductile strength, These zones may provide detachment horizons during the deformation. Which detachment zone is selected depends on the temperature gradient; higher detachment levels are favoured in warmer lithosphere with a young thermal age, whereas lower levels are favoured in cooler lithosphere with an older thermal age.

Figure 2.30 (A) Stress plotted against depth for a three-layer crustal model with a range of surface heat flows, showing the development of low-strength zones in the middle and lower crust. The stress-depth distributions correspond to 1 Ma after the application of a tensile force of $10^{12} N m^{-1}$. The low-strength regions represent the probable sites of detachment horizons. WQ, wet quartz; DQ, dry quartz, An, anorthosite; Ol, olivine. Numerical model; from Kusznir and Park (1987). (B) Model of the MOIST profile across north Scotland, after Blundell et al. (1985), showing the three main crustal layers recognized in the LISPB profile by Bamford et al. (1979). 1, lower crust (7 km s^{-1}, ?basic granulite); 2, middle crust (6.4 km s^{-1}, intermediate, granulite-facies Lewisian); 3, upper crust (6.15 km s^{-1}, amphibolite-facies Caledonian metamorphics). The uppermost (unornamented) layer corresponds with unmetamorphosed sediments from Torridonian to Mesozoic in age. *FF*, Flannan 'fault'; *OIF*, Outer Isles fault. Note that most of the major low-angle extensional faults sole out along the top of layer 1 at about 20 km depth, which therefore represents a key detachment horizon during post-Caledonian extension. The faults in the west, however, including a branch of the Outer Isles fault, appear to detach along the Moho. From Kusznir and Park (1987).

3 Plate movement and plate boundaries

3.1 Kinematic behaviour of plates

A major implication of the plate tectonic model is that crustal deformation is ultimately controlled by the relative movements of the lithosphere plates. We have seen that the plates have considerable lateral strength and suffer little lateral distortion over time periods of tens or even hundreds of Ma, and that they can transmit horizontal stresses through long distances in a regular fashion. We observe that large relative movements are confined to plate boundaries, where crustal deformation is therefore concentrated. The determination, both qualitatively and quantitatively, of the relative motion between two adjoining plates at their common boundary is therefore essential in understanding the deformation that takes place there. The converse is equally important: a knowledge of the deformational structure at a plate boundary can provide useful information about relative plate movements, especially in pre-Mesozoic time when evidence from ocean-floor stratigraphy, so vital in interpreting more recent plate tectonic history, is lacking.

The principles governing the motion of plates over a spherical surface were first explained by McKenzie and Parker (1967), Morgan (1968), Le Pichon (1968) and Isacks *et al.* (1968). By using the orientation of transform faults, slip vectors from earthquake focal-mechanism solutions and spreading rates on ridges, relative movement vectors may be found for the five major plates in relation to an arbitrarily stationary Antarctic plate (Figure 3.1). The magnitude of these velocity vectors varies from 2 to 17 cm/year.

Using known relative movement vectors, the direction and amount of relative movement at any plate boundary can be found by constructing a vector triangle (Figure 3.2). Although the direction and speed of angular rotation is uniform, as measured on a spherical surface, map projection indicates differences in the linear (i.e. tangential) velocity vectors, which can be considerable over large areas. The first step therefore in analysing the structures at any active or recent plate boundary is to determine this relative velocity vector.

It is convenient in general terms to distinguish three main types of relative motion: *divergent*, *convergent* and *strike-slip*, and these types correspond in turn to the main types of tectonic regime arising respectively from constructive, destructive and conservative plate boundaries. However we must recognize that the relative movement vector may make any angle with the plate boundary and that, in general terms, most vectors will be oblique.

Thus, in practice, a divergent or a convergent boundary may also exhibit a component of strike-slip motion which will impose a simple-shear stress across the boundary. Because new plate is created at divergent boundaries, and old plate destroyed at convergent boundaries, oblique relative motion can easily be accommodated. However, since plate is conserved at transform faults, the movement there should theoretically be pure strike-slip with no oblique component. This condition is violated when a change in relative plate motion occurs, in which case old transform faults may suffer extensional or compressional movements. Good examples of such changes in plate motion have been studied in the Indian and eastern Pacific oceans (see Figures 3.6, 3.8).

Migration of plate boundaries

One of the consequences of relative plate motion is that plate boundaries themselves may migrate in relation to one another. The present plate boundary network as shown in Figure 3.1 is a transient one, and many of the boundaries shown are moving at a determinable rate. The simplest example which demonstrates this is the boundary of the

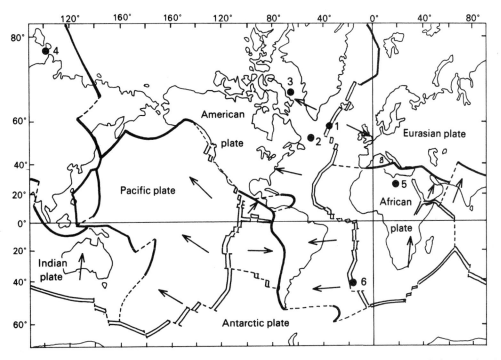

Figure 3.1 The six major plates and their boundaries, with approximate velocity vectors relative to the Antarctic plate, and poles of rotation for six plate pairs: 1, America-Africa; 2, America-Pacific; 3, Antarctica-Pacific; 4, America-India; 5, Africa-India; 6, Antarctica-Africa. Ridges, double lines; trenches, single lines; transform faults, dashed lines. After Vine and Hess (1970).

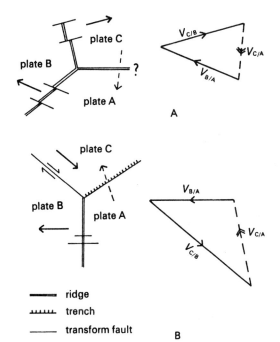

Figure 3.2 Determination of the relative velocity vectors of three plates meeting at a triple junction. (A) Three plates A, B and C are bounded by ridges. The velocity of B relative to A ($V_{B/A}$) is parallel to transform faults on the A/B boundary, and the magnitude can be determined from the spreading rate and is represented by OP in the vector diagram. Similarly $V_{C/B}$ is represented by PQ. The unknown vector $V_{A/C}$ is represented by the line QO. (B) This triple junction involves a transform fault, a ridge and a trench. $V_{B/A}$ can be determined in the same way as in (A), and the direction of the C/B vector is given by the orientation of the transform fault. $V_{A/C}$ can be determined if the orientation is known (by the orientation of transform faults cutting the trench) or if the magnitude of $V_{C/B}$ can be determined.

Antarctic plate. The movement vectors of Figure 3.1 were constructed assuming a stationary Antarctic plate. However this plate is completely surrounded by ridges offset by transform faults, and new plate material is being created at each. It follows that the Antarctic plate is growing in size and that, with reference to a fixed Antarctic continent, all the constructive boundaries surrounding the Antarctic plate must be moving outwards. A similar argument may be applied to the American and Eurasian plates. Since both these plates are growing by the addition of oceanic material along the mid-Atlantic ridge, either the destructive west Pacific boundary must migrate eastwards, or the destructive/transform east Pacific boundary must migrate westwards, or both. In other words, the Pacific plate must be shrinking, and the destructive boundaries on its NW and SE sides are approaching each other.

Stable and unstable triple junctions

It was recognized early in the evolution of plate tectonic theory by McKenzie and Morgan (1969) that there must be points on the Earth's surface where three plates meet. Such points were termed *triple junctions*. They divided triple junctions into types according to the nature of the boundaries involved. Thus if R symbolizes ridge, T trench, and F transform fault, an RTF junction is one involving the meeting of all three types of boundary. Similarly we may have RRR, TTT, TTR junctions, and so on. McKenzie and Morgan recognize 16 possible types of triple junction and discuss the stability of each. A junction is stable if it maintains its shape through time (disregarding its absolute motion). Some examples are immediately obvious: an RRR junction will always be stable. A good example of an RRR junction is where the Galapagos ridge meets the E. Pacific ridge west of Central America (Figure 3.8A), separating the small Cocos plate in the northeast from the Pacific plate in the west and the Nazca plate in the south. It is clear that continued spreading on all three ridges will not affect the basic geometry of the triple junction. In contrast, FFF and RRF

junctions will always be unstable. The stability of the remaining types is dependent on the geometry.

The kinematic behaviour and stability of a given triple junction may be determined by drawing the appropriate vector triangle (see Figure 3.2). The sum of the relative velocities of the three plates must be zero (i.e. $V_{B/A} + V_{C/B} + V_{A/C} = 0$, where $V_{B/A}$ is the velocity of B relative to A, etc.) provided that the plates are rigid. The lengths OP, PQ, and QO of the vector triangle represent the velocities $V_{B/A}$, $V_{C/B}$ and $V_{A/C}$ respectively. If we know $V_{B/A}$ and $V_{C/B}$, $V_{A/C}$ can be determined. Now consider the movement of a point on a boundary. If we take the RRR case (Figure 3.3A), a point P on the ridge axis AB will move with respect to A and is represented by the mid-point of AB in the velocity triangle. Similarly points Q and R on ridge axes BC and CA will be represented by the mid-points of BC and CA respectively. The velocity of all points P on the A/B ridge axis is represented by the dashed line ab parallel to the A/B ridge axis, that of all points Q by bc parallel to the B/C ridge axis, and that of all points R by ca parallel to the C/A ridge axis. The condition for stability is that ab, bc and ca meet at a point. This point represents the velocity of the triple junction, which for convenience we shall assume is stationary. In the RRR case, although the geometry is always stable, the triple junction will migrate if the spreading rates on the ridges are not equal (compare Figures 3.3A and B).

Consider now the TTT example given by McKenzie and Morgan (Figure 3.3C). Here, plate A is stationary, plate B is subducting below plates A and C, and plate C is subducting below plate A. The relative motion vectors are shown, and are of course different for each boundary. Figure 3.3C(2) shows the position of plates B and C at some later time, as if they were allowed to proceed horizontally instead of being subducted. The trench BC has migrated up the A/B boundary because it is consuming plate B and not plate C; that is, the margin of plate C with B is fixed to plate C and must migrate with it. One consequence of this is the obvious geometric change of the triple

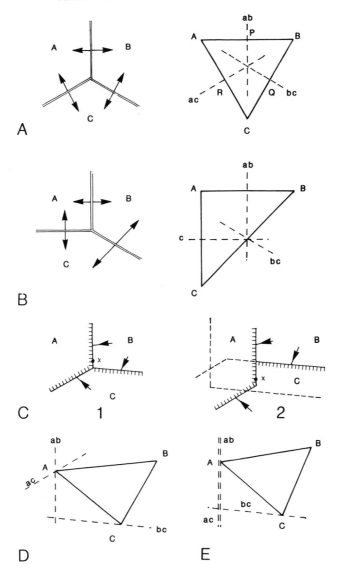

Figure 3.3 Stable and unstable triple junctions. (*A*, B) RRR junctions. The velocity of a point *P* on the ridge axis *AB* moves with respect to *A* at half the velocity of plate *B*. It is therefore represented by the mid-point of *AB*. Similarly, *Q* and *R* represent the velocities of points on ridge axes *BC* and *AC* respectively. *ab*, *bc* and *ac* represent the loci of all such points, and are parallel to the appropriate ridge axes. If these loci meet at a point, the junction is stable. If the spreading rates on the three ridges are equal (*A*), *ab*, *bc* and *ac* meet at the centroid of the vector triangle *ABC*, that is, the RRR junction is stationary with respect to the three plates. In (*B*), the spreading rate on the *BC* ridge is greater than on the other two, and *ab*, *bc* and *ac* must meet at a point, that is further from *A* than from *B* and *C*, indicating that the triple junction is moving towards *A*. The geometry of the junction is, however, stable. (*C*,*D*) TTT junctions. The positions of plates *B* and *C* after a given time are shown in (2) by the dashed line, assuming that no subduction takes place. Since the *BC* trench is fixed to plate *C*, and the *AC* trench is fixed to plate *A*, the *BC* trench must migrate with plate *C*, and the triple junction must consequently move along the *AB* boundary. The TTT junction of *C*(1) is therefore unstable, as shown in the vector triangle (*D*). Such a junction would only be stable if the vector *AC* were parallel to the trench *BC*. The new triple-junction configuration of *C*(2) is, however, stable, as shown in (*E*).

junction itself; another is the relative movement vector at a point such as x, originally on the A/B boundary. The motion perceived on plate A undergoes an abrupt change as the BC boundary moves past the point x. This change is diachronous and progressively reaches all points on the AB boundary given enough time. It is easy now to visualize the condition for stability of this type of triple junction (see Figure 3.3D,E). Since points on the AB boundary do not move relative to A, line ab is drawn through A in the velocity triangle. Line ac is drawn through A for the same reason, and line bc must be drawn through C. Therefore ab, bc and ca will only meet if bc is parallel to AC (Figure 3.3D), that is, if the CA movement vector is parallel to the DC boundary. In that case, the boundary will not migrate and the triple junction will be stable. Note that the new configuration of the triple junction in Figure 3.3C(2) is actually stable (see Figure 3.3E) although it is migrating with reference to its original position.

In Figure 3.4 an example given by McKenzie and Parker (1967) from the NE Pacific is shown. The present plate geometry is shown in Figure 3.4A(4). Plate velocities are shown with reference to a stationary American plate. There are two triple junctions: an FFT in the north (Figure 3.4B) and an RFT in the south (Figure 3.4C). Both junctions involve three plates, American, Pacific and Farallon (the Farallon plate is now divided into separate parts and partly incorporated into other plates). The evolution of the plate system is shown in Figure 3.4A(1–4). Since the Pacific plate is moving northwestwards and the Farallon plate northeast, any material point on the ridge (such as the intersection with the Murray transform) is moving approximately northwards. As the eastern ridge–transform junction travels north, it encounters the trench on the northeast side of the Farallon plate to form a triple junction. Immediately the Pacific plate comes into contact with the American plate, the relative movement vector becomes northwest, changing the boundary to a transform fault. Thus the southern RTF triple junction is formed. Since the ridge axis is still moving north, plates F and P together with the ridge itself are being consumed by the trench. However, the position of the triple junction is stationary with respect to the American plate. As more of the Pacific plate is consumed, the transform fault lengthens and the northern FFT triple junction migrates along the margin of the American plate, progressively converting trench to transform. Thus both triple junctions possess stable geometry but are moving relative to each other. If the San Andreas fault were not parallel to the Pacific/American movement vector, the junctions would not be stable (see Figure 3.4B).

A very common type of intersection occurs where a fault meets a ridge, initially forming an RRF junction. McKenzie and Morgan show that such a triple junction is always unstable and evolves into an RFF type. A good example occurs in the central Atlantic, at the junction of the American, Eurasian and African plates, where the mid-Atlantic ridge meets the Azores–Gibraltar fracture zone (Figure 3.1). Another example is the junction of the Chile fracture zone with the East Pacific ridge, at the SW corner of the Nazca plate (Figure 3.8A). A third is the young triple junction formed at the meeting of the Carlsberg ridge and the Owen fracture zone in the NW Indian Ocean, due to the breakaway of the Arabian plate (Figure 3.6B). This type of junction is stable if the two faults lie on the same small circle (i.e. are effectively a single fault) as is the case in these examples.

The evolution of both the Pacific and the Indian oceans through Mesozoic–Tertiary time provides useful illustrations of the kinematic evolution of plate structure. In broad terms, the evolution of world-wide plate structure since the early Mesozoic is brought about by means of the break-up of Pangaea through the opening of the Atlantic and Indian oceans, and the consequential shrinking of the Pacific (Figure 3.5). The opening of the Atlantic involved a relatively simple sequence of movements that resulted in a net convergence of the American and Eurasian plates across the Pacific.

The initial stage took place during the

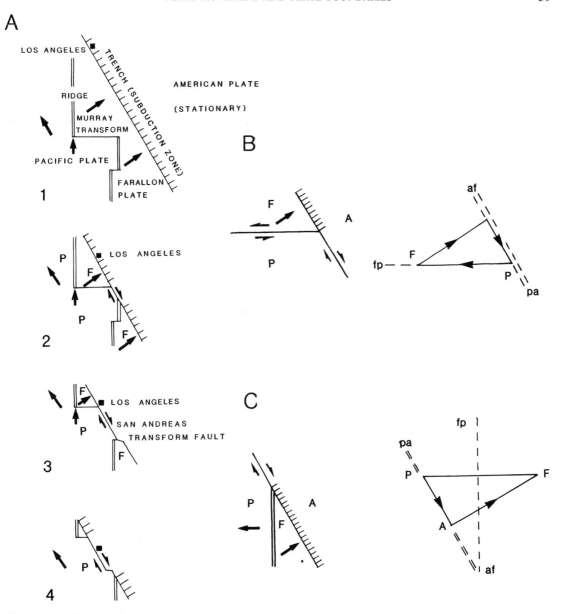

Figure 3.4 Initiation and evolution of the triple junctions of the San Andreas fault. (*A*, 1–4), four stages in the evolution of the plate boundary system in the region of the San Andreas fault at 53 Ma, 30 Ma, 10 Ma BP, and the present, assuming a stationary American plate, after Atwater (1970). When the Pacific plate meets the American plate (2), two triple junctions are formed, joined by the San Andreas transform fault, which gradually lengthens as more of the Pacific plate is subducted. The northern is a TFF junction (*B*), which is stable because the fault and trench are parallel, and are fixed to the American plate. The position of the junction, however, migrates along the boundary of the American plate. The southern junction is a RTF type (*C*). This also is stable for the same reason, but is stationary relative to the American plate.

Jurassic, between 200 and 170 Ma BP, with the opening of the Central Atlantic (Figure 3.5*A*, *B*). This movement was bounded on both sides by major transform faults, between Newfound-land and Gibraltar in the north, and between the Bahamas and Guinea in the south. Thus while South America and Africa remained unified to the south, and North America and

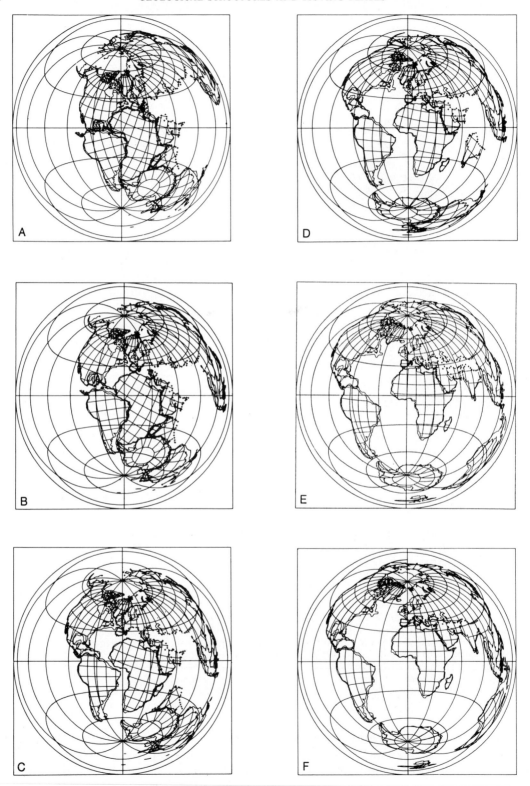

Europe to the north, there was a clockwise rotation of the latter away from Gondwanaland that resulted in a sinistral displacement through the Mediterranean and a convergence across Tethys.

The second stage took place mainly during the Cretaceous, between 170 and 65 Ma, when the South Atlantic opened about a rotation axis in the Central Atlantic, in addition to continued spreading there. These movements brought about a further rotation of North America–Eurasia and a consequential convergence across Tethys (Figure 3.5C). By the mid-Cretaceous, a new rift appeared through the Labrador Sea and across the Arctic Ocean, splitting Laurasia into two separate plates (Figure 3.5D). In the early Cenozoic (65 Ma BP), this spreading axis was replaced by the presently active ridge through Iceland, separating Greenland from northwest Europe (Figure 3.5E). This third stage was accompanied by the complete closure of the Tethys ocean.

The evolution of the Indian Ocean is more complex (Figure 3.6). A model for the plate tectonic evolution of the region was put forward by McKenzie and Sclater (1971). They suggest that the earliest movements took place along a rift separating the Indian from the African continents (Figure 3.5C). However, the magnetic record of this movement has been largely obliterated. At around 90 Ma ago, in the late Cretaceous, the movement of India changed dramatically, and a northwards movement commenced due to a fast-spreading ridge between the Indian and Australia–Antarctica continents (then joined) (Figure 3.5D).

This movement was accommodated to the north by subduction of Tethys ocean plate along the southern margin of Eurasia. Much of this northward movement was constrained between two major N–S transform faults, the Chagos and Ninety-East ridges. At about 45 Ma ago, the spreading ridge extended eastwards, causing the separation of Australia from Antarctica (Figure 3.5E). The record of this phase of movement, which lasted until about 35 Ma ago, can be seen in the NW Indian Ocean as a set of magnetic stripes and inactive transform faults that are oblique to the presently active ridge and transform system (Figure 3.6). The third major change in kinematic pattern was related to a change in spreading direction of about 45°, accompanied by a further extension at the NW end of the old ridge into the Gulf of Aden. This marked kinematic change appears to be diachronous, commencing around 35 Ma BP at anomaly 12 in the central area and progressing northwestwards until about 10 Ma BP when the Gulf of Aden opening commenced. The change in direction is reflected in the ocean floor in a discordance that represents a vertical 'unconformity' with new stripes cutting old stripes and old transforms (Figure 3.6B). It has been suggested that the kinematic change is related to the collision between the Indian and Eurasian continents. The effect of this collision may have been to bring about a change in the relative movement vector between the Indian and Eurasian plates.

The evolution of the Pacific Ocean can be reconstructed by examining the Mesozoic magnetic stripe pattern. Larson and Pitman (1972) present an interpretation of the plate structure within the Pacific in early Cretaceous time (Figure 3.7A). At that time, the ridge system appears to have been much more symmetrically arranged, with two RRR triple junctions separating four oceanic plates: the Kula plate in the north, the Farallon plate in the northeast, the Pacific plate in the southwest and the Phoenix plate in the southeast. Subduction zones formed an almost continuous rim around the Pacific Ocean of that time (the 'proto-Pacific', we might call it). According to Pitman and Hayes (1968), there has been a continuous northwards movement of the northern parts of this proto-Pacific Ocean since the early

Figure 3.5 Palaeomagnetic reconstruction of the continents illustrating the break-up of Pangaea in Mesozoic-Cenozoic time. Lambert equal-area S-polar stereographic projections at: (A) 200 Ma BP (late Triassic); (B) 140 Ma BP (late Jurassic); (C), 100 Ma BP (mid-Cretaceous); (D), 60 Ma BP (Palaeocene); (E), 20 Ma BP (early Miocene); and (F), Present. From Smith and Briden (1977)

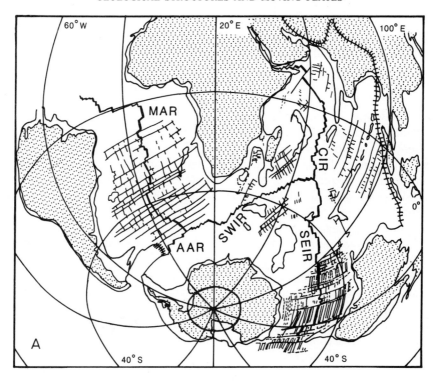

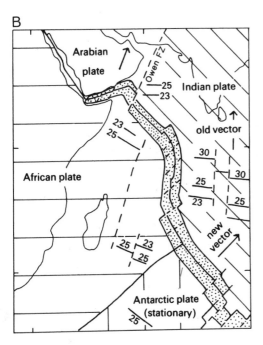

Figure 3.6 (*A*) Present-day positions of the Gondwanaland continents, with simplified distribution of ocean ridges, magnetic stripe patterns and transform faults. MAR, mid-Atlantic ridge; AAR, Atlantic–Antarctic ridge; SWIR, SW Indian Ocean ridge; SEIR, SE Indian Ocean ridge; CIR, central Indian Ocean ridge. (*B*) Simplified map of the western Indian Ocean region showing the discordance in magnetic anomaly patterns and transform directions at anomaly 5 (33 Ma BP) caused by a change in the Indian–Antarctic plate vector. After Laughton *et al.* (1973).

Mesozoic. With spreading occurring on all five ridges, the northern RRR junction would migrate north relative to the southern. When subduction ceased along the Antarctic margin, northward movement of the southern RRR junction would take place relative to a fixed Antarctica, causing the whole proto-Pacific system to migrate northwards. The Kula plate was rapidly subducted throughout Cretaceous time around the north and northwest Pacific rim, according to the Pitman and Hayes reconstruction. They show four stages in this process in the Alaskan region (Figure 3.7*B*). By early Palaeocene time (60 Ma ago), the Kula plate

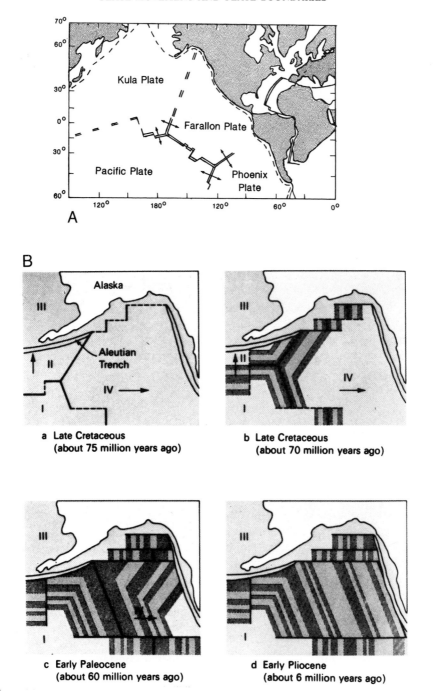

Figure 3.7 (*A*) Arrangement of plates in the Pacific Ocean at around 110 Ma BP (early Cretaceous time). Note the two RRR triple junctions and the projected Farallon–Kula ridge (see *B*). The Central Atlantic has commenced opening. After Larson and Pitman (1972). (*B*) Schematic diagrams showing four stages in the evolution of the magnetic stripe pattern in the NE Pacific Ocean; I, Pacific plate; II, Kula plate; III, American plate; IV, Farallon plate. From Uyeda (1978), after Pitman and Hayes (1968).

had almost entirely disappeared, and the Aleutian subduction zone was consuming progressively older material belonging to the Pacific plate. It is difficult to interpret Figure 3.7B(4) by itself, but it becomes clear once the stages of evolution are followed through. The ridge between the Pacific and Farallon plates has migrated northeastwards, and is about to be consumed by the Aleutian trench.

The other important effect on the Pacific region was the progressive westwards encroachment of the American plate brought about by the opening of the Atlantic. Coney (1973) has calculated that the western margin of North America has moved about 3700 Km westwards from the early Mesozoic until the formation of the San Andreas fault about 10 Ma ago. Almost all of the Farallon plate has been consumed by this movement, along the western American subduction zone. The later stages of this process are discussed above (see Figure 3.4).

Two major changes in movement direction can be traced in the magnetic anomaly pattern of the Pacific floor. The first, around the beginning of the Cretaceous, is marked by a discontinuity in transform directions in the Hawaii region (Figure 3.10). At 45 Ma BP, the East Pacific ridge extended westwards to connect with the Indian Ocean, by splitting Australia from Antarctica, but no corresponding change can be seen in the Pacific floor at this time. The second major change occurred 10 Ma ago when the plate boundary system in the east-central Pacific was completely reorganized (Herron, 1972). Two new plates were formed, the Cocos plate west of Central America, and the Nasca plate west of the Peru–Chile trench (Figure 3.8A). A new section of the East Pacific ridge was formed between latitude 45°S and the Gulf of California, breaking through the old ridge at the Galapagos triple junction (Figure 3.8B). As in the Indian Ocean, the change in relative movement vectors is clearly marked by an abrupt change in the orientation of the transform faults, which run approximately E–W until anomaly 5, where they are replaced by a NW–SE set. The heavily fragmented but still active Chile ridge represents the southwest boundary of the old Farallon plate, of which relics occur in both the Cocos and Nasca plates.

The reason for this change cannot be established with certainty, but several factors may have contributed. The creation of the new ridge represents a change in relative movement direction between the Pacific and American plates and corresponds in time with (i) a decrease in spreading rate in the Atlantic, (ii) the subduction of the northern part of the East Pacific ridge below North America (and the consequential initiation of the San Andreas fault), and (iii) the change in plate structure in the Indian Ocean already discussed. A major factor which may have influenced world-wide plate structure is, of course, the collision of India with Asia, although the exact time of this event is uncertain (see 5.4).

The structure of the western Pacific, which was the site of destruction of Pacific plate throughout Mesozoic and Tertiary time, is dominated by effects created by the northwards advance of Australia (on the Indian plate) and by the creation of numerous back-arc spreading basins (see 4.4). The structure of the complex Indonesian region is discussed in detail in 5.5.

3.2 The influence of plate geometry on the kinematic pattern

In the preceding section, we have assumed that the relative plate movement vectors were constant, and that they, together with the shape of the plate boundary network, controlled the subsequent geometry of the network. Plate tectonic theory, to a first approximation, assumes a constant network geometry. However, under certain circumstances, plate boundaries may be deformed as a result of plate movements, thus violating the principle of 'rigid' plates. Such deformation is largely confined to destructive boundaries, and particularly affects continental lithosphere which, as we have seen, is softer and weaker than oceanic.

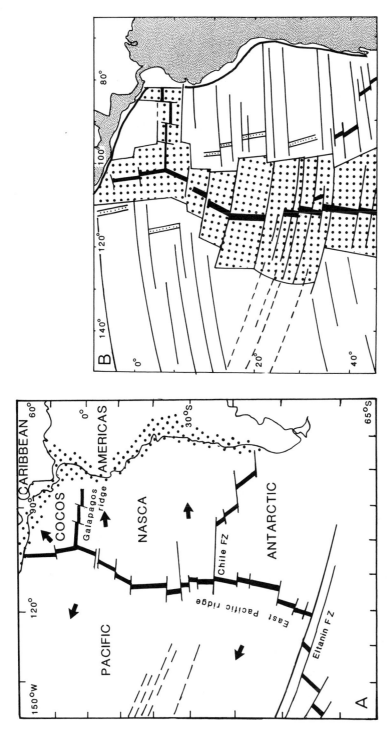

Figure 3.8 Change in plate structure in the East-Central Pacific Ocean. (*A*) Present plate boundary network. The earthquake zone marking the destructive western boundary of the American plate is stippled. (*B*) Discordance in magnetic stripe and transform fault pattern at anomaly 6 (approx. 10 Ma BP). Pre-anomaly 6 ridge segments are dotted, active ridge segments in black. New ocean crust since anomaly 6 is stippled. After Herron (1972).

Vogt (1973) suggested that the cuspate shape of island arcs and trenches could, in many cases, be ascribed to the deformation of a destructive plate boundary brought about by the interaction of aseismic ridges on the ocean floor of the subducting plate (Figure 3.9A). He claimed that subduction of such ridges was inhibited by their topographic relief, and by the greater buoyancy of the locally thickened oceanic plate. Thus when the trench moves oceanwards, driven by back-arc spreading (Figure 3.9B), the movement will be con-

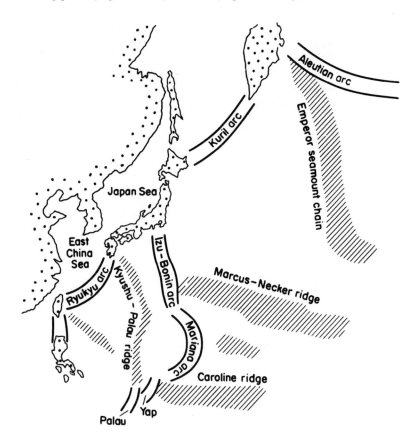

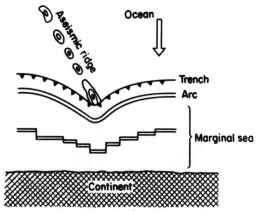

Figure 3.9 (A) Aseismic ridges and island arcs of the NW Pacific Ocean. After Miyashiro *et al.* (1982). (B) Schematic diagram showing how back-arc spreading may be inhibited at the cusps of the arcs by the presence of an aseismic ridge. From Miyashiro *et al.* (1982), with permission.

strained at the positions of the ridge intersections. The rate of plate convergence will be at a minimum there, but will increase to a maximum between the intersections, producing an arcuate pattern. Good examples of this are the Emperor seamount chain at the intersection of the Aleutian and Kurile arcs, and the Marcus-Necker and Caroline ridges at each end of the Marianas arc (Figure 3.9A). Such deformation is consistent with rigid-plate theory because it results from a progressive change in the position of down-bending and does not involve active lateral distortion at the surface.

Deformation of continental plates appears to result mainly from collision, and may involve major changes in geometry both of the plate boundary and of the plate interior. The best example of such deformation at the present day occurs in the Central Asian region described in 5.4. The deformation results from the collision of the continental part of the Indian plate with the southern margin of the (continental) Eurasian plate. Once the intervening oceanic plate had been consumed, further convergence would have been inhibited by the buoyancy of the continental part of the Indian plate. The processes of subduction and collision are considered in detail in Chapter 5. It is important to recognize that collision is the most effective way of altering plate kinematic patterns, often in a quite dramatic and worldwide fashion. Envisage the collision of two opposing continental margins, both typically irregular in shape, and oblique to each other and to the convergence direction. At the first point of contact between the two opposing margins, resistance to convergence will be introduced which may act either to change the convergence vector, or to deform the boundary geometry, or both. The tectonic effects of wedge-shaped protrusions of one plate as it meets another at a collision boundary are discussed in the 'indentation' model of Tapponnier and Molnar (1976) and applied to the India–Asia collision. The model is of general application and involves a protrusion or 'indenter' which causes local stress concentrations in the

indented plate sufficient to overcome its strength and to produce widespread distortions.

'Absolute' plate motion

The methods of analysing plate motion developed by McKenzie and Parker (1967) give vectors for relative motion only, and Figure 3.1 is based on the assumption of a stationary Antarctic plate. A method for determining 'absolute' plate motion was suggested by Wilson (1965). Wilson noted that, at a number of locations scattered over the Earth's surface, volcanic activity appears to have been concentrated over long periods of time. Wilson called these areas 'hot spots' and identified several, including Hawaii and Iceland. He showed that the motion of an oceanic plate over one of these hot spots would result in a linear chain of volcanic islands becoming progressively older from the currently active volcanic centre. Figure 3.10 shows the Hawaii–Emperor chain of volcanic islands and sea-mounts in the Pacific interpreted according to the Wilson model. The ages of the vulcanicity range from 70 Ma at the distal end of the chain, adjacent to the Aleutian trench, to the present hot-spot location in the Hawaii islands at the southern end. The bend in the ridge is interpreted as a change in plate velocity vector, occurring at c.35 Ma BP (see above). Wilson also identified lateral chains on either side of the mid-Atlantic ridge, such as the Tristan da Cunha–Walvis ridge off SW Africa (Figure 3.11). In this case, he showed that the present ridge axis is offset from the hot-spot site by about 400–500 km, and suggested that the ridge originally lay over the hot spot, but had been moved westwards over the last 25 Ma as a result of a change in the pole of rotation for the America–Africa separation.

Morgan (1972) developed Wilson's ideas further and reconstructed a set of 'absolute' plate movement vectors with reference to the hot-spot frame of reference (Figure 3.12). He found that the relative movement between the hot spots has been very much less than that

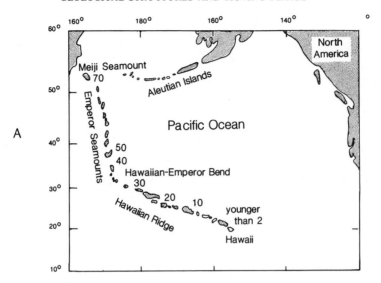

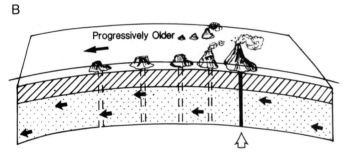

A

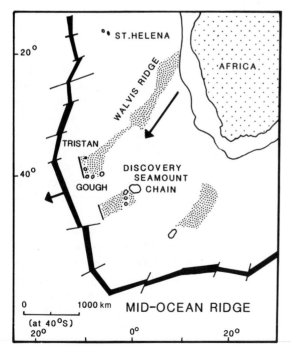

B

Figure 3.10 (*A*) Location of the Emperor and Hawaiian volcanic island and sea-mount chains in the northern Pacific Ocean. (*B*) Model to illustrate the formation of the volcanic chains by movement of the oceanic lithosphere over a 'fixed' hot-spot. After Wilson (1963).

Figure 3.11 Map of the southern Atlantic Ocean showing volcanic centres (Tristan, Gough etc.) offset from the present position of the mid-ocean ridge. According to Wilson (1973) the line through the southwest ends of the volcanic chains shows the position of the ridge 25 Ma ago, since which time the ridge has migrated westwards. After Wilson (1973).

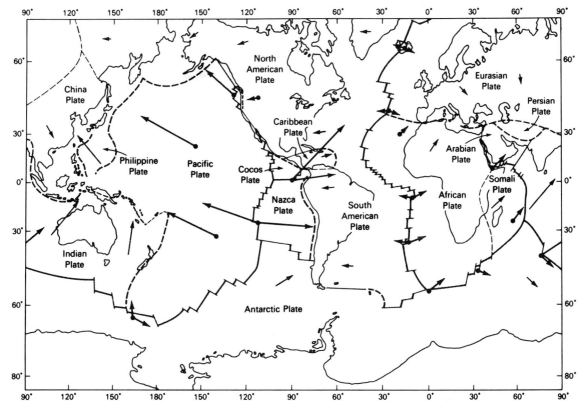

Figure 3.12 Plate movement vectors relative to a fixed hot-spot frame of reference. Lengths of arrows are proportional to plate velocities. From Uyeda (1978), after Morgan (1972).

between the plate boundaries over the last 200 Ma or so. Although it is likely that the hot spots also migrate with time relative to a notional fixed mantle frame, this migration is probably an order of magnitude slower than that of the plates. Since the hot spots are formed by the upward migration of hot mantle material, they must be related to the mantle convective system. We have seen (2.4) that the convective circulation in the mantle must be complex, and constantly changing because of the constraints imposed on the surface mobility of the plates by their strength and geometry. The lateral movement of ridges away from their likely locus of creation is a good illustration of this. It is thought that the hot spots represent the sites of ascent of deep mantle 'plumes' that form the rising columns of a convective circulation pattern, separate from but linked with, the plate movement circulation (see Figure 2.11).

An interesting feature of the 'absolute' plate movement vector pattern is the much greater velocities of the wholly or mainly oceanic plates, such as the Pacific, Cocos and Nasca plates, compared with those plates carrying large continental masses such as the Eurasian, American and Antarctic plates. It might be thought that the presence of large pieces of continental lithosphere acts as a resistance to motion. However, the fast-moving Indian plate argues against this notion. It is clear from Figure 3.13 that the significant factor affecting the velocity of a given plate is the length of attached subduction zone, indicating that the subduction process (i.e. slab pull) is a major kinematic control, as well as an important dynamic control, as shown in 2.5.

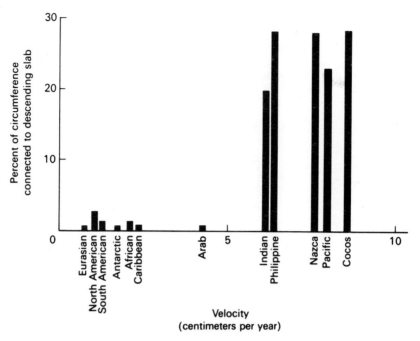

Figure 3.13 Proportion of plate circumference (%) connected to subducting slabs *v*. 'absolute' velocity of plates. Note that all the plates with long subduction zones have velocities greater than 5 cm/year. From Forsyth and Uyeda (1975)

3.3 The effects of relative plate motion at plate boundaries

It is clear from Figures 3.1 and 3.12 that the movement vector across plate boundaries is generally oblique, and that only in the case of transform faults is the relative motion constrained in a particular direction (parallel to the boundary). The importance of oblique relative movements in orogenic belts was originally highlighted by Harland (1971), who introduced the terms 'transpression' and 'transtension' to describe tectonic regimes exhibiting elements of both strike-slip motion and either convergence (compression) or divergence (extension) respectively. In the intervening years, these ideas have been comparatively neglected by structural geologists, who have been applying them extensively to orogenic belts only quite recently.

Assuming initially that we are dealing with movements that are wholly in the horizontal plane (i.e. tangential to the Earth's surface),

there are eight possible categories of relative motion at plate boundaries (Figure 3.14*A*). These are: normal convergent, sinistral convergent, dextral convergent, normal divergent, sinistral divergent, dextral divergent, sinistral strike-slip and dextral strike-slip. However where the plate boundary is inclined, as it typically is in subduction zones, the relative movement takes place with reference to an inclined plane, introducing the vertical dimension. It is useful to employ fault displacement terminology for such movements. All the possible categories of relative movement are listed in Figure 3.14*B*. Thus we see that convergent movements lead to thrust displacements on inclined boundaries, and divergent movements to normal displacements. At subduction zones (or more generally, destructive boundaries) therefore, the following possibilities exist: dip-slip thrust and oblique sinistral or dextral thrust. At constructive boundaries (ridges or rifts), we may have dip-slip normal, and oblique-slip sinistral or dextral normal, in

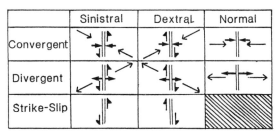

A. Movements in the horizontal plane

Dip-Slip Movement	Strike-Slip Movement		
	Sinistral	Dextral	None
Thrust	x / •	• / x	/
Normal	x / •	• / x	/
None	x / •	• / x	

x = Away from observer   • = Towards observer

B Movements in three dimensions
(vertical profile) on inclined boundaries.

Figure 3.14 Categories of movement across plate boundaries: (A), in the horizontal plane; (B), in three dimensions (vertical profile). In each case, the heavy arrows mark the movement vectors and the light arrows the components of motion along and across the boundary.

addition to the categories of divergent motion listed above applied to a vertical boundary. At conservative boundaries, sinistral or dextral strike-slip motion applies either to vertical or inclined planes.

Movements across a deformable boundary

The above analysis ignores one of the most important aspects of plate boundaries, which is that they do not represent a discrete plane, but a volume of deformable material. Orogenic belts are, to a large extent, an expression of this deformation. We therefore have to take into account relative movements *across* the boundary as well as along it. This is particularly obvious in the case of constructive boundaries, where new plate material is created to accommodate the divergent movement. It is therefore necessary to consider the plate boundary

as a deformable sheet rather than a plane, in order to determine the relationship between plate movements and deformation. To each of the categories of relative movement listed in Figure 3.14, must be added a component of either convergent or divergent movement across the sheet, resulting in either compression or extension of the sheet in that direction. These movements are accompanied by stresses on the walls of the sheet that give rise to strain within the sheet. These stresses and strains possess components of *simple shear* and also components of *compression* or *extension*. The combination of simple shear and compression is termed *transpression*, and the combination of simple shear and extension is termed *transtension* (Harland, 1971; Sanderson and Marchini, 1984; see Figure 3.15*A*). It is important to recognize that simple shear strain itself involves both compression and extension, but that these balance such that there is no net volume change in the sheet. Typically, of course, volume changes do occur: for example magmatic material adds to volume in extensional regimes, and movement of volatiles and metamorphic reactions create volume changes in compressional regimes. The process of mountain building in a convergent regime reflects crustal thickening, which is an expression of compressive strain across the belt.

We may illustrate the general case of kinematic behaviour at a plate boundary by considering two examples: oblique convergence and oblique divergence (Figure 3.15*A*,*B*). Case *A* produces transpression across the sheet, and case *B* transtension across the sheet. However, in *A* there is also extension along the sheet, and in *B*, compression along the sheet. The general case is illustrated in Figure 3.15*C*.

In attempting to understand the geometry of deformational structures, it is important therefore to be able to visualize, in three dimensions, the components of movement at the plate boundary resulting from the relative movement vector. Many attempts to apply plate-tectonic theory to individual structural case histories have been two-dimensional or otherwise over-simplified, thus restricting their

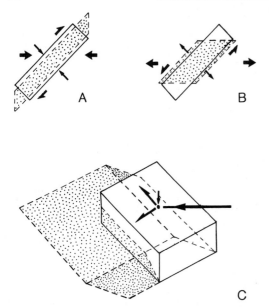

Figure 3.15 (A) transpression, and (B), transtension (plan views) produced by a movement vector (heavy arrows) oblique to a plate boundary represented by a deformable sheet. The light arrows represent the components of movement across and along the boundary. (C) A three-dimensional diagram illustrating transpression. In each case, the deformed shape is stippled.

usefulness. In Chapters 8 and 9, where individual examples of orogenic belts are consi-

dered, we shall see how far these principles have been put into practice.

Summary

In applying plate tectonic theory to the study of geological structures in orogenic belts, the primary object is to relate relative plate movements at plate boundaries to strain within the belt. Since oblique movements across plate boundaries are the norm rather than the exception, oblique convergence and divergence, both in the horizontal plane and in three dimensions, must be considered. Eight categories of relative movement across a plate boundary are recognized. By considering the plate boundary as a deformable sheet rather than a discrete plane, the further possibility arises of compression or extension taking place normal to the walls of the sheet, causing volume changes of the kind expressed in orogenic belts by crustal thickening, and at constructive boundaries by the creation of new lithosphere. At a typical plate boundary, oblique-slip extensional or compressional movements are expressed in terms of transtensional or transpressional strains within the material adjacent to the boundary.

4 Divergent (extensional) tectonic regimes

4.1 Types of extensional regime

The main sources of extensional stress in the lithosphere arise from the density imbalances produced by ocean ridges, continental margins and plateau uplifts, and by the forces arising from subduction (see 2.5). Plateau uplifts are relatively common in the continental crust, and even quite small uplifts of the order of 100 km across and 1 km or so high will produce an appreciable extensional stress (see Bott and Kusznir, 1979). Extensional tectonic regimes are therefore primarily associated with divergent (i.e. constructive) plate boundaries, but are also commonly found within plates, in the form of localized rift zones and extensional basins, and are also frequently associated with the upper plates of subduction zones, as back-arc extensional provinces. Strike-slip regimes typically involve local extensional provinces. Extensional regimes are much commoner than compressional ones, simply because the crust is considerably weaker in extension than compression, as explained in 2.7.

The main types of regime that will be discussed are therefore as follows:

A. Plate boundary regimes
(i)   Ocean ridges
(ii)   Continental rifts of type I (plate boundary)
(iii) Continental extensional provinces at convergent boundaries
(iv)   Back-arc extensional provinces
(v)   Strike-slip extensional provinces

B. Intraplate regimes
(vi)   Continental rifts of type II (intraplate)
(vii) Intraplate extensional basins.

In this chapter, we shall discuss examples of types (i)–(iv). Type (v) will be discussed in Chapter 6, and intraplate regimes in Chapter 7.

4.2 Ocean ridges

The world-wide network of ocean ridges constitutes the most significant topographic feature on the Earth's surface, surpassing even the great mountain ranges in scale. A typical section of ridge is about 1000–2000 km wide and 2–3 km high. The mid-Atlantic ridge occupies about one-third of the surface area of the Atlantic Ocean (Figure 4.1*A*). The tectonically active central rift is marked by a zone of concentrated earthquake activity. The wide topographic swell with central rift is also characteristic of the continental rift zones (Figure 4.2*A*) which display approximately the same dimensions. According to Menard and Smith (1966) the ridge system as a whole makes up 32.7% of the surface area of the oceans, or 23.2% of that of the Earth.

The geophysical structure of ridges is now known in considerable detail. The very large excess topographic mass of the ridge is almost exactly compensated by a mass deficiency

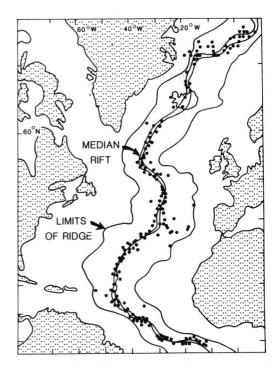

Figure 4.1 The mid-Atlantic ridge, showing central rift and distribution of earthquake epicentres. After Heezen (1962).

73

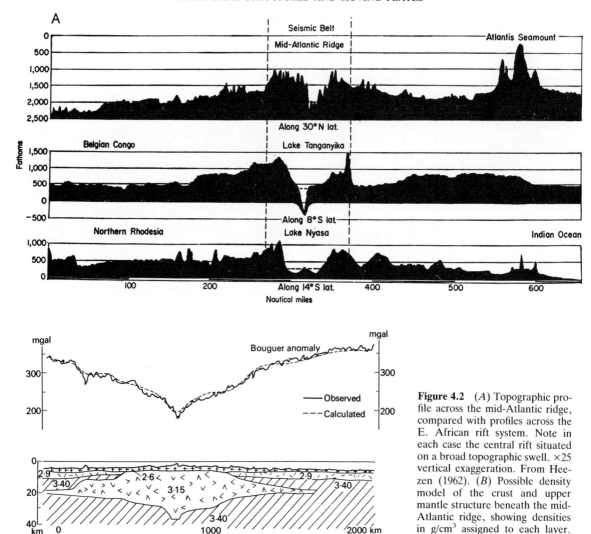

Figure 4.2 (*A*) Topographic profile across the mid-Atlantic ridge, compared with profiles across the E. African rift system. Note in each case the central rift situated on a broad topographic swell. ×25 vertical exaggeration. From Heezen (1962). (*B*) Possible density model of the crust and upper mantle structure beneath the mid-Atlantic ridge, showing densities in g/cm^3 assigned to each layer. From Bott (1971), after Talwani *et al.* (1965).

caused by a volume of less dense mantle material below the ridge. Figure 4.2*B* shows profiles combining gravity information with seismic refraction data that indicate anomalously low seismic velocities in the region occupied by the low-density material (see 2.2). The most satisfactory explanation of this anomalous structure is that the asthenosphere effectively rises much closer to the surface, and that the lithosphere thickens away from the ridge crest as it cools (see 2.1). The presence of

material with the properties of asthenospheric mantle has been confirmed by other geophysical evidence: low electrical conductivity, high attenuation of seismic waves, high surface wave dispersion, and inefficient propagation of S_n waves, which correlate with the zone of high heat flow along the ridges (see 2.3).

The active tectonic zones occupy a narrow central rift valley about 100 km wide, in which the earthquake activity and vulcanicity are concentrated. Rather low heat flows have been measured from areas adjacent to the high heat-flow zone along the ridge axis. This pattern is thought to be due to convective circulation of fluids, which gives rise to intense hydrothermal activity.

Detailed information about ocean-ridge morphology and its relationship to volcanic and tectonic activity was obtained in the FAMOUS project (Heirtzler and van Andel, 1977) which employed manned submersible dives along a section of the mid-Atlantic ridge west of the Azores islands, between latitudes 36°30′ and 37°N. The manned dive programme was supplemented by a number of techniques including seismic refraction, side-scan sonar, surface sampling and deep drilling. This enabled a detailed picture to be obtained of the morphology of the ridge and its tectonic construction (Ballard and van Andel, 1977).

The ridge in this sector is topographically complex and is offset by several fracture zones (transform faults). Bathymetric profiles show ranges of rift mountains with peaks at a depth of 1300 m on each side of the rift, separated by 30–32 km of valley floor which slopes gently towards the rift axis. The axis is offset towards the western margin of the rift. The asymmetry of the central rift in relation to the margins of the first magnetic anomaly suggests asymmetric spreading of 0.7–1.0 cm/year on the west side and 1.2–1.4 cm/year on the east side of the rift. The fracture zones are located in deep U-shaped valleys up to 3 km deep and 10 km wide. Seismically active scarps of the order of 100 m high mark the tectonically active portion of the fracture zone, and sheared rocks were dredged up to 10 km on either side of the southern fracture zone.

The rift floor contains an inner rift valley (Figure 4.3) with a floor 1–3 km in width, bordered by sloping terraces. The centre of this inner rift valley floor consists of a line of elongated hills up to 1 km wide and 100–200 m high, lying at a depth of about 2500 m. The morphology of the rift valley was found to result mainly from volcanic activity modified by tectonic effects. Vertical tension fractures occur on the inner rift floor and dip-slip

displacements characterize the flanking fault scarps, which dip at about 60° inwards, forming the walls of the rift. These features indicate extension normal to the rift axis but oblique to the plate divergence vector.

The vertical displacement on the extensional fissures of the axial zone is always less than 1 m but, in the marginal tectonic provinces, the throws increase to between 1 and 3 m, with the formation of tilted blocks. On the inward-facing walls of the rift, major vertical scarps occur, with throws of 5–100 m, on planes dipping around 60°, bounding fault blocks tilted away from the rift at 5–7°.

The horizontal extension measured across the inner rift on these faults and fissures amounted to 5.7% on the western side and 8% on the eastern. The difference is proportional to the difference in spreading rate. New rift axis positions appear to occur along the lower slopes of the volcanic edifices rather than through the middle. Consequently, short transient transform faults are required to accommodate the spreading. The width of this central zone of offset rifts is about 1 km. The central axial ridge contains volcanic material with dates of 20 000 to 35 000 years, but the marginal ridges contain material 100 000–160 000 years old (Figure 4.3A). Thus the inner rift contains the youngest volcanic rock and is interpreted as the result of axial volcanic activity in the form of small piles or cones of lava 5–7 m high and 10–15 m wide draped by radiating lava tubes. This vulcanicity is accompanied by pure extension, in contrast to the faulting in the rift walls, which results from shear. It is suggested that, shortly after formation, the volcanic edifice in the axial zone collapses vertically along the boundaries of the central ridge, whereas in the outer portions of the valley, older volcanic blocks are uplifted to form the walls of the inner rift (Figure 4.3C).

Since the distance from the boundary wall to the axis is correlated with spreading rate, it seems likely that the uplift of the walls is related to the thickening of the lithosphere as it moves laterally away from the locus of new volcanic activity and cools. The crustal thickness in the inner rift valley is only about 3 km, based on thermal models (Sleep, 1975). About 10–20% of this thickness is made up of pillow basalts, overlying oceanic crust of layer 2 type (presumably sheeted dykes), which in turn overlies a magma chamber that appears to extend the whole width of the inner rift.

The horizontal tensional stress exerted across the whole width of the ridge, due to its density contrast with the adjoining ocean floor, is therefore transmitted across a very thin, weak, brittle layer in the central rift zone. This must cause repeated extensional failure, enabling the spreading process to continue.

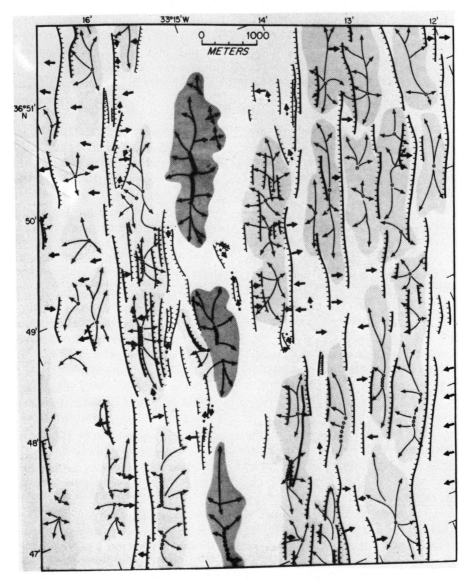

Figure 4.3 (*A*) Distribution of faults (hachured lines) and volcanic bodies (shaded) in the inner rift zone of the mid-Atlantic ridge at 36°N. The darker shading indicates the youngest volcanism; dots, vents and crest lines of volcanoes; thick arrows, dip of fault blocks; thin curved arrows, volcanic flow lobes. From Ballard and Van Andel (1977)

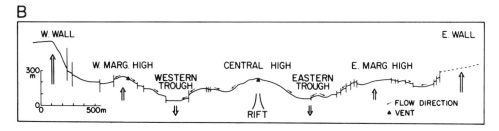

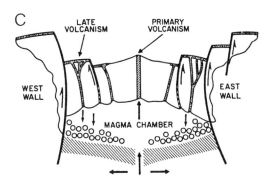

Figure 4.3 (*B*) Diagrammatic profile of the inner rift zone showing structural subdivisions. (*C*) Diagrammatic crustal cross-section of the inner rift valley, showing magma chamber in which lateral magmatic differentiation and cumulate deposition are taking place. Later volcanism on the flanks draws on differentiated magma, and crustal thickness increases towards the main boundary faults. *B* and *C* from Ballard and Van Andel (1977)

The type of topography described in the FAMOUS area appears to be characteristic of ridges with rather slow spreading rates. Ridges with moderate to high spreading rates (e.g. the East Pacific ridge between 21°N and 20°S) are characterized by central ridges with narrow, shallow axial depressions (Figure 4.4). Francheteau and Ballard (1983) describe the results of a detailed survey of three small sections of the East Pacific ridge at 21°N, 13°N and 20°S respectively, and compare these with the FAMOUS results from the Atlantic. They show that at moderate spreading rates (e.g. at 21°N) there is a shallow axial depression comparable in width to the rift valley on the Atlantic ridge (3–5 km) and with similar volcanic edifices 10–50 m in height. The fault-controlled walls of this rift average 100 m in height. The axial trough is situated on a central swell up to 7 km in width. However the highest relief occurs on the uplifted blocks immediately flanking the axial trough. At faster spreading rates (e.g. 13°N) there is only a very narrow axial graben, 200 m in width and 50 m deep, situated on the central swell. At the fastest spreading rates, the graben is absent, and there is only a prominent axial ridge, 2 km wide and 200 m high, situated on the central swell. This swell is still about 7 km wide, but is now 500 m high. These features characterize the East Pacific ridge at 20°S, and are superimposed on a general ridge topography that is much smoother and flatter than that of the mid-Atlantic ridge. In this section of the East Pacific ridge, the spreading rate is about 17 cm/year. The ridge is almost aseismic and displays no signs of major faulting on the axial ridge, which appears to be of volcanic construction. Fluid lavas forming sheet flows are more prominent than pillow lavas in the axial region, and would tend to cover the fissures as soon as they opened up. This would account for their relative scarcity. Hydrothermal activity is evident from the existence of numerous vents and, locally, massive sulphide deposits along the axial zone. Faults and fissures are comparatively uncommon.

Francheteau and Ballard propose a model for an ideal accreting segment of ocean ridge

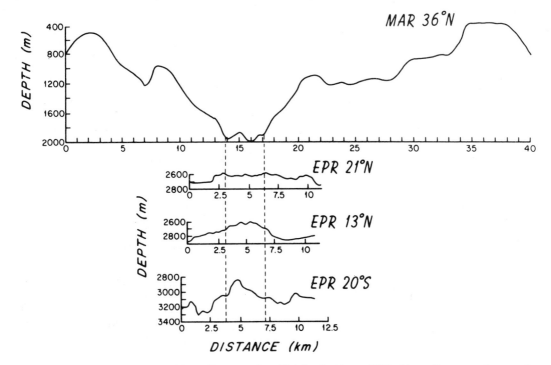

Figure 4.4 Comparison of topographic profile across the mid-Atlantic ridge at 36°N with profiles across three sections of the East Pacific ridge characterized by different spreading rates, at 21°N (6 cm/year), 13° N (10.2 cm/year) and 20°S (16 cm/year). The position of the inner rift zone in the mid-Atlantic ridge is projected onto the other profiles for comparison, to highlight the differences in topography. Vertical exaggeration of ×6.45. From Francheteau and Ballard (1983)

(Figure 4.5). The segment is bounded by active transform faults which are associated with topographic depressions. They suggest that each segment is underlain by a separate magma reservoir that lenses out due to cooling at the transform-bounded ends of the segment. The magma chamber will thus be thicker and the overlying crustal 'lid' thinner, in the central part of the segment, resulting in isostatic uplift and to the observed topographic highs. These hotter regions where the lid is thinnest will produce surface-fed fluid lavas rather than pillow lavas, which will occur distally, around the flanks of the highs.

Lichtman and Eissen (1983) suggest that eruptive phases migrate in a probably random fashion along the ridge crest. The initial, highly active phase is characterized by magmatic inflation, increase in instantaneous spreading rate, and eruption of sheet-flow lavas, and lasts for about 10 years. Then, as the pulse of replenishment moves to another point on the ridge axis, the previous site is characterized by pillow-flow eruptions and decreasing hydrothermal activity.

The explanation for the differences in topography between fast and slow-spreading ridges appears to lie in the differing effects of broad and narrow axial magma chambers. Sleep and Rosendahl (1979) constructed numerical fluid dynamical models for fast and slow-spreading ridge sections, and produce calculated profiles which closely match the observed topography. They conclude that ridges can be classified into slow-spreading (half-spreading rates less than 4 cm/year) with axial rifts about 1 km deep; fast-spreading (half-spreading rates greater than 3 cm/year) with axial ridges several hundred metres high; and hot-spot ridges with spreading half-rates less than 4 cm/year, pro-

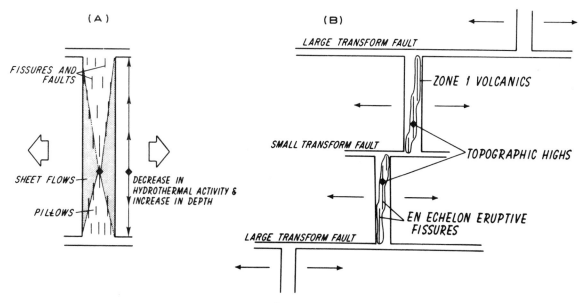

Figure 4.5 Model for an idealized accreting segment of ocean ridge, showing topographic, volcanic, tectonic and hydrothermal variation as a function of distance from bounding transform faults (A), and the predicted arrangement of small scale transform faults and en-echelon eruptive fissures. From Francheteau and Ballard (1983)

nounced axial ridges and higher general elevation. In the fast-spreading ridges, the topography is considered to be the consequence of isostatically compensated thermal expansion due to a large magma chamber. In the slow-spreading ridges, the rifts result from viscous head loss in the upwelling material ascending through a narrow conduit. The hotspot type of ridge is associated with a thick crust overlying a thick magma chamber and produces abnormally elevated topography, as seen for example in the Reykjanes ridge near Iceland, and in the Azores section of the mid-Atlantic ridge, where axial peaks reach nearly to sea-level.

The relationship between rifting and spreading or divergence direction is not as simple as might at first be assumed. Theoretically, extension may occur across a plane making an angle with the plate divergence, or spreading direction (see Figure 3.14). However, in practice, the axial rift fissures appear to be generated by pure extension, with no component of lateral shear. Moreover, the normal faults in the axial region are also generated by pure dip-slip

motion. The 'normal' pattern of a spreading ridge will therefore be a *crenellate* one (see Figure 4.6A) with rift sections offset by short transforms similar to the structure found in the FAMOUS area. Once formed, these transforms will be of constant length if spreading is uniform on both sides. However there is evidence that ridges can either 'straighten' or form more exaggerated crenellations with time, by asymmetric spreading on either side of the axis, balanced by alternations of faster and slower spreading across the transforms bounding each section (Figure 4.6A). This process is believed by Menard (1984) to account for the crenellated Cretaceous anomaly pattern in the Pacific, which is sandwiched between earlier and later straight sections. If the amount of spreading along a rift varies, a different type of stepped pattern results, characterized by wedge-shaped segments (Figure 4.6B). Menard shows how a stepped ridge may evolve from a straight one by alternations of fast and slow sections across the boundary transforms, combined with variations between fast and slow within the segments. This process explains

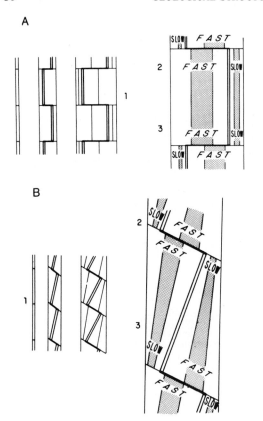

Figure 4.6 Models illustrating the creation of crenellated and stepped ridge patterns by asymmetric spreading. (*A*) Evolution of a crenelate ridge from a straight one by creation and lengthening of transform faults separating segments of faster and slower spreading. (*B*) Evolution of a stepped ridge from a straight one due to a change in the direction of spreading: spreading rate varies from fast to slow at opposite ends of each segment, creating wedge-shaped segments; ridges eventually become orthogonal to the transform faults and to the new spreading direction. From Menard (1984)

how a change in plate divergence direction can be accommodated. For example, the post-10 Ma stepped pattern of the Carlsberg ridge in the NW Indian Ocean (Figure 3.6) is a response to a change in relative movement vector between the African and Indian plates that resulted in a discordance between the old and new sets of magnetic stripes and transforms.

Iceland

This area is the only part of the currently active ocean ridge system that is exposed on land and

has been subjected to detailed examination. Iceland lies on the site of one of the major hot spots (Wilson, 1963; Morgan, 1971) and has been created as a result of continuous volcanism from Miocene times (*c.*16 Ma BP) to the present. The currently active rift runs through the middle of the island in a complex pattern dictated by offsets along three transform faults (Figure 4.7*B*). The active rift coincides with a zone of high heat flow and active high-temperature steam fields. Each new section of rift appears to have formed first as a fault-bounded topographic depression in which accumulated thick sequences of volcaniclastic sediments prior to the first fissure eruptions. The rift zone changes orientation south of the central transform from N–S to NE–SW, and again south of the southern transform to ENE–WSW. According to Saemundsson (1974), this pattern evolved from a much simpler pattern about 4.5 Ma ago (Figure 4.7*A*) when a single spreading axis ran southwards until the junction with the present central transform, whereupon it changed direction to NE–SW. Spreading rates at this time were low, about 2 cm/year divergent motion.

The explanation for the eastward shift was considered by Saemundsson to be the westward drift of Iceland (and of the old ridge axis) in relation to the 'stationary' Iceland hot spot.

The axial rift on the island is a 70 km-wide zone of flood basalts younger than 0.7 Ma, and has been studied in detail by Saemundsson in the northern part of the island (Figure 4.8). The currently active belt contains a series of fissure swarms parallel to the general rift orientation, together with N–S normal faults and a number of calderas and central volcanoes. The older basalts forming the flanks of the rift are gently tilted towards the rift axis at angles of 5–10°, increasing to steeper dips (20–35°) near the rift margins. Further west, the dip changes to around 25°W indicating a broad flexure along the rift flanks.

Within the short ENE–WSW section, on the Reykjanes peninsula, the individual faults and fissures trend 030–035° and have an en-echelon arrangement. A detailed analysis of the fractures in this southern area (Jefferis and

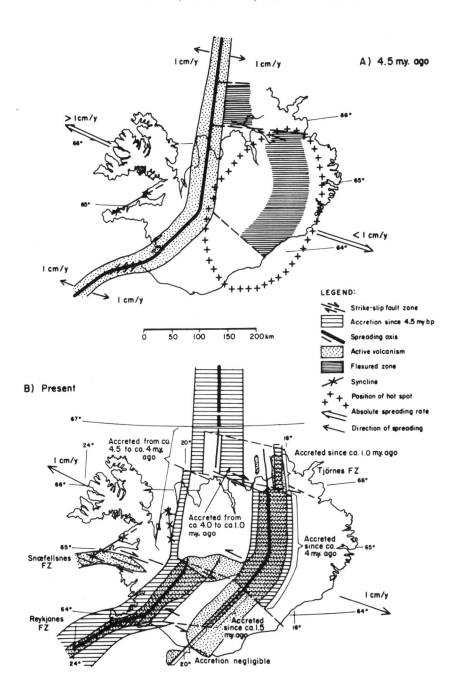

Figure 4.7 Plate tectonic interpretation of the Iceland rift. (*A*) Reconstruction of the rift of 4.5 Ma BP before the eastwards shift of the central segment. New eastern rift zones are beginning to develop, marked by flexural troughs where thick sediments are accumulating. (*B*) Present pattern, after the shift was completed. The new rifts were initiated at different times: the central zone at 4 Ma BP, the southern segment at 1.5 Ma and the northern at 1.0 Ma. The position of the abandoned western rift is indicated by a synclinal structure in the basalts. From Saemundsson (1974)

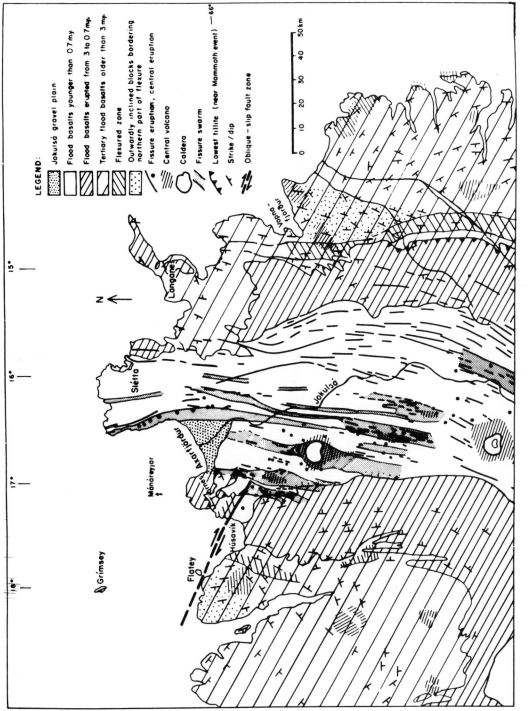

Figure 4.8 Structural map of NE Iceland showing positions of fissures, fissure eruptions and central volcanoes within the active rift zone and the flexure of the older basalts in the flanking regions. From Saemundsson (1974)

Voight, 1981) indicates two main orientations, one NNE–SSW parallel to the local trend of dykes and faults, and the other a broad E–W system ranging from 070° to 130° in trend. The NNE–SSW system is interpreted as extensional fractures (indicating about 0.4% extension) arising from the stress field responsible for the normal faulting and dyke injection. The E–W set occurs near the edges of the active zone and indicates a reorientation of the active stress field as the crust moves away from the active zone. These fractures are interpreted as the result of thermoelastic stresses due to cooling, and indicate extension of the order of 0.1%.

Focal mechanism solutions of earthquakes in this NE–SW zone give uniform horizontal least stress axes oriented NW–SE. Since the plate boundary here has an ENE–WSW strike, these extensional zones are en-echelon zones of normal extension defining an oblique spreading ridge on the regional scale. North and south of this section, the spreading direction coincides with the direction of extension in the normal way.

4.3 Continental rifts

The classification and origin of continental rifts has been a subject of great debate and controversy for many decades. A number of symposia have been devoted to the topic in recent years (e.g. Neumann and Ramberg, 1978; Illies, 1981; Palmason, 1982; and Morgan and Baker, 1983). Some well-known rift systems, for example the African–Red Sea–Gulf of Aden system, the Rhine–Ruhr system, the Oslo graben, the Baikal rift, and the Rio Grande rift have been studied in great detail.

A primary classification of currently active rifts would attempt to subdivide them into those representing constructive plate boundaries, and those that are intraplate. However this subdivision is not as clearcut as it seems, since many apparently intraplate rifts have been interpreted as resulting directly or indirectly from plate boundary processes, and some are directly connected with plate boun-

daries although they do not themselves form plate boundaries. Certain of the latter type of continental rifts provide good evidence as to the nature of the processes of initiation of constructive boundaries in continental lithosphere. The Red Sea–Gulf of Aden rift is a currently active constructive/transform plate boundary which links with the Indian Ocean ridge system and provides a modern analogue of the rift system which resulted in the break-up of Pangaea. This rift system is discussed in detail below. Examination of the passive continental margins of the Atlantic and Indian Oceans provides supporting evidence for this process.

Equally clearly, many apparently intraplate rifts have existed over long periods of geological time without opening to form oceans. The Baikal and African rifts are examples of such structures that are currently active. The Baikal rift originated in Cretaceous times (Logatchev et al., 1978) and the East African rifts in the Miocene (Baker et al., 1972). Presently inactive rifts such as the Permian Oslo graben and the North Sea basin obviously never made the transition to constructive boundaries, although appreciable extension took place.

The *failed arms* of rift triple junctions are an important class of rift first recognized by Burke and Dewey (1973), who suggested that continental splitting took place by the joining together of pairs of rifts from adjacent triple junctions to form a continuous but irregular constructive boundary (Figure 4.9). The rifts that were not developed in the eventual spreading became 'failed arm' graben, or *aulacogens* (Shatsky, 1955). Such rifts, although structurally and genetically linked to the plate boundary, are not part of it. A good example of an aulacogen is the Benue trough in West Africa (Figure 4.11; Burke and Dewey, 1973). The Niger triple junction developed in the Cretaceous prior to the opening of the South Atlantic. The other two rifts developed into the Atlantic Ocean while the Benue trough became a failed arm.

Structurally, a rift is essentially an elongate downfaulted block or graben. However, major

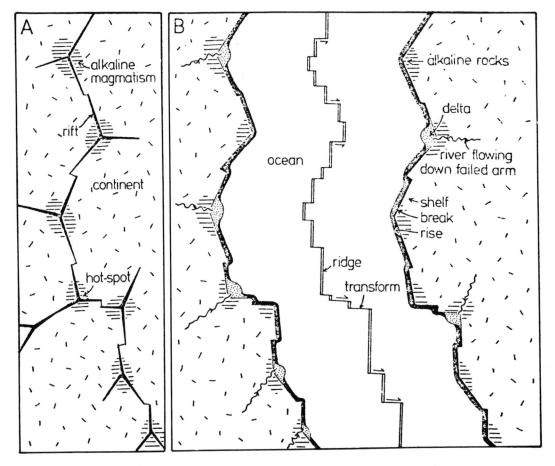

Figure 4.9 Diagram illustrating the early stages of break-up of continental lithosphere to form an ocean. The failure of the crust is attributed to linkage of triple-junction rift systems situated over hot spots. Irregularities in the passive continental margin so formed are perpetuated as transform offsets in the ocean rift system. 'Failed-arm' rifts are preserved on the continental margins. From Burke and Dewey (1973)

rifts comprise complex systems of extensional faults, and exhibit seismic, volcanic and sedimentary features that are directly related to the primary extensional structure. Horizontal extension is therefore fundamental to all rifts, although in many cases the amount of extension may be quite small (around 10% for the Rhine and Baikal rifts for example). All rifts exhibit anomalous crustal and upper mantle profiles, usually interpreted as the result of crustal and/or lithosphere thinning and extension, or asthenospheric diapirism. The geophysical characteristics are very similar to those of ocean ridges and are explained in terms of a region of low-density mantle material, which has been termed the *rift pillow*, correlating with high heat flow. This pillow underlies the rift and often supports an uplifted plateau or dome in the flanking regions. Associated vulcanicity is highly variable both in type and amount. Some rifts exhibit very little vulcanicity while in others volcanic rocks are abundant. Magmatism is typically bimodal, with both alkali basalt–phonolite and tholeiite–rhyolite magma series being represented. In many rifts, the volcanic rocks become increasingly alkaline away from the rift axis. There is usually a complex sequence of events involving several episodes of faulting and vulcanism.

Many rifts are associated with broad domal uplifts as suggested by the Burke and Dewey model, but there are arguments as to whether these are a consequence or a cause of the rifting, or whether both are aspects of a more fundamental driving mechanism. According to Burke and Dewey (1973), rifts are initiated as a result of mantle plumes or diapirs, which produce both the domes and the vulcanicity. In this model, the rifts are a consequence of the initial mantle upwelling. Such rifts have been termed 'mantle-activated' by Condie (1982). However many rifts do not fit this model. An alternative mode of origin is where the rift is produced entirely as a result of lithosphere extension. The conditions necessary for complete extensional failure of the lithosphere were discussed in 2.7. It appears that such conditions can readily be met, particularly at times of unusual plate configurations when continental intraplate extensional stress would be at a maximum. Such a period must have preceded the break-up of Pangaea, when that supercontinent was surrounded by subduction zones, thus maximizing the potential effect of the subduction-suction force (see 2.5). Extra mantle heat sources may not have been necessary for failure to occur, although failure would preferentially occur in warmer, younger orogenic crust rather than in the cooler shield areas. Rifts originating in this way are termed 'lithosphere-activated' by Condie (1982). Clearly such rifts, once formed, will produce lithosphere thinning and volcanism and might

evolve into oceans if the stress conditions were appropriate (see 2.7).

These two fundamentally different types of rift may be distinguished most easily in their initial stages. Plume-generated rifts ought to commence with doming, and should exhibit abundant vulcanicity from the earliest stages. Stress-generated rifts, on the other hand, should commence with graben and basins of sedimentation, and develop vulcanicity at a later stage (Figure 4.10). Some authors have used the terms 'active' and 'passive' to describe the mantle-generated and lithosphere-generated types respectively (Şengör and Burke, 1978). However, this terminology begs the question of which mechanism is more 'active', and is therefore confusing. Furthermore, the classification cuts across the more widely used subdivision between currently active and currently inactive or 'fossil' rifts that have been active in the past.

Some active rifts fall easily into one or other of these categories. The East African rift system for example, with its abundant vulcanicity, although exhibiting extensional stress locally, exists within a continental plate in a state of general compression (see Figure 2.21). At first sight, therefore, this rift is a good candidate for the mantle-generated category. However, as we shall see below, the East African rift is connected via the Afar triple junction with the Red Sea–Gulf of Aden constructive plate boundary, and the whole network needs to be considered in its entirety. Moreover, the state of stress at the time of initiation of the rift system may be very different to the present one.

Mantle-activated rifting requires either sub-

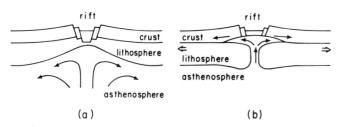

(a) (b)

Figure 4.10 Models illustrating two opposed mechanisms for generating rifts. (a) The lithosphere thinned by mantle convection, with resulting crustal doming and rifting. (b) 'Passive' injection of asthenospheric material to the base of the crust following extensional failure of the continental lithosphere. From Turcotte and Emerman (1983)

lithospheric thinning ('erosion') or the emplacement of an asthenospheric diapir within the lithosphere, preferably spreading along the base of the crust for maximum effect (Figure 4.10*b*). It has been argued that the diapir mechanism is more efficient and that the thinning mechanism involves too long a time-scale (Turcotte and Emerman, 1983; Mareschal, 1983). Neugebauer (1983) has summarized the stages in the development of a rift generated by such a mechanism. The initiation of the rift requires a small perturbation in the inverted density interface at the base of the lithosphere. This perturbation causes a surface uplift with a wavelength of around 500 km for reasonable values of crustal and asthenospheric viscosities. When the amplitude of the perturbations reaches a critical level, the overlying layers are deformed and the crust is thinned. At this stage, the evolution of the rift becomes more rapid. The diapir, now formed, rises at its maximum rate (*c.*5 km/Ma) and is accompanied by more rapid and extensive crustal thinning and high-level rifting. Volcanic activity may now commence. Eventually the diapir loses its buoyancy due to cooling, when the supply of new low-density material diminishes. At this stage the influence of the mass excess above the diapir, caused by the volume of solidified volcanics, prevails, causing basin subsidence and a cessation of vulcanism.

The Baikal and Rhine rifts have been considered as active examples of the lithosphere-generated type. They are intraplate, exhibit only minor vulcanicity and are associated in their early stages with sedimentary basins (Logatchev *et al.*, 1978; Illies, 1978).

The state of stress within the Eurasian plate at the time of initiation of these rifts can only be guessed at, but it is not unreasonable to envisage a state of general E–W extension that would be replaced in the west by compression when the North Atlantic ridge broke through between Britain and Greenland in the early Cenozoic. These rifts therefore could possibly be ascribed to stress generation, since they were initiated in late Mesozoic (Baikal) to early Cenozoic time (Rhine). However several

authors argue for special kinematic factors in the initiation of these rifts. Illies considers that the north–south compression resulting from the Alpine collision was a major contributory factor in the origin of the Rhine rift, and early Eocene collision between India and the Lhasa block may have had a similar effect in the Baikal initiation. The Rhine rift is described in detail below.

Many rifts show evidence of partial control by lines of structural weakness. This is particularly evident in the African rift system, where the East African rift commonly parallels the trend of Precambrian mobile belts and follows the boundaries of the older cratons. Similar examples can be seen in the line of opening of the Atlantic, which in several places follows the trend of older orogenic belts (e.g. the Appalachian belt in N. America and the Pan-African belts in Africa and South America. This structural control is never complete however, and rifts also locally cross-cut previous structures, showing the influence of their own stress field. Failure along planes of weakness with suitable orientations would be expected with either of the principal mechanisms of formation.

Other types of rift develop at convergent plate boundaries on continental crust, and are therefore analogous to the back-arc extensional provinces found on oceanic crust. The best-known example of this type of province is the Basin-and-Range province of the western USA, which is discussed later. Rifting is also associated with collisional boundaries, although in this case the rifts are usually aligned perpendicular to the plate boundary. Examples of such structures north of the Himalayas are described by Molnar and Tapponnier (1978) who ascribe their formation to the stress field generated by the collision (see 5.4).

Two examples of currently active intra-continental rift systems will now be examined in more detail: the Rhine and Afro-Arabian rift systems. These exhibit contrasting properties in some respects but neither can be convincingly classified into either mantle-generated or lithosphere-generated types. In this, as in other respects, they can probably be regarded as fairly typical.

The Afro-Arabian rift system

The well-known rifts of East Africa are part of a much larger regional system that extends across Central Africa to the west to link up with the Atlantic Ocean on one side, and embraces the Red Sea–Gulf of Aden plate boundary on the other (Figure 4.11). To the south, the two main branches of the East African system join and continue southwards to meet the Indian Ocean at Beira, in Mozambique. Associated with these rifts are three major domal uplifts that have been major centres of alkaline vulcanicity since the Miocene, in Ethiopia, Kenya and Darfur respectively. Northwest of Darfur are two other centres, at Tibesti and Hoggar, that do not appear to be directly linked with the main rift network. According to Fairhead (1976), the Ethiopian and Kenya domal uplifts are the focal points for the volcanism of the rift system. It is in these areas that the lithosphere has undergone the greatest amount of thinning. Geophysical evidence suggests that within the domal uplifts, the crust away from the rifts is of normal thickness but is underlain by hot, low-density mantle material with anomalously low seismic velocities within the upper

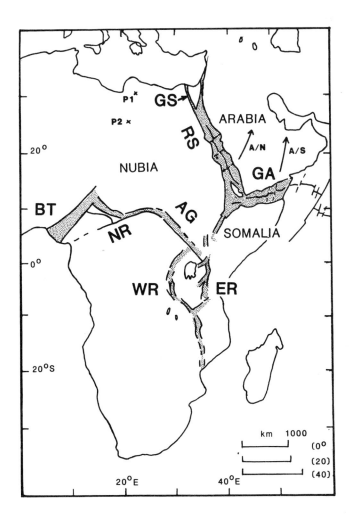

Figure 4.11 Main elements of the Afro-Arabian rift system. BT, Benue trough: NR, Ngaoundere: AG, Abu Gabra; WR, western E. African; ER, eastern E. African; GA, Gulf of Aden; RS, Red Sea and GS, Gulf of Suez rifts. P1 and P2 are the poles of rotation for the Arabian/Nubian and Arabian/Somalian plate movements respectively. After Girdler and Darracott (1972).

50 km of the mantle (Gass *et al.*, 1978), indicating that asthenospheric material has been emplaced at relatively shallow depths.

The Kenya or Eastern Rift

The structure of this rift is summarized by Baker and Wohlenberg (1971). The rift extends from Tanzania, where it joins the Western Rift, to the Red Sea at the Afar triple junction in Ethiopia. It crosses the Kenya domal uplift, which is elliptical in plan and about 1000 km in length along its major axis parallel to the rift. It had a maximum height of 1400 m during the late Pliocene to mid-Pleistocene.

The major faults of the Kenya rift zone (Figure 4.12A) define a complex branching graben structure with an overall N–S trend, although individual faults and graben segments generally strike NNW–SSE or NNE–SSW. The well-defined central graben (the 'Gregory rift') traverses the elliptical uplift and, at its northern and southern ends, is replaced by less

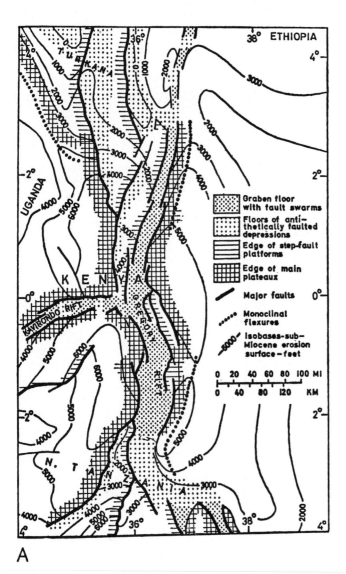

Figure 4.12 (*A*) Structure of the Kenya rift zone showing fault pattern and flexural margins to the rift. Contours indicate amount of crustal uplift since the mid-Cenozoic in feet. From Baker and Wohlenberg (1971)

A

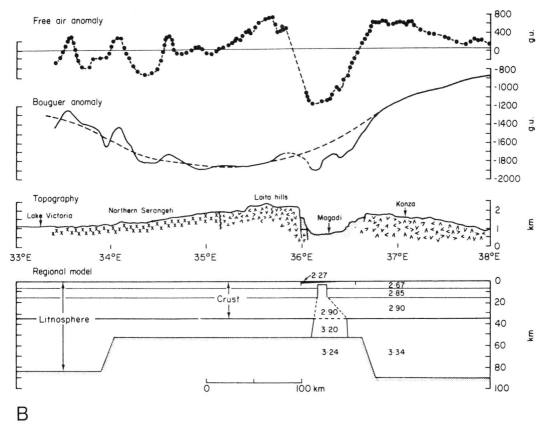

Figure 4.12 (*B*) Free-air and Bouguer gravity anomaly profiles across the eastern E. African rift in Kenya, together with an interpretative model satisfying the gravity data. Note the wide region of thin lithosphere underlying the regional dome and the narrow region of anomalous density immediately underlying the rift. From Darracott *et al.* (1973)

well-defined broad depressions marked by splay faults. The main graben is 60–70 km wide and 750 km long, and is bounded by normal faults arranged in an en-echelon pattern. Between the ends of adjacent en-echelon faults, sloping ramps descend from the marginal plateaux to the rift floor. Major fault escarpments are preserved, ranging up to 2000 m in height. It is estimated from the thickness of the rift fill that the total throws on the major faults may be 3–4 km. The floor of the graben is cut by abundant young minor faults subparallel to the graben walls.

The earliest rift-related structure is a mid-Cenozoic (pre-Miocene) monoclinal flexure along the western rift flank. During the Miocene, extensive vulcanicity occurred, and the

first extensive rift faults developed in the early Pliocene. Voluminous vulcanicity, of predominately alkaline affinity, characterized the period from the Pliocene to the present day, accompanied by periodic fault movements.

At least 5 km of crustal extension is required to explain the displacements on the visible faults, and this figure should perhaps be doubled to take account of the concealed faults beneath the graben floor. The amount of extension is too great to be explained as the result of crustal doming alone.

The rift closely follows the axis of a regional negative Bouguer anomaly (Figure 4.12*B*) interpreted as a zone of thin lithosphere. The gravity data also suggest the presence of a narrow shallow crustal body of dense material thought to represent a wedge-shaped basic intrusive mass about 10 km wide at its top, reaching to 1500 m below sea-level. Focal mechanism solutions of earthquakes along the rift (see Figure 2.22) indicate mainly WNW–ESE to NW–SE extension.

If we assume 10 km of crustal extension across the Kenya rift since the Miocene, this represents a strain-rate of about 10^{-15}/s. Further north, in Ethiopia, the rift system widens and the apparent rate of extension is 3–5 mm/year, corresponding to a strain-rate of about 3×10^{-14}/s across the 25 km-wide zone (Tryggvason, 1982). This difference suggests an increasing rate of widening towards the triple junction. The progressively more alkaline trend of the vulcanicity from Ethiopia to Kenya suggests that tectonic widening has progressed southwards (Mohr, 1982).

The East African rift is linked across Central Africa to two other domal uplifts, at Darfur and Adamaoua, by means of two rifts: the NW–SE Abu Gabra rift and the NE–SW: Ngaoundere rift (Figure 4.13; Browne and Fairhead, 1983). The latter parallels and is linked with the Benue trough, which, according to Burke and Dewey (1973), forms the failed arm of a triple junction with the proto-Atlantic rift.

The Ngaoundere and Abu Gabra rifts are subsiding sediment-filled troughs which have been active since the Cretaceous. The south-west end of the Ngaoundere rift crosses basement over the Adamaoua uplift, where the basement structure is seen to be a dextral shear zone of Pan-African age (De Almeida and Black, 1967). This zone can be traced into Brazil as the Pernambuca lineament. This Precambrian structure was reactivated in the Lower Cretaceous during the initial separation of South America from Africa. Deposition occurred until 80 Ma ago when it was terminated by compressive folding. The lineament was re-activated again during the Cenozoic, when the basement uplifts were formed with their associated vulcanicity. Faults parallel to the Ngaoundere rift show dextral movements with a total displacement of 40 km.

The NW–SE Abu Gabra rift consists of a 150 km-wide trough filled with Cretaceous and Cenozoic sediments, locally more than 4500 m thick. Faulting is parallel to the trend of the rift, which is considered to be purely extensional. The Cenozoic tectonic movements on the Central African rift system are summarized in Figure 4.13. Both rifts are currently aseismic.

To the north, the East African rift is linked with the Red Sea and Gulf of Aden rifts at the Afar triple junction (Figure 4.11). This junction separates three distinct plates: the Nubian (or main African), Arabian and Somalian plates, that are moving independently.

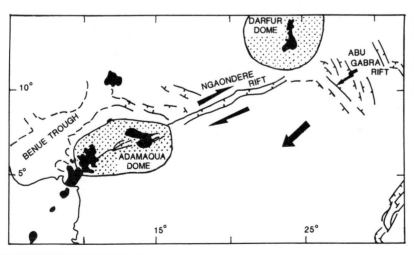

Figure 4.13 Arrangement of rifts and volcanic domes in West and Central Africa. The arrow indicates the inferred direction of opening of the Abu Gabra rift and related dextral shear along the Ngaoundere rift. After Browne and Fairhead (1983).

The Gulf of Aden (Figure 4.14) is composed of oceanic crust, and commenced opening 10 Ma ago in the late Miocene (Laughton *et al.*, 1970). About 260 km of continental separation has taken place at an average rate of about 2 cm/year about a pole situated in NE Africa (Figure 4.11). This rift represents an extension of the Carlsberg ridge in the NW Indian Ocean discussed in 3.1.

The Red Sea rift is also floored by oceanic crust (Girdler, 1969) and the magnetic anomaly pattern indicates an average divergence rate of 2 cm/year for the last 3–4 Ma. Both the Gulf of Aden and the Red Sea have undergone subsidence and volcanism since the Cretaceous.

The Gulf of Suez

Angelier (1985) provides a detailed structural analysis of the Gulf of Suez segment of the Red Sea rift system (Figure 4.11). This area lies at the northwest end of the rift between the Mediterranean Sea in the north and the Gulf of Aqaba–Dead Sea (transform) rift in the south. The main structures trend NW–SE parallel to the orientation of the rift, and to the axis of the main Red Sea rift. The flanks of the rift consist of blocks tilted at 5–35° away from the rift axis, and bounded by large normal faults marking the margins of the main graben (Figure 4.15). There is no evidence of the rift prior to the Oligocene. Nor is there any evidence of pre-rift doming in the plateaux bordering the rift. Immediately pre-rift strata of Eocene age are widespread in the axial zone.

Fault movements and limited tilting appear to have taken place throughout the Miocene and continue to the present. Analysis of the late Cenozoic fault geometry indicates a main set of faults with a mean orientation of 135–140° with minor sets at 080°, 120° and 155°. Fault displacements are predominantly normal dip-slip with a small dextral component. Frac-

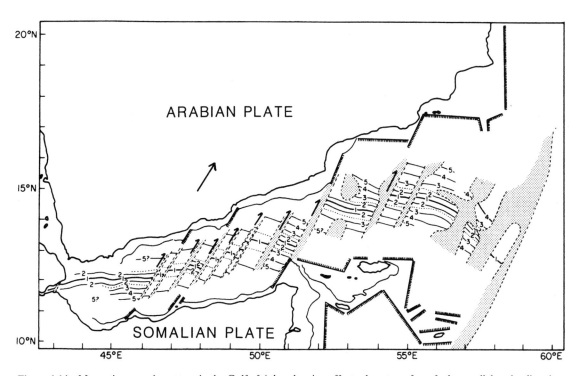

Figure 4.14 Magnetic anomaly pattern in the Gulf of Aden showing offsets along transform faults parallel to the direction of relative motion between the Arabian and Somalian plates. Stippled areas show no clear pattern. After Laughton *et al.* (1970).

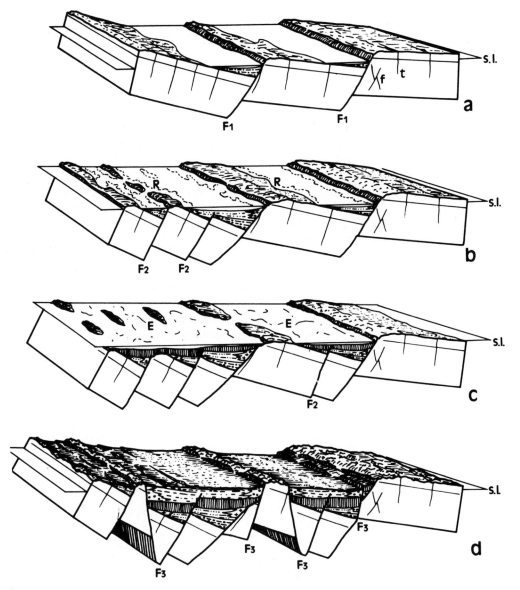

Figure 4.15 Structure of the Gulf of Suez rift. Block diagrams *a–d* show the evolution of a system of tilted blocks accommodating to gradually increased extension: (*a*) late Oligocene–early Miocene; (*b*) early Miocene; (*c*) mid-Miocene; (*d*) Present. F1–3 indicate successive generations of faults; *f*, early conjugate normal fault system; *t*, early tension fractures; *R*, reefs; *E*, evaporite basins (vertical ruling in section); *sl*, sea level. From Angelier (1985)

tures appear to have developed perpendicular to bedding by pure extension and to have been rotated during block tilting, when the dip-slip movements took place (Figure 4.15). The fault geometry indicates an extensional horizontal stress oriented at 045°. The amount of exten-sion is estimated at 20–30%. Estimates based on the amount of subsidence (cf. McKenzie, 1978) yield higher values of 45–50% over the entire width of the 80 km rift section. It is likely that the fault reconstruction method under-estimates the extension.

Origin of the rift system

The Afro-Arabian rift system consists of a network of rather disparate structures connected with the current plate boundary system along the Gulf of Aden–Red Sea line. Although most of the network displays extensional structures, either well-defined graben or swarms of extensional faults, not all of these are or were active simultaneously, and some rifts exhibit important strike-slip components of movement. Parts of the rift network are intimately associated with major domal uplifts that have been a focus of vulcanicity. The most important of these are situated in Kenya, Ethiopia, Darfur and Adamaoua. Other volcanic domal uplifts are located at Tibesti, Hoggar and Air, in Central Africa.

There appear to have been three main phases of movement on the rift network. The initiation of the system seems to be early Cretaceous in age, and is related to the opening of the Atlantic Ocean via the Benue trough failed arm. Cretaceous sedimentary basins on the sites of the Ngaoundere and Abu Gabra rifts, and in the Red Sea and Gulf of Aden, suggests that a continuous network of extensional basins may have existed then, linking the opening Atlantic and Indian Oceans with the Tethys Ocean. Ethiopia and Kenya were characterized by uplift at that time.

The second major phase of activity commenced in the late Eocene with the initiation of the Red Sea rift. This was followed in the late Oligocene–early Miocene by the Gulf of Aden and Gulf of Suez rifts, and in the late Miocene by the East African rifts. The latter developed on the site of the major volcanic domes of Ethiopia and Kenya. At this time also (10 Ma BP), ocean-floor spreading commenced in the Red Sea and Gulf of Aden, and the present plate movement pattern was initiated. This kinematic pattern produces extension across the East African rifts with important seismic activity, but no seismically detectable movement on the Abu Gabra or Ngaoundere rifts.

In the argument as to whether the rift system is lithosphere- or stress-generated, it is important to observe that the basis of the extensional network was in existence in the Cretaceous, before the major volcanic centres were formed. Undoubtedly in East Africa the rift is genetically related to, and post-dates, the volcanic dome. This rift may thus be said to be lithosphere-generated. However in the context of the network as a whole, the location of the domal uplifts is likely to be controlled by the earlier, probably stress-generated, zones of failure and weakness.

The Rhine rift

The Rhine rift system (Figure 4.16) is one of the best-known and intensively studied examples of currently active extensional rifts. A study by Illies and Greiner (1978) summarizes the tectonic evolution of the Rhine graben and relates it to the development of the Alpine orogeny. Tectonic activity in the Rhine system appears to have commenced in the late Cretaceous, about 80 Ma ago, which coincided with the initiation of compressional deformation in the Alps. Rifting and subsidence occurred during the Eocene and Oligocene as the Alpine compressional deformation continued, reaching its climax at the Eocene–Oligocene boundary. In the mid-Miocene to early Pliocene, extensional activity appears to have decreased as the Alpine convergence slowed down, to be replaced by regional epeirogenic uplift. In mid-Pliocene times, the rift became reactivated, but the sense of movement changed from extensional to sinistral shear. This phase is thought to have lasted for the last 4 Ma.

The present stress field has been thoroughly investigated using the overcoring technique for in-situ stress determination (see 2.6). Stress magnitude is calculated as excess maximum horizontal compressive stress over the hydrostatic value, and varies from -0.3 to about 36 MPa. Low values ($c.2$ MPa) characterize the foreland (i.e. the rift region), and high values the Alpine belt. The pattern of stress orienta-

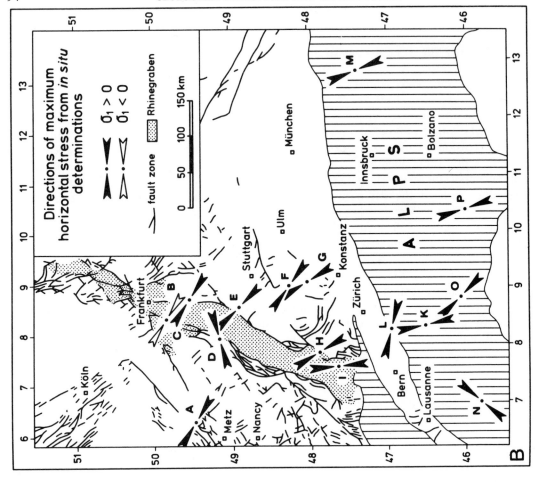

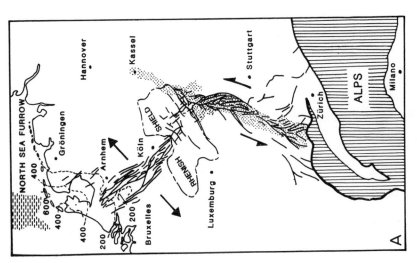

Figure 4.16 Structure of the Rhine–Ruhr rift system. (*A*) Map showing the extent of the rift system and its relationship with the Alps to the south, the North Sea furrow in the north, and the Rhenish shield. Note the differences in fault patterns between the northern sector with orthogonal extension and the southern sector with a sinistral strike-slip component of movement. After Illies (1981). (*B*) Directions of maximum horizontal stress derived from in-situ measurements in the southern sector of the rift. Letters A–P refer to measurement locations. From Illies and Greiner (1978)

tion (Figure 4.16*B*) indicates a general NW–SE-directed σ_1 both in the Alps and across the Rhine rift. This direction is oblique to the rift, which trends at 015° in its southern sector, and is clearly responsible for the current sinistral movements.

The fault pattern is complex (Figure 4.16*A*). Most of the active faults in the graben strike about 170°. A relatively simple orthogonal set of NW–SE and NE–SW normal faults outside the rift change to a complex pattern of branching faults, curving from NE–SW, parallel to the rift, into a N–S trend, indicating a sinistral shear component. Dip-slip displacements occur on both synthetic and antithetic fault sets, indicating extension in an ENE–WSW direction consistent with the trend of σ_1. At many localities, former dip-slip faults parallel to the graben have been overprinted by horizontal slickenside striations arising from the more recent sinistral movements (Mulleried, 1921). Fault-plane solutions of recent earthquakes indicate an oblique sinistral shear mechanism with an overthrust component (Ahorner, 1975). Ahorner calculates the seismic slip rate parallel to the graben axis to be about 0.05 mm/year.

The northern segment of the rift system, northwest of Frankfurt, is characterized by a NNW–SSE trend. Here abundant neotectonic normal faults parallel to the trend of the rift show dip-slip movements but there is no continuous graben feature. This belt extends as far as Arnhem, where it is obscured by a late Quaternary depression bordering the North Sea (Figure 4.16*A*). This depression links with the buried central North Sea graben which was active in the Mesozoic (see 7.3).

The rift system has undergone considerable vertical movement. Depression during the Pleistocene, for example, reached a maximum of 380 m. However the rift is also associated with a regional upwarp, forming raised borders to the graben. Pebble analysis of conglomerates from the rift valley fill indicates that upwarping evolved simultaneously with the main stages of graben subsidence.

The geophysical evidence indicates a thinned crust (*c*.24 km) within the graben. The deep structure, together with gravity and heat-flow data, are interpreted as an expression of a mantle diapir which first produced the early volcanic phase of the system, beginning about 100 Ma BP, and was followed by the tectonic effects associated with the extension.

Kusznir and Park (1984) have discussed the initiation of the Rhine graben in terms of the extensional strength of the lithosphere during the initiation of the rift. Using their mathematical model, and the heat flow of 73 mW m^{-2} measured for the rift flanks (see Table 2.2) they calculate that an extensional stress of *c*. 10 MPa would be required to initiate the rifting. The origin of the extension is likely to be related to the general late Cretaceous–Palaeocene extension preceding the break-up of Laurasia (cf. Bott and Kusznir, 1984) at a time when the supercontinent was subjected to extensional subduction-suction forces on all sides (see Figure 2.15). The effect of the N–S Alpine collision is suggested by Illies and Greiner as an important factor in the initiation of the rift. This event, however, is much later (in Eocene-Oligocene time) than the initiation of the crustal extension, but may have contributed to the initiation of the actual rift feature.

4.4 Extensional provinces at convergent boundaries

It may seem at first paradoxical that extensional regimes should be associated with convergent plate boundaries. However we have seen (see 2.5) that the effect of the subduction process, under certain circumstances, is to exert a tensional stress on the plates on either side of the subduction zone, as first suggested by Elsasser (1971).

Extensional tectonic regimes formed in this way are found on both continental and oceanic crust on the upper slab of the subduction zone. It has been suggested (see 5.1) that the angle of subduction is important in determining the state of stress in the upper slab, and that shallow-dipping slabs are associated with compressional, and steep with extensional, stress. It is important to remember that the upper slabs of subduction zones are the sites of enhanced heat flow associated with the production of volcanic arcs, and that much smaller extensional stresses will be required to produce lithosphere failure in the upper slab under these conditions. Thus, although the extensional stress generated by the subduction process is applied equally to both plates, only the thermally weakened upper plate will fail. The most intensively studied example of a continental extensional tectonic regime related to a convergent plate boundary is undoubtedly the Basin-and-Range province of the western USA, which we shall now examine in some detail.

The Basin-and-Range province

In the south, this extensional tectonic province (Figures 4.17, 4.18) lies immediately east of the San Andreas fault, which marks the American plate boundary. In northern California and Oregon, the province lies east of the Cascades volcanic arc. The Cascades chain is related to the subduction zone marking the boundary between the Juan de Fuca and American plates in the north. The province is about 1000 km wide at its maximum in the north, but narrows southwards as it bends around the margin of the Colorado plateau to link with the Rio Grande rift. The province is characterized by normal faulting, seismicity, high heat flow (c. 90 mW m^{-2} (see 2.3) and a high regional elevation.

The modern extensional structure is summarized by Zoback et al. (1981). It consists of a linear topography of elongate ranges separated by basins filled with Cenozoic and Quaternary terrestrial sediments. The range blocks are spaced about 25–35 km apart from crest to crest, with intervening basins 10–20 km in width. Thus horsts and graben are of approximately equal dimensions. This block structure is controlled by normal faulting in response to a stress field with a WNW–ESE minimum horizontal stress, as indicated by earthquake

focal mechanism solutions and in-situ stress measurements (Zoback and Zoback, 1980). Alignment of volcanic feeder dykes and detailed fault slip data indicate that the present stress system was also characteristic of the late Cenozoic (Figure 4.17B). This stress field appears to be fairly constant throughout the northern and southern sectors of the province, and continues into the Rio Grande rift. Zoback et al. (1981) suggest that this modern extensional structure developed earlier (c. 13 Ma BP) in the south than in the north (c. 10 Ma BP). A comparison of the stress orientation data with the fault and topographic trends shows that the latter are approximately perpendicular to the direction of extension in the north, but oblique in the south. Eaton (1980) notes that the southern part of the province indicates low levels of seismicity compared with the northern, and suggests that the former sector is now largely quiescent structurally. It is likely therefore that the orientation of the stress field has changed since the structure of the southern sector was initiated.

Seismicity is concentrated in broad belts 100–150 km wide along the margins of the northern sector, delimiting a relatively aseismic region in the central part of the Great Basin. Seismic activity is concentrated in the depth range 5–16 km. Much discussion has centred on the problem of how this high-level normal fault extension is accommodated at depth. It now seems clear that, although a few faults exhibit listric geometry (i.e. are concave upwards) at shallow depths of 4–5 km, most steep faults whose geometry has been investigated at depth continue to dip uniformly. The prevalence of tilted blocks and the regional consistency of tilt (Stewart, 1980) suggests that the extension has been accomplished mainly by a tilted block mechanism in the upper crust, detaching on major low-angle displacements at depths of 5–15 km and thus separating the brittle upper-crustal extension from more ductile deformation in the lower crust (Wernicke, 1981).

The amount of extension of the modern phase of deformation is estimated by Zoback et

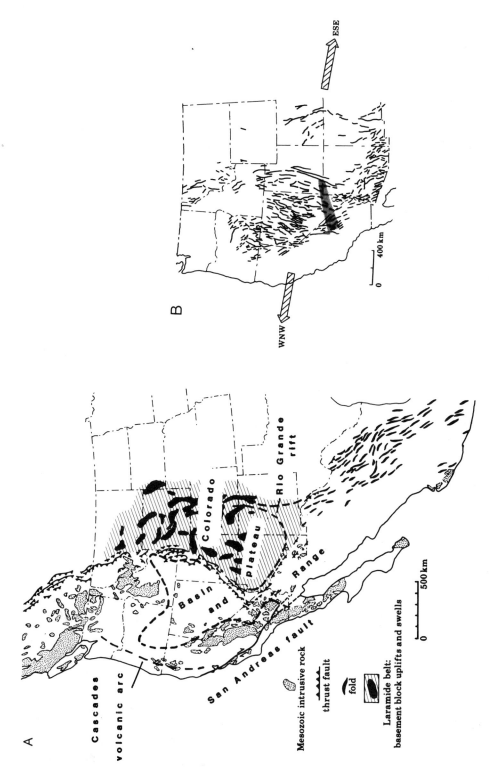

Figure 4.17 Regional tectonic setting of the Basin-and-Range province. (*A*) Map showing the location of the province and of other important tectonic features: the Cascades volcanic arc, the San Andreas fault, the Colorado plateau and the Rio Grande rift, in relation to the belt of intrusive rocks and structures of the Sevier and Laramide orogenies. After Zoback *et al.* (1981). (*B*) Late Cenozoic normal fault trends and present extensional stress orientation derived from earthquake focal-mechanism solutions and in-situ stress determinations; from Zoback *et al.* (1981)

97

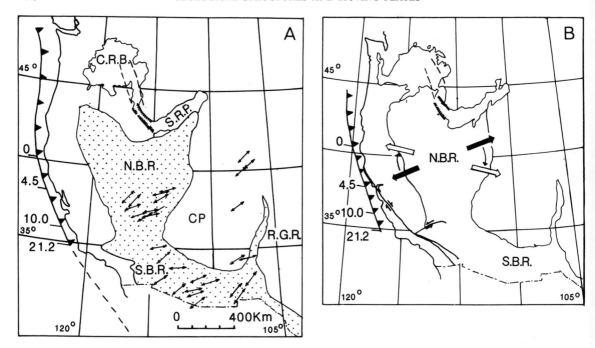

Figure 4.18 (A) Map showing directions of Miocene extension derived from fault and tilted block orientations within the Basin-and-Range province (stippled). Numbers along the subduction zone (toothed line) indicated times (in Ma BP) of cessation of arc volcanism, indicating the contemporary end of the subduction zone. NBR, North, and SBR, South, Basin-and-Range province; CRB, Columbia River basalts; SRP, Snake River plateau; CP, Colorado plateau; RGR, Rio Grande rift. (B) Simplified map based on Figures 4.17B and 4.18A showing the clockwise change in least horizontal principal stress orientation at c.10 Ma ago within the Basin-and-Range province. (A) and (B) after Zoback *et al.* (1981).

al. (1981) as 17–23% from fault analysis. Much larger estimates, up to 100–300% (Hamilton, 1975; Wernicke, 1981), are attributed by Zoback *et al.* to the effects of the earlier phase of extension (see below).

The present crustal structure and tectonic pattern of the Basin-and-Range province cannot be considered in isolation, as they are the products of a tectonic history extending back to Mesozoic and late Palaeozoic times. During the early Mesozoic, a subduction zone extended along the whole coastal belt of western America with a volcanic arc to the east represented by the Cordilleran Mesozoic batholith belt (Figure 4.17A). At about 80 Ma BP, magmatism ceased and was replaced by compressional deformation of the Laramide and Sevier orogenies in late Cretaceous to early Cenozoic time (c.80–40 Ma BP). The fold-thrust belt resulting from this compressional phase affected a broad region up to 1500 km wide, marked by a series of basement uplifts resting on thrust faults. These structures resulted from NE–SW to ENE–WSW compression. This pattern of compressional deformation has been attributed by Coney (1973) to the effects of low-angle subduction along the plate boundary and nearly perpendicular to it.

The compressional regime ended between 40 and 50 Ma ago, probably as a result of the reorganization in Pacific plate movement patterns discussed earlier (3.1). It was replaced by a phase of calc-alkaline volcanism that lasted until the collision of the Farallon–Pacific ridge changed the western boundary of the American plate from a subduction zone to a transform fault. As we saw earlier (Figure 3.4A), this change was diachronous, and the ridge–fault–trench triple junction marking the northern end of the San Andreas fault migrated northwestwards from about 25 Ma ago until the present day. The migration was accompanied by the progressive cessation of the subduction-related calc-alkaline volcanism which ended at successively younger dates northwards (Figure 4.18A).

Extension-related basaltic volcanism occurred in the south between 20 and 30 Ma ago, and in the north around 17 Ma BP. There is evidence over large areas for faulting and tilting associated with vulcanicity, where the faults are cut by modern faults at a distinct angle. In the 'metamorphic core complexes' which form a semi-continuous belt along the western Cordillera, lineated mylonitic gneisses occur, interpreted by Davis and Coney (1979) as low-angle shear zones. These exhibit normal dip-slip displacements and are thought to represent the mid-crustal detachment horizons for the extensional deformation. Some of the low-angle normal displacement zones may have originated as Laramide thrusts.

Considerable support is given to this interpretation by COCORP seismic reflection data from the eastern part of the province (Allmendinger et al., 1983). A series of continuous low-angle reflectors extending more than 120 km across the strike can be traced to a depth of 15–20 km (Figure 4.19). None of these low-angle reflectors appears to be cut by the steep normal faults that appear at the surface. A major detachment, the Sevier Desert detachment, can be traced from a surface zone of normal faulting to a depth of 12–15 km. The authors suggest that 30–60 km of extension has taken place on this fault.

Various estimates have been made of the extensional strain during this early phase. Locally high strains of 50–100% have been reported (e.g. by Anderson, 1971) but it is difficult to estimate the overall strain. The high strains imply average strain rates of around 3×10^{-14}/s compared with the modern deformation which yields a strain rate of $c.5 \times 10^{-15}$/s. The extensional deformation would thus appear to be slowing down. The direction during the early phase was NE–SW (Figure 4.18A), markedly oblique to the present and recent extension. The change in σ_1 direction, illustrated in Figure 4.18B, may be related to the northward migration of the southern end of the subduction zone, the probable source of the extensional stress.

Given the present crustal thickness of about 30 km, a net extension of around 80–100% may be calculated if we assume that the crust attained a thickness of 45–50 km (comparable to that of the Colorado plateau) as a result of the Mesozoic compressional phase, as suggested by Wernicke et al. (1982). The present plateau uplift appears to be compensated by warm low-density asthenospheric mantle replacing normal lithosphere.

The commencement of back-arc extension in the province is linked by Zoback et al. to the last phase of subduction, occurring eastwards of the calc-alkaline volcanic arc. According to their model, the enhanced heat supply, volcanism and extension are secondary effects related to the subduction process. However an alternative model has been suggested by Gough (1984) who argues that the East Pacific ridge may have been underlain by a convective mantle upwelling or plume that became overridden by the American continent as the ridge was subducted during the Cenozoic (Figure 4.20). He thus attributes the formation of the province to a 'primary' mantle source rather than to a 'secondary' diapir as implied by the alternative model.

Résumé

The Basin-and-Range province is taken as a typical example of a continental back-arc extensional regime. Extension commenced in the early Cenozoic when Laramide compressional deformation, attributed to shallow subduction along the western America trench, was replaced by a volcanic arc behind which extension occurred around 440–50 Ma ago. An early phase of extension with a WSW–ENE orientation produced large strains, for which estimates of around 100% have been made, implying strain rates of $c.3 \times 10^{-14}$/s. This early extensional phase was progressively terminated and

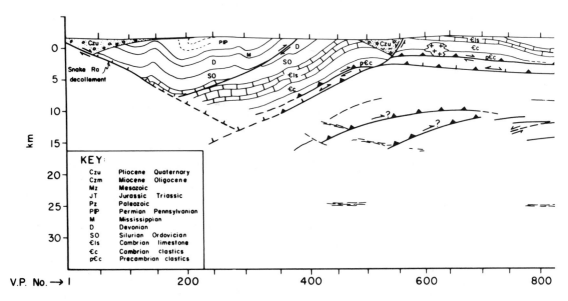

Figure 4.19 Structural interpretation of COCORP deep seismic reflection line across part of the Basin-and-Range province. Solid-toothed lines indicate thrust faults; ticked lines, normal faults; open-toothed ticked lines, low-angle normal faults; solid-toothed ticked lines, thrusts re-activated as normal faults. From Allmendinger *et al.* (1983)

replaced by the current extensional phase with a σ_1 orientation of WNW–ESE between *c*.20–10 Ma ago.

Extension was accomplished by the formation of tilted blocks and by rotation on listric faults detaching on low-angle normal faults/shear zones.

The clockwise rotation of extensional strain axes has been attributed to the northward migration of the end of the subduction zone as it is replaced by the northwards-extending San Andreas fault.

Back-arc spreading basins

A series of oceanic basins (often termed 'marginal basins') are situated 'behind' the island arc/trench systems of the northern and western Pacific rim (Figure 4.21), and also the Caribbean and Scotia arcs of the Atlantic Ocean. The basins occur in the upper slab of the subduction zones, on the concave side of the arcs, and between the arcs and the continental margins. The concept of 'back-arc

spreading' to explain these marginal basins is due to Karig (1971). In Karig's model (shown in Figure 4.23*A*), a new spreading axis is developed above the descending slab because of the diapiric rise of hot mantle material released by the subduction process. An alternative way of viewing this process is as a secondary convective cell of the kind predicted by the thermal model of Elder (see Figure 2.13).

A study of the magnetic anomaly pattern of the basins of the west Pacific (Figure 4.22) shows that they are all relatively young and short-lived in comparison with the Pacific plate itself. The oldest basins in this region commenced spreading about 60 Ma BP and the inactive basins generally had a life of only about 10 Ma. Among the basins that are still active are the Bonin trough, the Marianas trough, and the Lau trough/Havre basin. The Japan, Parece–Vela and South Fiji basins are examples of inactive systems (Figure 4.22*B*).

Weissel (1981) shows that the complex magnetic stripe patterns of these basins cannot be

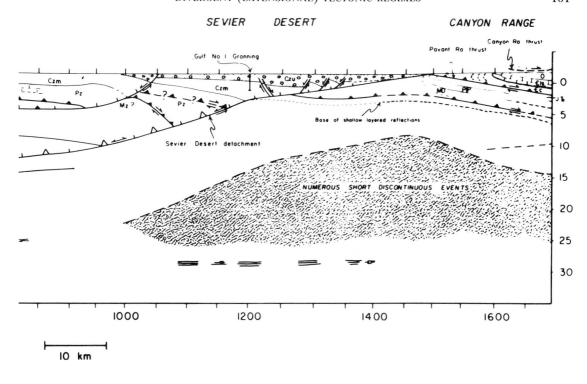

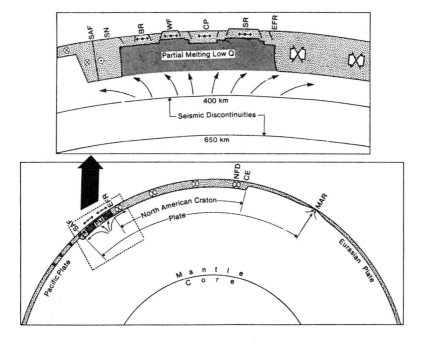

Figure 4.20 Cartoon section illustrating a possible interpretation of the formation of the Basin-and-Range province. SAF, San Andreas fault; SN, Sierra Nevada; BR, Basin-and-Range province; WF, Wasatch Front; CP, Colorado plateau; SR Southern Rocky mountains; EFR, East front of the Rocky mountains; NFD, Newfoundland; CE, continental edge; MAR, mid-Atlantic ridge; PM, zone of partial melting. From Gough (1984)

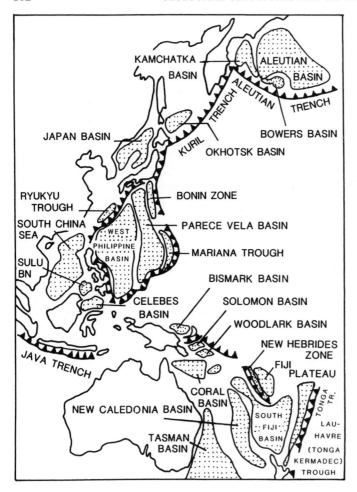

Figure 4.21 Distribution of marginal or back-arc basins in the north and west Pacific Ocean. Subduction zones indicated by toothed lines. After Karig (1974).

interpreted by a simple back-arc extension mechanism. Only some of the most recent spreading patterns can be simply related to the present arc geometry (Figure 4.22A). Most of the basins in this region show patterns of magnetic lineations that are repeated across active or extinct spreading ridges, although the pattern is often complicated by the super-imposition of an active system on an inactive system with a quite different trend. For example, NNE-SSW spreading in the West Philippine basin changes to E-W spreading in the Parece–Vela basin, across the Kyushu–Palau ridge (Figure 4.22–B). In several basins, the spreading axis is offset towards the volcanic arc. This is particularly evident in the Tonga–

Kermadec and Marianas arcs (Figure 4.22B) and may be because the flank region of the volcanic arc is the hottest and weakest part of the basin.

The magnetic anomaly patterns in general indicate that back-arc spreading rates are similar to those on the main ocean ridges, although the duration of the spreading episodes is much shorter. It would appear that tectonic conditions favourable for the generation of back-arc basins are either relaxed relatively quickly, or are easily interrupted, for instance by buoyant ocean floor material on the descending slab arriving at the trench.

The South Fiji and Lau basins appear to have formed as a result of the evolution of

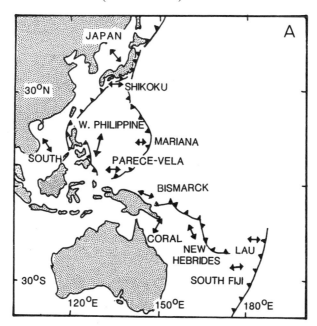

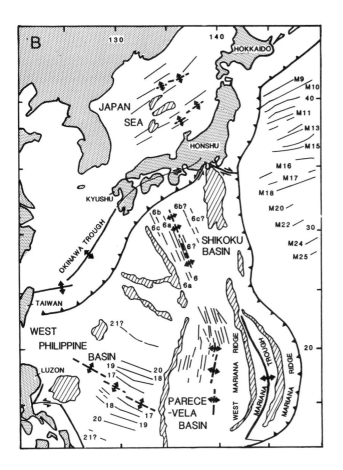

Figure 4.22 (*A*) Spreading directions in marginal basins in the western Pacific Ocean. (*B*) Magnetic stripe patterns, aseismic ridges (hachured) and spreading axes (heavy lines with arrows) in part of the western Pacific ocean. Dashed lines indicate inactive, continuous lines active, spreading axes. Based on Weissel (1981).

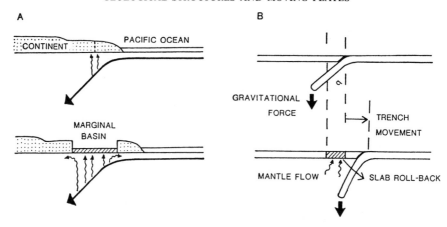

Figure 4.23 Models to explain marginal or back-arc basins. (*A*) The Karig model: 'back-arc spreading'. After Uyeda (1978). (*B*) The 'trench roll-back' model.

three, rather than two, plates, although the details are not very clear.

Although many of the back-arc basins in the Pacific can be explained by the back-arc spreading mechanism of Karig (Figure 4.23*A*), it has been suggested by several authors (see Chase, 1978*a*,*b*; Molnar and Atwater, 1978) that back-arc basin formation may be related more plausibly to global plate motions. In this model, back-arc extension occurs when the resultant of the velocities of the overriding plate and the trench 'roll-back' motion has a component directed away from the trench, when viewed in a hot-spot frame of reference. *Trench roll-back* (Figure 4.23*B*) was defined by Dewey (1980) as the gradual seaward migration of a trench caused by the gravitational pull of the descending slab. This motion is greater for old, cold ocean plate than for younger, warmer ocean plate, which accounts for its importance in the western Pacific, where trenches are subducting old Mesozoic ocean crust. When the component of motion is directed from the trench outwards, towards the ocean, tensional failure will occur in the over-riding plate along the volcanic axis. The spreading direction will be governed, according to this model, by the direction of the resultant velocity vector, and spreading occurs passively just as in the ocean ridges. According to the world

map of 'absolute' plate motions (Chase, 1978; Minster and Jordan, 1978) several of the basins discussed above (e.g. the Marianas and Lau–Havre basins) are opening in accordance with the predictions of this model.

Hynes and Mott (1985), in a study of the Tonga–Kermadec and Marianas arcs, have argued that the roll-back mechanism may not be viable because it cannot account for the variability in back-arc spreading along a specific margin, and in particular cannot explain its recent initiation. They suggest that a more attractive explanation for trench migration is a change in profile of the descending slab, due to changes in the relative rates of motion of the converging plates — a mechanism proposed by Furlong *et al.* (1982) on the basis of thermal modelling. The mechanism proposes that an increase in velocity of comparatively old, cool, and strong oceanic lithosphere will produce a decrease in slab inclination at shallow depths. However, due to the strength of the slab, the line of commencement of down-bending migrates oceanwards, pulling the trench away from the upper plate. In other respects, however, Hynes and Mott confirm that the global plate motion model is applicable to these two arcs in that they both indicate absolute trench motion towards the Pacific.

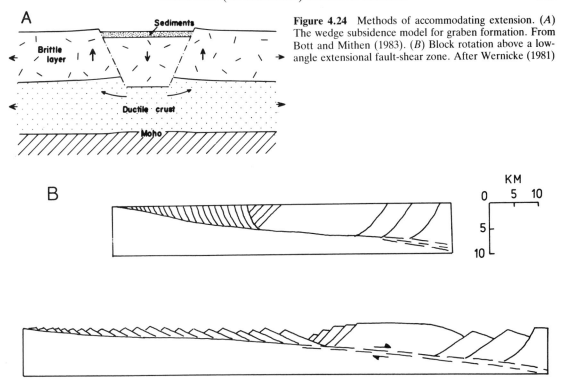

Figure 4.24 Methods of accommodating extension. (*A*) The wedge subsidence model for graben formation. From Bott and Mithen (1983). (*B*) Block rotation above a low-angle extensional fault-shear zone. After Wernicke (1981)

4.5 Structures associated with extensional regimes

Classical views on faulting (e.g. Anderson, 1951) and on the formation of graben and rifts (Vening Meinesz, 1950) envisaged that extension was accommodated primarily by dip-slip movements on steep faults, or by filling of extensional fissures by magma. However the amount of extension achieved by displacement on a steep normal fault is limited by the depth to which a crustal block can sink, and begs the question of how the displacement is accommodated at depth — see for example the keystone or wedge model for rifting (Figure 4.24*A*) proposed by Bott (1976). It is clear that some additional mechanism is required to produce extensions larger than a few km.

The importance of alternative fault mechanisms in achieving large extensions has only been discovered relatively recently, mainly by field workers in the Basin-and-Range province (e.g. Davis *et al.*, 1981; Wernicke, 1981) and by the analysis of seismic records obtained in petroleum exploration of marine basins, for example the North Sea (e.g. Gibbs, 1984). Wernicke and Burchfiel (1982) present a geometric and kinematic analysis of extensional fault systems which highlights three principal methods of achieving large horizontal displacements. These are: (i) the use of low-angle normal faults or detachment horizons; (ii) the rotation of fault blocks, and (iii) the use of curved or '*listric*' fault planes. The demonstration by Wernicke (1981) that large displacements could be accommodated by a combination of rotated imbricate blocks detaching on a low-angle fault or ductile shear zone (Figure 4.24*B*) was of great importance in the study of extensional regimes.

If fault blocks are allowed to rotate (Figure 4.25*A*), the amount of displacement is cumulative, each new fault displacement adding an increment of extension without necessarily causing any additional depression of the surface or additional crustal thinning. The listric

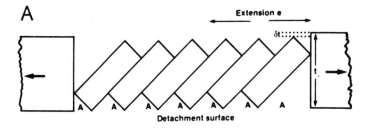

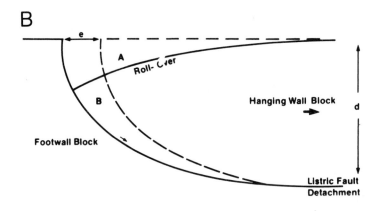

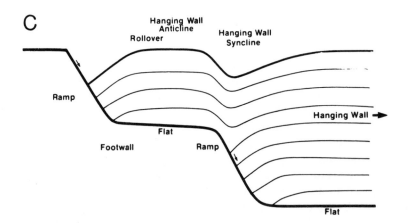

Figure 4.25 Geometry of extension. (*A*) Block rotation above a detachment. Triangular gaps *A* are left. (*B*) Listric fault with hangingwall roll-over anticline. Areas *A* and *B* are equal. The adjustment in hangingwall shape implies internal strain. (*C*) Flat/ramp geometry of fault produces geometrically necessary folding in hangingwall. (*A*), (*B*) and (*C*) from Gibbs (1984)

fault (Figure 4.25*B*) achieves a rotation in the hanging wall merely by its displacement, and is accompanied by an accommodation anticline termed a '*rollover*'. The role of the low-angle detachment fault is critical in both structures. This detachment horizon will generally possess a ramp-flat geometry (Figure 4.25*C*) similar to that found in compressional thrust belts (Dahlstrom, 1970; Boyer and Elliott, 1982). Both these mechanisms require accompanying ductile deformation to alleviate space problems. These problems can be minimized by the introduction of antithetic listric faults, producing listric 'fans', which have the effect of thinning

a

A

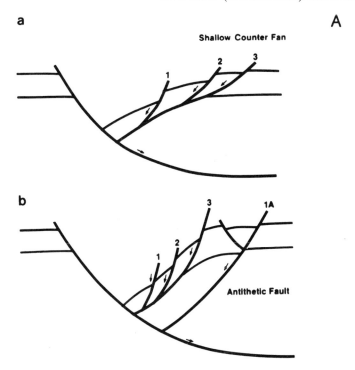

Shallow Counter Fan

b

Antithetic Fault

c

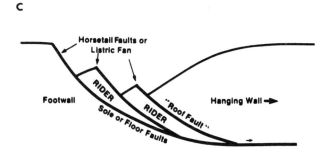

Horsetail Faults or
Listric Fan

Footwall

RIDER

RIDER

"Roof Fault"

Hanging Wall →

Sole or Floor Faults

d

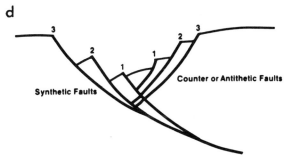

Synthetic Faults

Counter or Antithetic Faults

Figure 4.26 More complex extensional geometry. *A(a,b)* Thinning and extension in the hangingwall by antithetic fault systems. *(c,d)* Horsetail or listric fan producing a series of riders on the sole fault, formed by the migration of the sole fault into the footwall. From Gibbs (1984)

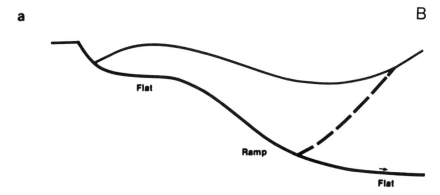

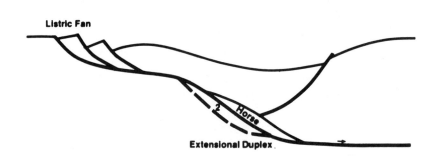

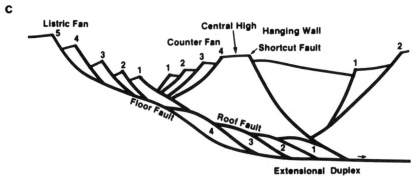

Figure 4.26 *B(a–c)* Evolution of an extensional duplex with associated listric fan, counter fan and elevated central high. From Gibbs (1985)

the rollover in the hangingwall. The addition of synthetic fault sets in the footwall produces complex graben structures (Figure 4.26A,B). Further complications are produced if the basal detachment has a ramp-flat geometry, in which case the detachment may migrate into the footwall to produce extensional duplexes analog-ous to those found in thrust zones (Figure 4.26B).

Gibbs (1984) presents an interpretation of the central graben of the North Sea basin (see Figure 7.10) based on both seismic and well control. His section illustrates well how the conventional model of a symmetrical 'keystone' graben has been replaced by an asymmetrical structure where the key element is a low-angle extensional fault. Although the structure as a whole as seen at the surface retains an overall symmetry, individual elements in the structure are asymmetrical, typically forming V-shaped 'half-graben' with an inclined fault on one side, and inclined rotated bedding on the other.

An important consequence of the low-angle detachment is that displacements may be transferred considerable distances laterally. Extensional structures at the surface may be accommodated by ductile displacements in the lower crust, but the precise form which these take cannot usually be discovered. Lower crustal extension could be accommodated largely by pure shear extensional thinning, or alternatively by simple-shear displacements on low-angle shear zones, or by a combination of both mechanisms. The importance of mid-crustal shear zones in the deformation of high-grade Precambrian complexes is stressed in 9.4.

Deep seismic profiling across the Mesozoic basins off the coast of northern Scotland (the MOIST line), combined with data on deep-crustal structure obtained from the on-land deep seismic profile LISPB (Bamford *et al.*, 1977) indicate that the fault displacements forming the basins (mostly of half-graben type) are transferred along major mid-crustal detachment horizons using a network of pre-existing faults and shear zones used as thrusts during the Caledonian orogeny (see Figure 2.30). It appears likely that in this case the extensional displacements cut through the whole thickness of the crust.

These results have important implications for continental separation and the structure of passive continental margins. Gibbs (1984) illustrates a typical passive continental margin structure based on a section in the Bay of Biscay (after De Charpal *et al.*, 1978) which demonstrates the importance of a basal detachment horizon. The detachment effectively transfers the lower crustal displacement away from the continent towards the ocean. This

mechanism allows broad zones of upper crustal thinning to take place by extensional faulting as shown above, leaving the lower crust unaffected. The extensional structures of passive continental margins are thus likely to be allochthonous, and the basal detachment plane will transfer the displacements to the original site of the rift, where the whole lithosphere was broken through (see Figure 2.29).

Extensional fault systems frequently contain many steep faults with strike-slip displacements. These faults appear to be integral to the displacement pattern and are termed *transfer faults* by Gibbs. They play the same role as oceanic transform faults in transferring displacements from one dip-slip movement plane to another, but differ in that all the movement planes normally detach on the sole fault, rather than at the base of the lithosphere. Transfer faults may separate imbricate fault-fold packages that are quite distinct and uncorrelateable across the transfer faults (Figure 4.27).

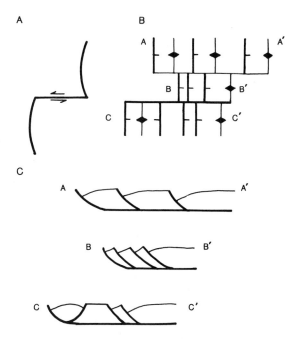

Figure 4.27 Geometry of transfer faults. (*A*) Plan of a simple transfer fault connecting two normal faults. (*B*) plan of two transfer faults separating three zones with different arrangements of normal listric faults, all detaching on the same basal plane. (*C*) Sections *AA'*, *BB'* and *CC'* across (*B*). After Gibbs (1984)

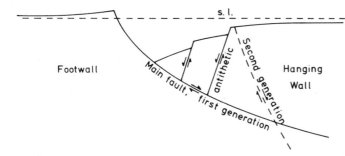

Figure 4.28 Diagrammatic profile illustrating elevation of the footwall, and the development of antithetic faults and of a second-generation synthetic fault in the hangingwall of an extensional listric fault system. From Jackson and McKenzie (1983)

The problem of how listric faults evolve is central in the understanding of the extensional faulting mechanism (Figure 4.28). This problem is addressed by Jackson and McKenzie (1983) in the light of observations in a zone of active seismic faulting in Greece, the Aegean and western Turkey. Seismological observations of the earthquake source are used to determine the fault orientation and slip vector at the origin of fault failure, typically at a depth of 8–12 km. The great majority of large normal-fault earthquakes in this region have dips of 40–50° at their foci, and the dips appear to correspond with observed surface dips in at least two cases. Jackson and McKenzie therefore suggest that faults normally propagate to

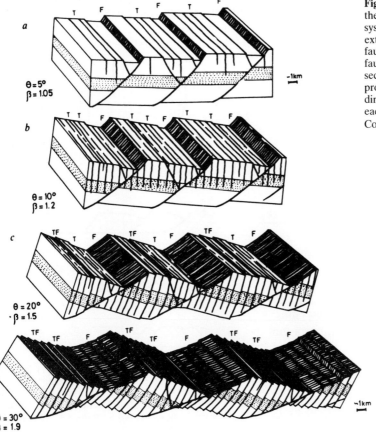

Figure 4.29 (a–d) Model showing the evolution of an extensional fault system with increasing amounts of extension ($\beta = 1.05–1.9$). F, normal fault oblique to bedding (first-order faults); T, tensional fractures; TF, second- and third-order faults, approximately perpendicular to bedding. θ = average tilt of blocks in each case. From Angelier and Colletta (1983)

the surface as planes, and that the listric geometry is developed in a semi-ductile zone below the depth of fault initiation, that is, between 10 and 15 km depth. Increasing extension in the brittle layer is produced by rotation of the initially planar faults as shown in Figure 4.25A. The rotation produces a gradual increase in dip of the initial flat surface or bedding, accompanied by the elevation of the hangingwalls and the depression of the footwalls (Figure 4.28). These vertical movements are the isostatic response to any dip-slip fault displacement and may be clearly seen for recent fault movements in the Aegean, because sea-level is a convenient datum.

It is geometrically necessary for a concave-upwards curved fault to form in order that voids do not occur as displacement takes place. Even if the fault in the ductile layer is initially planar and meets the brittle fault at an angle, the result of movement on the fault system will cause accommodation strains and secondary faulting in the region of the change in dip. This process will eventually result in a listric geometry by abrasion of the hangingwall angle.

According to Jackson and McKenzie, antithetic faults are nucleated at this zone of strain concentration where the abrupt change in dip occurs. After the initial fault set has been rotated to much lower angles of dip, new sets of faults may form at steeper angles. As the earlier faults are cut by the new faults, the former lock, and movement is transferred to the younger set.

Angelier and Colletta (1983) demonstrate this principle in a study of the evolution of extensional fault geometry, comparing the Gulf of Suez (with 10–50% extension, see Figure 4.15), the western Gulf of California (50–100%) and the southern Basin-and-Range province (up to 200%). They consider that at small extensions (10–20%), blocks are gently tilted up to 10° between parallel, widely-spaced faults dipping at 60–65° (Figure 4.29a,b). Many closely-spaced vertical tension fractures are developed approximately perpendicular to bedding during the early stages of extension, and these are used for the later dip-slip displacements when they have been rotated into a suitable attitude (Figure 4.29c,d).

5 Convergent tectonic regimes

5.1 Subduction

The concept of a special type of orogeny related to the subduction process is implied by the plate tectonic model, and was first clearly explained by Dewey and Bird (1970) in their classic paper 'Mountain belts and the new global tectonics'. They demonstrated that the process of subduction at a destructive plate boundary inevitably produces a characteristic association of rocks and structures that, in pre-plate tectonic terminology, would have been regarded as a type of orogenic belt.

These ideas were foreshadowed to some extent by R.S. Dietz (1963) who proposed an 'actualistic' model to explain the formation of 'eugeosynclines', which were so familiar in the literature on orogenic belts. In his model, an accretionary sedimentary wedge is formed on the continental slope, on isostatically depressed oceanic crust, and subsequently deformed and accreted to the continental margin. At about the same time, the essential link was made between trench formation and subduction (see Hess, 1962) that led directly to the ocean-floor spreading hypothesis and to the plate tectonic theory.

The two essential features of the subduction zone are the volcanic or magmatic arc and the trench (Figure 5.1). The trench typically contains a thick prism of sediments overlying the volcanic rocks of the oceanic crust. These sediments are usually undeformed on the outer flank and floor of the trench, but become deformed at the foot of the inner trench wall. This inner trench region, characterized by high pressures and low geothermal gradients (see 2.3), was suggested by Takeuchi and Uyeda (1965) as the site of formation of the blue-schist metamorphic belts characteristic of the circum-Pacific region. These belts form a paired set with the high-temperature, low-pressure metamorphic belts found on the inner side of island arcs and mountain belts. The latter correspond to the zone of high geothermal gradient associated with the volcanic arcs, on the down-dip side of the subduction zone. The presently-exposed, dissected orogenic belts of the circum-Pacific region, with their paired metamorphic belts, thus represent the uplifted products of originally active arc-trench systems.

Dewey and Bird visualized the sequence of events in the evolution of a subduction zone as follows. As the subducting oceanic plate begins to descend, forming a trench, thrusting occurs at its inner wall forming a series of thrust wedges of ocean-floor material such as chert, argillite, carbonates and even basic and ultra-basic igneous material. The thrust wedging process results in a tectonic elevation on the inner side of the trench to form a ridge. This

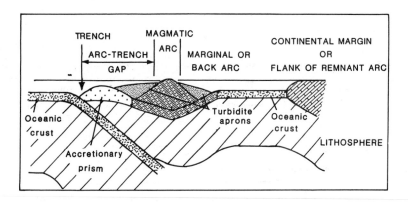

Figure 5.1 Schematic profile across an island-arc subduction zone, showing the main tectonic features. After Windley (1977).

process may be accompanied by gravity sliding of part of this material back into the trench.

Geometry of trench systems

The present trench network is shown in Figure 3.1. Trenches occur on the ocean side of volcanic island arcs and active continental margins. Some trenches are remarkably continuous over great distances. The Peru–Chile trench is 4500 km long and the Tonga trench about 700 km. Typical widths are around 100 km and depths between 2 and 3 km below the ocean basin floor, or up to 11 km below sea-level, although up to 2 km of sedimentary trench fill may be present. The position of trenches bordering active continental margins is obviously determined by the location and shape of the margin. Trenches bordering island chains, however, are typically arcuate. The explanation for this curvature (discussed in 3.2) is thought to be due to back-arc spreading of ocean crust which is inhibited at the cusps of the arcs by aseismic ridges or other obstructions (see Figure 3.9).

Morphology and structure of island arcs

A typical gravity anomaly profile is shown in Figure 5.2 together with a crustal structure profile based on combined gravity and seismic refraction data. Trenches are obviously not in isostatic equilibrium, and represent a mass deficiency which is largely explained by the depth of water in the trench. The crustal thickness does not appear to vary across the trench. It may be concluded, therefore, that the gravity anomaly may be satisfactorily explained by a downward force due to the subduction process depressing the crust; that is, the trench topography is dynamically maintained. The variable sediment infill (some trenches have very little sediment, others are virtually full) appears to have little influence on the amount of crustal depression and depends purely on sediment supply.

The negative gravity anomaly associated with the trench is flanked by positive anomalies: a large positive anomaly follows the volcanic arc, and a smaller broad anomaly occurs on the oceanward side of the trench. The latter anomaly has been explained by Watts and Talwani (1974) as the result of upward bending of the oceanic plate as it approaches the trench, and is a consequence of the lateral strength and continuity of the plate. The positive anomaly associated with the volcanic arc is explained by the mass excess represented by the relatively dense volcanic rocks of the arc. The lower crustal layer is greatly increased in thickness below the arc (Figure 5.2), which possesses a crustal structure similar. to that of the continents. The elevated topography of the volcanic pile is thus only partly compensated isostatically.

Seismicity and the mechanism of subduction

The seismicity associated with subduction zones is one of their most characteristic features. The dipping zone of earthquake foci widely known as the *Benioff zone* constitutes one of the most important pieces of evidence for the hypothesis of subduction of oceanic lithosphere. We have already discussed the seismic evidence for stress orientation and distribution in subducting slabs (see 2.6). The temporal and spatial distribution of earthquakes in subduction zones was investigated by Mogi (1973) who demonstrates, firstly, a progression in time of earthquakes from shallower to deeper levels on the slab, and secondly, that major shallow earthquakes are always preceded by a marked increase in deep seismic activity. He suggests that strain is accumulated gradually near the surface by continued convergence, and that a large earthquake occurs when the accumulated strain reaches a critical value. However, before the main shock occurs, the region experiences numerous smaller shocks that indicate a slight movement along a restricted sector of the slab (Figure 5.3). This movement is transmitted progressively down the slab at a rate of about 50 km/year and terminates at the lower end of the slab in a large shock. The sudden downwards move-

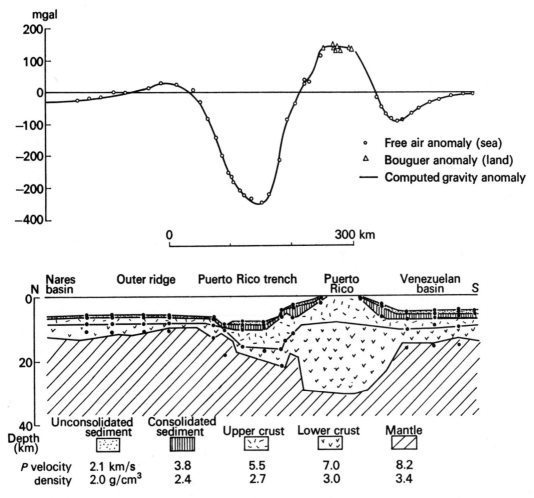

Figure 5.2 Gravity-anomaly and crustal-structure profile across a typical island arc (Puerto Rico). From Bott (1971), after Talwani *et al.* (1959).

ment at the end of the slab rapidly propagates upwards to trigger the large shallow earthquake at the top of the slab. The large shallow earthquakes may have either compressional or extensional focal-plane solutions (see Figure 2.23) and both may be explained by the above mechanism (Figure 5.3). Compressional solutions are explained by underthrust faulting (Figure 5.3A). Compressional solutions are explained by underthrust faulting (Figure. 5.3A) and extensional solutions by normal faulting (Figure 5.3B). Both types of fault have the same effect of releasing the upper portion of the slab, previously 'stuck' at the

trench, and allowing it to move downwards.

A comparison of the seismicity of a large number of subduction zones led Ruff and Kanamori (1983) to suggest that the extent of 'seismic coupling' is an important control on earthquake magnitude. The degree of seismic coupling is controlled by the number and size of 'asperities' on the slab, that is, of strong regions that resist the motion of the slab and have to be broken through before motion can be continued. Earthquake magnitude also appears to correlate both with age of subducted lithosphere and with convergence rate (Figure 5.4). The largest earthquakes occur in

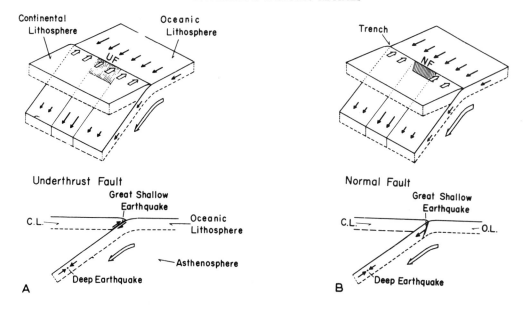

Figure 5.3 Schematic diagram illustrating the mechanism for shallow and deep earthquakes in subduction zones. (*A*) illustrates the underthrust fault (UF) case, and (*B*) the normal fault (NF) case for the production of great shallow earthquakes. Hatched areas denote the faulted segment. From Mogi (1973)

zones where young, warm lithosphere is subducting at fast convergence rates. Ruff and Kanamori note that most of the seismic slip on subduction zones occurs above a depth of about 40 km and that this depth corresponds to a sharp bend in some slabs. They suggest that below this level, seismic activity is related to

the basalt–eclogite phase change which produces superplastic deformation within the slab.

The various factors affecting the geometry of subduction zones (Figure 5.5) are analysed by Cross and Pilger (1982). They recognize four interdependent factors: (i) rate of relative plate convergence: (ii) velocity of absolute upper-

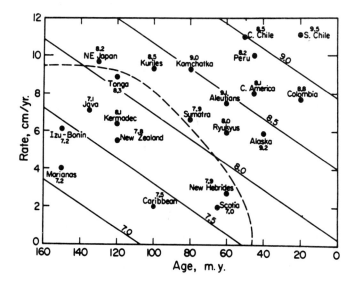

Figure 5.4 The correlation between earthquake magnitude, plate convergence rate, and lithosphere age for 21 subduction zones. The characteristic magnitude values are shown for each zone in a convergence rate *v*. age plot. Regression lines of constant magnitude are shown. The broken line encloses the subduction zones where back-arc spreading is thought to occur (in the lower left part of the diagram). The diagram shows that larger earthquakes are associated with faster convergence and younger age. From Ruff and Kanamori (1983)

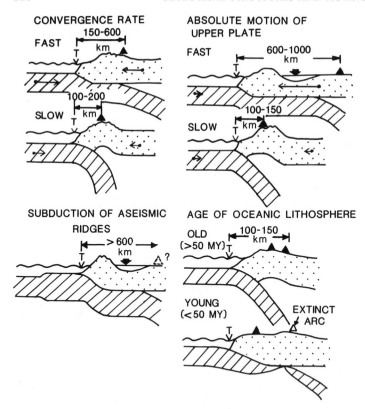

Figure 5.5 Principal controls on the geometry of subduction zones. Four controls are illustrated: convergence rate, absolute motion of upper plate, subduction of aseismic ridges, and age of oceanic lithosphere, each of which influences slab dip, and hence the size of the arc–trench gap. After Cross and Pilger (1982).

plate motion towards the trench; (iii) age of the oceanic lithosphere of the subducting plate; and (iv) presence or absence of intraplate 'obstacles' such as sea-mounts or oceanic plateaux. The effect of each of these factors was examined in examples where the effects of the other factors could be minimized. Cross and Pilger conclude that large convergence rates are associated with a low subduction angle (a relationship proposed earlier by Luyendyke, 1970), depressed isotherms, and a large arc–trench gap (150–600 km). An important consequence of the low subduction angle is the increased length of the inclined Benioff zone. A contemporary example is the Mexican subduction zone, where the Cocos plate is destroyed below the N. American plate. Slow convergence rates, in contrast, are associated with steep slab dips and short arc–trench gaps.

Absolute motion of the upper plate is an important factor independent of the conver-gence rate, since the trench is fixed to the upper plate and must move with it relative to ascending or descending mantle flow regimes. Thus rapid upper-plate motion may override the trench and reduce the influence of gravitational sinking on the slab. Fast absolute upper-plate motion towards the trench also produces a low subduction angle and a large arc–trench gap. The position of the arc is liable to change, and a new arc may develop 600–1000 km inland from the old. Again the Mexican subduction zone may be used as an example. Slow or retrograde absolute motion of the upper plate has the opposite effect of steepening the slab and causing a seaward migration of the trench. The Central American subduction zone, where the Cocos plate descends beneath the Caribbean plate, is an example. In contrast to the subduction zone further north, where the North American plate is rapidly overriding the trench, the Caribbean plate has a small component of motion away from the subduction zone and the arc–trench gap is much smaller.

A correlation between age of subducted oceanic lithosphere and slab length was noted by Farrar and Lowe (1978). The related correlation between age and slab inclination is a consequence of the increasing thickness, average density and decreasing sea-floor elevation produced by the gradual cooling of ocean basin lithosphere as it moves away from its site of formation at a ridge (see 2.3). Since most of the increase in thickness is produced in the mantle rather than in the crust, the mean density of the lithosphere increases, thus reducing buoyancy, increasing the slab dip, and hence potentially decreasing the arc–trench gap. However this effect is opposed by a tendency for the zone of magma generation to occur at higher levels in younger, warmer slabs, thus decreasing the arc–trench gap. A

test of this relationship is provided by the subduction of the Nazca plate along the Peru–Chile trench (Figure 5.6). Cross and Pilger (1982) demonstrate that the southwards decrease in age of Nazca plate being subducted along this trench correlates with a change to shallower and less well-defined seismicity, and with a decrease in the arc–trench gap.

The presence of aseismic ridges, sea-mount chains or oceanic plateaux on the subducting plate all involve regions of reduced mean density which increase the relative buoyancy of the slab and consequently decrease the angle of subduction. Very low angles of subduction are common, and the volcanic arc may be completely extinguished. Nur and Ben-Avraham (1983) note several examples of volcanic gaps in subduction zones which they relate to oblique consumption of aseismic ridges. Examples cited are the Cocos, Nazca and Juan

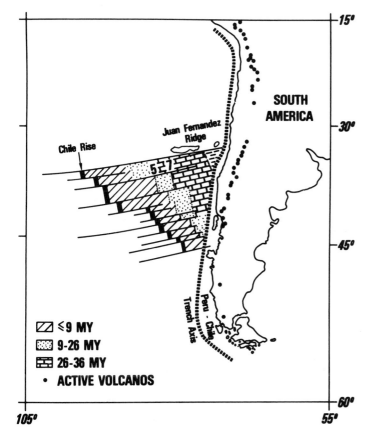

Figure 5.6 Relationship between the width of the arc–trench gap and the age of subducted lithosphere along the southen Peru–Chile trench. Note that the width of the gap decreases southwards with decreasing age of oceanic crust of the Nazca plate. From Cross and Pilger (1982)

Fernandez ridges on the Cocos and Nazca plates (Figure 5.7). The effect is most strongly marked in the case of the Nazca ridge which, because of its marked obliquity to the direction of subduction, has swept across a long sector of the subduction zone. This has caused a large gap in the vulcanicity in north and central Peru, that coincides with a very low subduction angle (< 10°) (Barazangi and Isacks, 1976).

Two additional effects were noted by Cross and Pilger. Accretion of sediment in trenches tends to flatten the slab inclination at shallow depths, because the weight of the accretionary prism depresses the plate prior to subduction (Karig *et al.*, 1976). Secondly, long-continued subduction may thicken the upper plate because of the cumulative effects of accretion and depression of the isotherms (James, 1972).

The four major factors described above all control the state of stress in the upper plate and consequently affect the structures produced there. Moderate to steep subduction is associated with extensional stress in the upper plate, and shallow subduction with compression. These relationships are a consequence of the relative size of the contributions of: (i) thermally-induced isostatic uplift of the upper plates, coupled with the subduction-suction force (producing tension); and (ii) the shear stress produced along the plate contact (producing compression). Oceanic plates are normally in compression (see 2.6). Low-angle subduction increases the area of mechanical coupling between the two plates, and allows shear stresses with a large component of horizontal compression to be transmitted into the upper plate. Steep subduction minimizes this effect and maximizes the effects of the extensional forces.

As discussed earlier (2.5) the state of stress

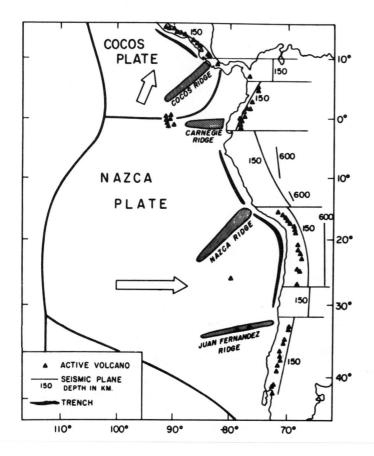

Figure 5.7 Map showing the relationship between the positions of aseismic ridges on the Nazca and Cocos plates and gaps in the volcanic arc. Note the wide gap associated with the oblique Nazca ridge, discussed in the text. From Nur and Ben-Avraham (1983)

across subduction zones is affected mainly by the opposing effects of negative buoyancy generated by the subducting slab (producing extension) and the compressive resistive forces opposing the convergent motion. As Solomon and Sleep (1980) pointed out, the general state of intraplate stress suggests that these two effects are broadly in balance, with some subduction zones exhibiting net extension and others net compression.

Bayly (1982) has pointed out, following Scholz and Page (1970) that the curvature of arcuate trenches, together with the curvature of the Earth's surface, must force a subducting slab into a more constricted space at depth. It is suggested that this constriction is taken up by buckling below a depth of about 100 km involving a range of dips from 35° to 55° in a slab initially dipping at 45° (Figure 5.8). There is some evidence that plate geometry as indicated by seismic foci distribution is compatible with this model.

Structure of accretionary complexes

A combination of seismic reflection, deep-sea drilling and side-scan sonar programmes in a number of active subduction zones has led to a significant increase in understanding of the structures and processes of deformation affecting accretionary sedimentary wedges at subduction zones. We shall examine in detail results from studies in the Peru trench in the West Pacific, the Barbados ridge in the Caribbean, the Makran complex in the Arabian Sea and the Hellenic trench in the Mediterranean. These examples cover the main types of active subduction zone currently recognized. These studies have suggested that the structures of active accretionary prisms are analogues of onland fold-thrust belts. The main processes and structures that have been inferred for accretionary complexes are shown schematically in Figure 5.9 (see Scholl *et al.*, 1980). The front or leading edge of the prism is dominated by frontal accretion whereby *offscraping* of the sedimentary cover (Figure 5.9*a*) is achieved by a synthetic imbricate thrust complex above a basal décollement plane that underthrusts the sedimentary prism. This basal plane and other major thrusts are thought to act as conduits for dewatering of the buried sediments. Further down the décollement plane, *underplating* or 'subcretion' may take

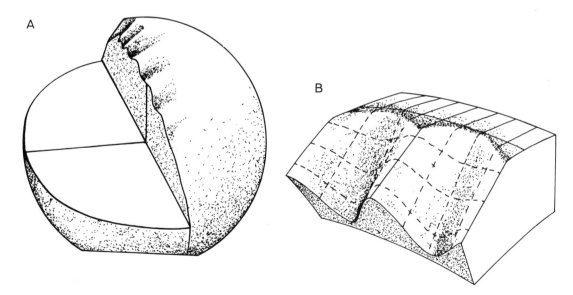

Figure 5.8 Schematic models to illustrate the buckling of a subducted slab due to the geometric shortening effect brought about by curvature of the Earth's surface (*A*), and by the arcuate outcrop of the zones (*B*). After Bayly (1982).

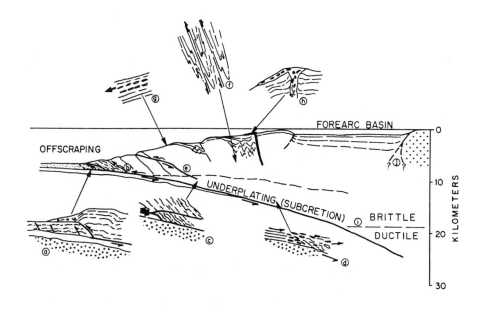

Figure 5.9 Structures and processes in an idealized accretionary prism. (*a*) Zone of frontal accretion by imbricate thrusting where the upper part of the incoming sedimentary sheet is *offscraped*; (*b*) décollement beneath which the deeper part of the incoming section is underthrust; the décollement plane and fault at (*e*) may serve as *dewatering* conduits, as shown by wiggly arrows; (*c*) *underplating* (or *subcretion*) by development of duplex structures; (*d*) underplating by diffusive mass transfer; (*e*) discrete fault cutting through the entire accretionary prism; this fault may enable the prism to adjust its shape to the 'critical taper', and may subsequently be used in strike-slip displacements; (*f*) deeply buried offscraped material, subsequently tightly folded, cleaved and rotated to a steep orientation; (*g*) gravitational spreading of slope sediment apron; (*h*) diapir of water- and perhaps gas-charged, disrupted sediment; (*i*) brittle-ductile transition below which viscous-flow models of deformation may be more appropriate; (*j*) basement terrain defining the inward boundary of the accretionary prism. From Moore *et al.* (1985)

place by the formation of thrust duplexes, which have the effect of thickening and raising the accretionary complex (Watkins *et al.*, 1981). More distal portions of the complex may be characterized by strongly deformed, steeply dipping material exhibiting tight folding and cleavage. Near-surface parts of the complex may show gravity spreading and slumping of the sediment pile.

The process of accretion at a subduction zone may thus be visualized as the progressive upward stacking of the sedimentary pile by the successive emplacement of new thrust wedges at the base, causing overall thickening and uplift of the prism. Eventually this process produces a topographic ridge or *trench–slope*

break, which is separated from the volcanic arc or continental margin by an undeformed *fore-arc basin*.

The sedimentary cover on the oceanic plate is not all accreted on to the upper plate. Hilde (1983) has highlighted the importance of graben in preserving pockets of sediment which are then subducted along with the oceanic crust (Figure 5.10), a mechanism first suggested by Isacks *et al.* (1968). Oceanic plates are typically scarred by normal faults that are caused by flexural extension as the slab bends down into the trench. According to Hilde's model, the graben may be empty on approaching the trench but become filled tectonically with material scraped from the upper slab. Another

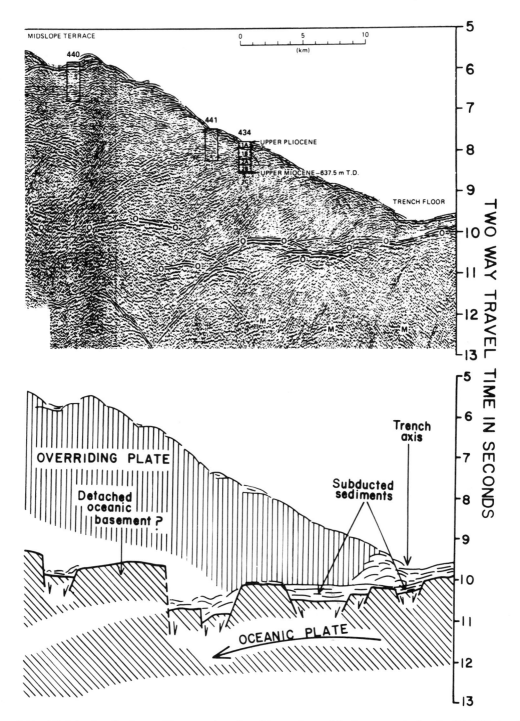

Figure 5.10 (*A*) Seismic reflection profile across the axis and lower slope of the Japan trench along lat.39°45′N, showing the geometry of the subducted oceanic plate (see lower sketch). Numbers refer to drill holes; *O*, surface of oceanic plate; *M*, Moho. From Hilde (1983)

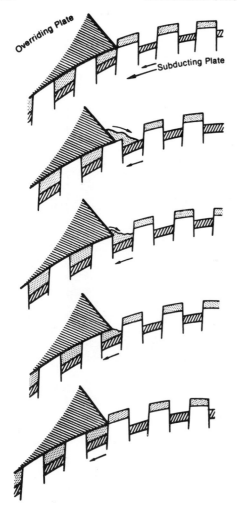

Figure 5.10 (*B*) Model illustrating the process of sediment subduction by tectonic displacement of sediment into graben, which preserve the sediment beneath the overriding plate. From Hilde (1983)

mechanism for removing sedimentary material from the inner trench slope is suggested by Cande and Leslie (1986) in a study of the south Chile trench. They point to the large basement relief associated with the transform faults, in which large pockets of sediment can be preserved, and also to the erosive effect of major fault scarps if they move obliquely across the base of the upper plate.

The importance of these mechanisms is that they provide answers to two problems. One is the apparent discrepancy between the rela-tively small sediment volume in trench complexes compared with the calculated accretion rates (Gilluly, 1971). The second is the crustal isotopic signature of certain mantle-derived magmas (see McKenzie, 1983).

5.2 Some active subduction zones

The Peru trench between 10° and 14°S

Warsi *et al.* (1983) report the results of a study, combining GLORIA side-scan sonography and seismic reflection data, on the Peru trench in the region of its intersection with the Mendano fracture zone (Figure 5.11*A*). This section of trench is interesting in that no obvious accre-tion of sediments seems to occur on the inner trench wall, and it has been supposed that sediment subduction is dominant and that the upper plate is being tectonically eroded (Kulm *et al.*, 1981; Hussong and Wipperman, 1981).

The Nazca plate in this region exhibits a NNW–SSE sea-floor spreading fabric (Figure 5.11*B*) consisting of fault blocks formed at the spreading axis. The blocks are generally tilted, and are bounded by normal faults with scarps facing in both directions. Individual blocks display 100–200 m relief and are draped by 100–150 m of pelagic sediments. Some blocks can be traced for over 100 km along strike.

The Mendano fracture zone consists of a series of parallel ridges and troughs with up to 1 km relief. It widens from less than 50 km across in the west to about 100 km near the trench. Several sea-floor volcanoes were found, about 10 km in diameter and rising to over 700 m from the sea-floor. At about 100 km west of the trench, the Nazca plate starts to bend downwards towards the trench. In this zone, the spreading fabric is overprinted by a set of new faults sub-parallel to the trench axis and to the spreading fabric in the north, but oblique south of 12°S. The new faults displace the pelagic sediments with maximum vertical throws of about 200 m in some cases, and have a spacing of 3–5 km. The faults form graben, some of which can be traced for more than 100 km.

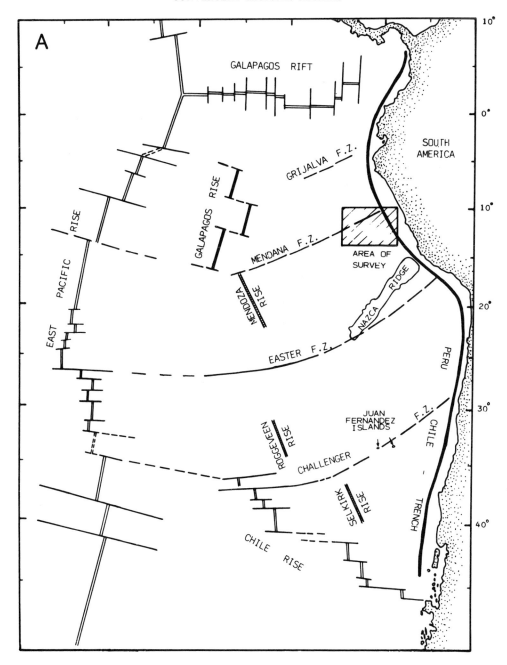

Figure 5.11 (*A*) Simplified tectonic map of the Nazca plate, showing the location of the area of Figure 5.11*B*. From Wars: *et al.* (1983)

The trench axis is turbidite-filled, with a flat floor at a maximum depth of 6300 m. The turbidites mask the structures on the subducting plate in the north, where they are folded against the base of the inner slope. The tectonic deformation front appears as a single continuous feature. It is believed that this tectonic accretion is temporary and that the sediments will eventually be emplaced in the graben by slumping and will be subducted.

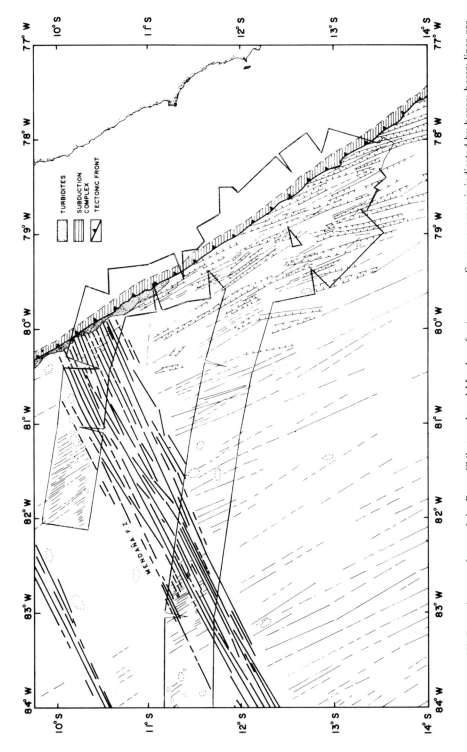

Figure 5.11 (B) Tectonic map of part of the Peru–Chile trench and Mendana fracture zone. Sonar coverage is indicated by boxes; heavy lines are fracture-zone lineaments; light lines indicate spreading fabric; ticked lines are faults on the subduction down-bend; oval features are sea-floor volcanoes. From Warsi et al. (1983)

This sedimentary fill thins southwards and terminates at about 11.5°S. Reflection profiles across a large partly-subducted graben showed two stages in the progressive filling of the graben by slumped sediments, the section nearest the trench axis being the more completely filled.

The Barbados ridge complex

This region, described by Westbrook (1982), has been interpreted as an example of a mature fore-arc complex. It probably represents an intermediate stage between the young, deep-water, accretionary prisms associated with well-defined trenches, such as the western American example already described, and the various uplifted, fossil complexes of Mesozoic/ early Cenozoic age forming inactive arc systems around the Pacific rim and in Indonesia.

The Barbados ridge is a N–S elevated structure lying 150 km east of the active volcanic arc of the Lesser Antilles (Figures 5.12, 5.13). This arc marks the site of the Lesser Antilles subduction zone where the Americas plate passes beneath the Caribbean plate at a rate of about 2 cm/year. The arc is bounded on both sides by major transform faults (Figure 6.12A). West of the volcanic arc lies the Grenada back-arc basin. The island of Barbados is the highest part of the ridge, which extends for over 500 km. The ridge is an accretionary complex made up of sedimentary rocks lying in a linear trough in the volcanic basement. The axis of this trough reaches a depth of 20 km beneath Barbados, and corresponds to an isostatic gravity low. The trough is interpreted as the site of the original trench marking the line of subduction. The eastern part of the ridge is marked by a positive isostatic gravity anomaly corresponding to the outer trench slope of other complexes.

The deformation front lies at the eastern margin of the complex, on the lower continental slope (Figure 5.12B). The complex becomes increasingly deformed westwards, and the island of Barbados exhibits strongly folded and faulted Eocene flysch overlain by

Upper Eocene to Miocene pelagic sediments. West of the ridge, and between it and the volcanic arc, lies the Tobago trough, containing about 3 km of undeformed sediments.

The style of deformation varies along the deformation front. In the south, gently asymmetric east-facing folds with amplitudes of 500 m and wavelengths of 8–9 km ride on thrusts which dip westwards at 20°. These thrusts are interpreted as listric faults detaching on a major décollement plane, parallel to basement, that can be traced at least 30 km westwards beneath the complex. Undeformed bedded sediments lie below this major décollement. In the north of the complex, the deformation is so intense and chaotic that no obvious structure can be discerned from the seismic profiles. This change corresponds to a northwards increase in slope gradient and to a decrease in width and height of the complex. These changes may relate to a northwards decrease in sediment supply. A series of steps in the ridge topography appear to be caused by the intersection of ridges in the oceanic basement. These are oblique to the line of subduction and would produce ponding and lateral sweeping of sediments as described above for the Peru trench.

The structure of the southern part of the central Barbados ridge and Barbados trough has been investigated by means of detailed SEABEAM-sonar bathymetric mapping and a high-resolution seismic survey (Biju-Duval et al., 1982). The seismic profiles (Figure 5.17) show clearly the asymmetric folds and reverse faults of the frontal overthrust region (Figure 5.14A). To the west, broad, km-wide synclines are evident, with syntectonic infilling in their upper parts, separated by narrow asymmetric anticlines associated with steep reverse faults dipping both east and west (Figure 5.14B). On the west side of the Barbados ridge, the sediments are deformed into gentle west-facing asymmetric folds associated in places with westwards-directed reverse faults (Figure 5.14C). The ridge complex thus displays a degree of overall structural symmetry.

In the north, thrusts have been proved in

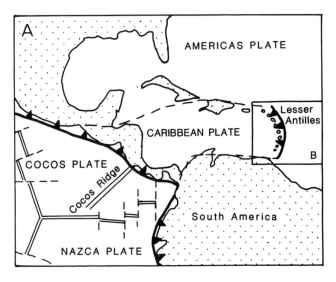

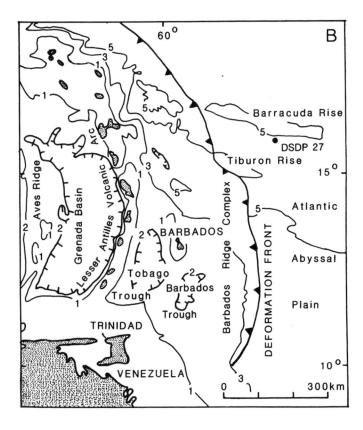

Figure 5.12 (*A*) The location of the Lesser Antilles subduction zone in the plate tectonic framework of Central America. (*B*) Main structural features of the Lesser Antilles subduction zone, showing the location of the Barbados ridge and Tobago trough, situated between the volcanic arc and the deformation front. After Moore *et al.* (1982).

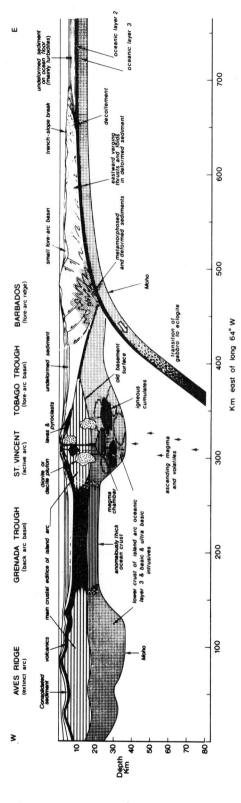

Figure 5.13 Interpretative cross-section of the Lesser Antilles subduction zone. Note the position of the Barbados ridge situated over the site of subduction. From Westbrook (1982)

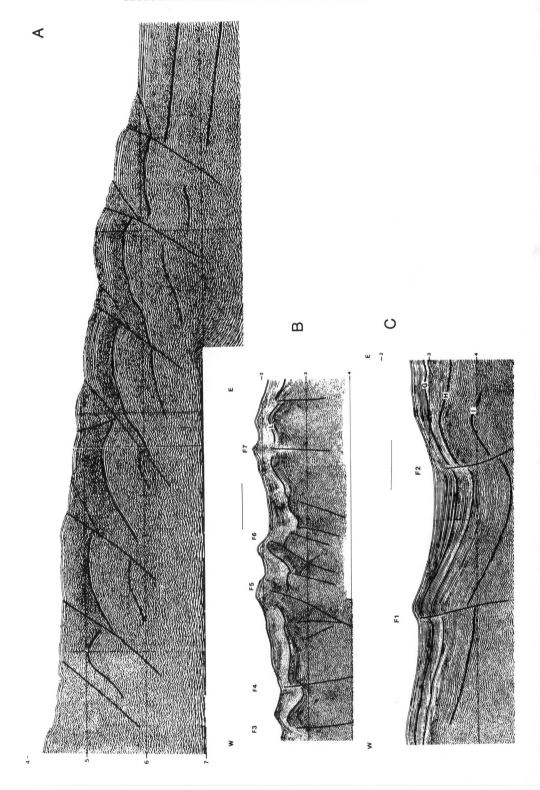

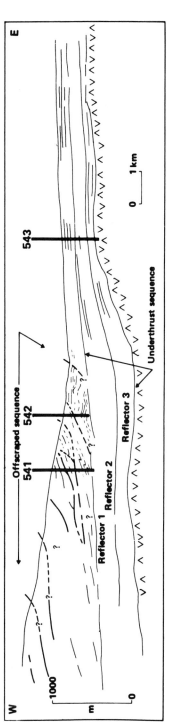

Figure 5.14 (*A–C*) W–E Seismic reflection profiles across the southern part of the Barbados ridge complex (scale bar 2.5 km) (*A*) asymmetric E-facing folds and reverse faults near the deformation front; (*B*) more symmetric pattern of E- and W-facing fold/fault structures on the central part of the ridge; (*C*) W-facing fold/fault pattern on the western side of the ridge; note rapid changes in stratal thickness in both (*B*) and (*C*) indicating the influence of deformation contemporaneous with sedimentation: for example, most faults die out below the surface, and synclines deepen towards their centres. From Biju-Duval *et al.* (1982)

Figure 5.14 (*D*) Tectonic summary profile illustrating the interpretation of the seismic and borehole data (true scale), from the Atlantic abyssal plain in the east to the centre of the ridge in the west. Thick, vertical, numbered lines are boreholes. From Moore *et al.* (1982)

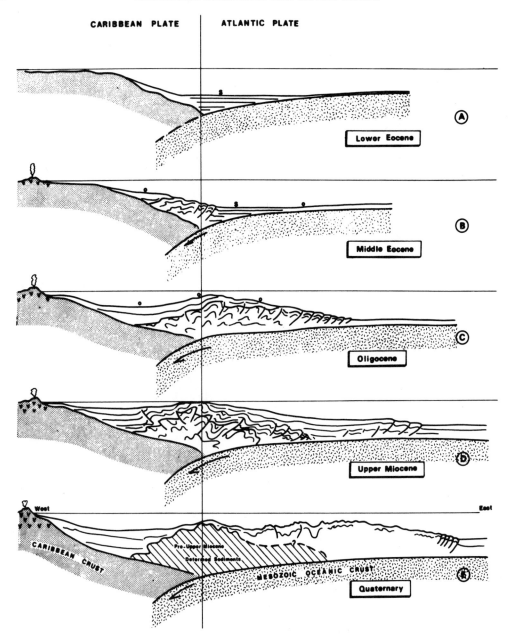

Figure 5.15 Sequence of cartoon model profiles illustrating the tectonic evolution of the southern part of the Lesser Antilles subduction zone. 'O' and 'S' refer to possible palaeo-environments for the deposition of the oceanic and Scotland formations, respectively. Thick sedimentary influxes in Eocene and Neogene times are inferred to have come mainly from the South American continent. From Biju-Duval *et al.* (1982)

drill cores in the DSDP drilling project leg 78A (Moore *et al.*, 1982). Near the postulated basal décollement plane (which was not penetrated by the drilling owing to technical problems), drilling revealed zones of intense deformation with fractured mudstone passing downwards into intensely foliated 'scaly' mudstone revealing slickensides, and ultimately to a tectonic breccia. Abnormally high fluid pressures were measured, roughly equivalent to lithostatic pressure. These high fluid pressures undoubtedly facilitated the underthrusting process as originally envisaged by Hubbert and Rubey (1959).

Correlation of drill core sections with seismic profiles enabled Moore *et al.* to reconstruct the stratigraphy of the accretionary wedge (Figure 5.14*D*). The offscraped sequence consists of Miocene and younger oceanic deposits. The layered sequence below consists of Upper Cretaceous to Lower Miocene pelagic clays, resting on oceanic basement, which are being underthrust below the younger deposits.

The Lesser Antilles subduction zone has been in existence since the early Eocene (about 50 Ma BP), much longer than most other active subduction zones. It therefore provides us with a useful model with which to compare supposed fossil examples in orogenic belts. During this time, the position of the trench migrated eastwards relative to the South American continent and to the mid-Atlantic ridge, owing to spreading within the Caribbean. An evolutionary model of the subduction zone (Figure 5.15) demonstrates how the trench has been first filled, then obliterated by the building of an Oligocene accretionary ridge (the present Barbados ridge), causing the Upper Miocene to Recent accretionary complex to migrate eastwards to its present position.

The Makran complex

The accretionary complex of the Makran (see White, 1982; Platt *et al.*, 1985) lies along the continental margin of Iran and Pakistan on the north side of the Gulf of Oman (Figures 5.16, 5.18). The complex is formed by the northwards subduction of the oceanic part of the Arabian plate beneath the Eurasian plate. The

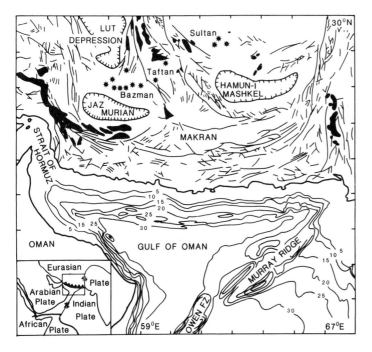

Figure 5.16 Location of the Makran accretionary prism in the Gulf of Oman. Note that the accretionary prism is situated at the subduction zone marking the boundary of the Arabian and Eurasian plates (inset), and is truncated on its eastern side by the Murray transform fault and its continental continuation. Stars mark volcanic centres along the active volcanic arc; black areas are ophiolite outcrops; thin lines on land are faults; ticked lines mark boundaries of major depressions. After White (1982)

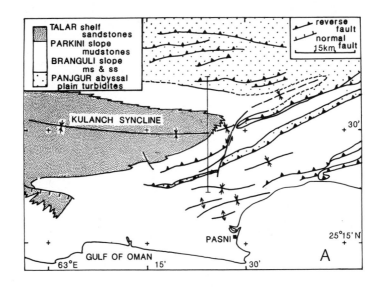

Legend:
- TALAR shelf sandstones
- PARKINI slope mudstones
- BRANGULI slope ms & ss
- PANJGUR abyssal plain turbidites

reverse fault
normal fault

15km

KULANCH SYNCLINE

30'

25°15'N

PASNI

GULF OF OMAN

63°E   15'   30'

A

B

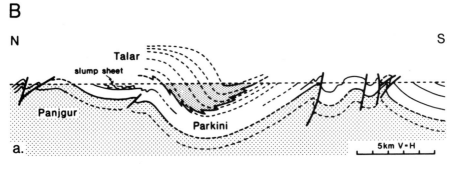

N   S

Talar

slump sheet

Panjgur

Parkini

5km V=H

a.

Present day

prism front

shelf   slope

abyssal plain

underthrust abyssal plain sequence

oceanic crust

10km V=H

b.

Early Pliocene

C

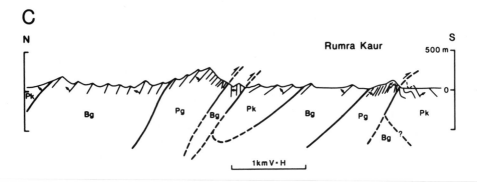

N   Rumra Kaur   S

500 m

Pk

Bg   Pg   Bg   Pk   Bg   Pg   Pk

Bg ?

1km V=H

subduction zone is terminated on its eastern side by the Murray transform fault that separates the Arabian and Indian plates (see Figure 3.6) and continues northwards on the continent as the Ornach Nal-Chaman fault system (see Figure 5.30). The subduction zone ends in the west at the Straits of Hormuz, where the Arabian and Eurasian continents are in contact. Sediment 6–7 km thick covers oceanic crust in the Gulf of Oman, which is thought to be between 70 and 120 Ma old.

The active volcanic arc consists of a chain of Cenozoic volcanoes situated 400–600 km north of the coast (Figure 5.16). There is no topographic trench, and the accretionary complex is unusually broad, about 300 km in width, more than half of which lies onshore. Seismic reflection profiles across the offshore part of the complex show a linear pattern of ridges with some intervening troughs (White, 1982). Folding appears to take place initially at the southernmost or frontal part of the prism which seems to have migrated southwards at a rate of 10 km/Ma. These frontal folds are then incorporated into the accretionary complex by uplift along a basal thrust. Little subsequent deformation appears to have occurred in this sector of the complex. However, 70 km to the north of the present front, a further uplift occurs which rises eventually above sea level 100 km north of the front to form the onshore Makran. Here a thick, faulted flysch sequence is exposed, extending about 200 km inland to the north.

The onshore structure, summarized in Figure 5.17, is described by Platt et al. (1985). A concordant sequence of marine sediments commences with Oligocene to mid-Miocene abyssal plain deposits, followed by Upper Miocene slope deposits, and by a late Miocene to Pliocene shallow-water shelf sequence, indicating rapid shoaling of the sedimentary prism in the mid-Miocene. There is apparently no field evidence for accretionary structure prior to the deposition of the slope and shelf sediments. Nor is there evidence for the progressive growth of structures during deposition, although the growth of gentle folds might be undetectable owing to the effects of the later deformation, which caused 25–30% shortening. This main deformation occurred after the early Pliocene (4 Ma BP) at a time when the accretionary front probably lay 70–100 km south of the present shore line, and has resulted in a series of E–W to ENE–WSW, asymmetric, S-facing folds and associated reverse faults (Figure 5.17B).

The uplift of the onshore Makran and the accompanying deformation are thought to have been accomplished by underplating at depth (see Figure 5.17B). The authors suggest that this process may have operated by the formation of a progressively widening duplex at a ramp in the basal thrust. Such a structure would cause tilting of the upper part of the sequence, leading to shoaling and possibly to syn-sedimentary deformation, but not to major folding or faulting.

Platt et al. attempt to apply mass balance calculations on the accretionary process by comparing the likely quantity of sediment input with the estimated volume of the prism. They show that a significant proportion of the sedimentary sequence must have been under-

Figure 5.17 Structure of the onshore Makran complex. (A) Simplified structural map of the coastal Makran, showing major folds and reverse faults. Note lateral facies change between Talar and Parkini formations. After Platt et al. (1985). (B) (a) Section across the Kulanch syncline (see line on Figure A) showing the relationship between structure and facies. Dashed lines represent time planes. (b) Interpretation of (a) during the Pliocene, suggesting that the shelf and slope sequences may have been deposited on an undisturbed abyssal plain sequence that was being uplifted along a major décollement surface. Note the suggested duplex structure causing contemporaneous uplift of the northern limb of the Kulanch syncline.
(C) Section across the faulted sequence east of section (A), illustrating the series of northward-younging sequences bounded by reverse faults characterizing the southern limb of the Kulanch syncline. Pg, Panjgur; Bg, Branguli; Pk, Parkini formations; arrows represent younging directions. (B) and (C) from Platt et al. (1985)

plated or subducted, enough for a layer around 6 km thick above the oceanic crust (see Figure 5.17B).

The Aegean arc

The complex Aegean region (Figure 5.18) has been intensively studied over the last decade. The basic plate boundary network and relative plate motions were established by McKenzie (1972), and subsequently refined in a comprehensive review of the neotectonic pattern of the region using earthquake data, Landsat photographs and seismic refraction records (McKenzie, 1978b). Several distinct tectonic units are apparent. The *Hellenic trench*, along which the African plate is descending below Eurasia, lies immediately southwest and south of the *Hellenic arc*, which is a non-volcanic island arc extending from mainland Greece west of the Peloponnesos to Crete and Rhodes.

On the south side of the trench is the broad *Mediterranean ridge* on the sea-floor between Greece and North Africa. North of Crete lies the *Cretan Sea basin*. The active volcanic arc runs from the eastern Peloponnesos through the southern Aegean Sea to the coast of Anatolia. A large tectonically active region north of the volcanic arc consists of the *Aegean Sea basin* and the adjoining land masses of mainland Greece in the west and Anatolia in the east. This Eurasian hinterland is divided into two separate parts by a major fault system connecting the north end of the subduction zone with the North Anatolian strike-slip fault running along the south side of the Black Sea. The region south of this fault is recognized as a separate small plate, the Anatolian plate, which is moving westwards as a result of the N–S convergence of the Arabian and Eurasian plates.

The Hellenic trench system is described by

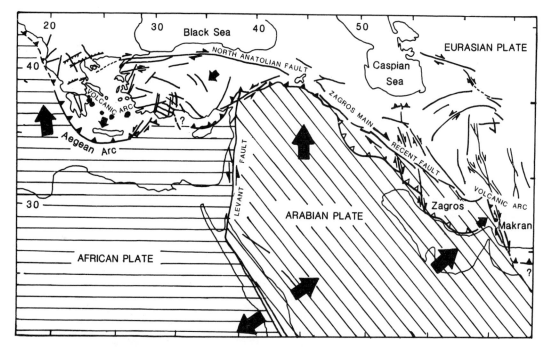

Figure 5.18 Map showing the location of the Aegean arc in the general tectonic setting of the Eastern Mediterranean region. Note the deformed southern edge of the Eurasian plate to the north, and the African and Arabian plates to the south. The large arrows denote the movement vectors of the southern plates in relation to each other and to the Eurasian plate, assumed to be stationary. Movement directions of the Anatolian and Aegean blocks relative to the main Eurasian plate are shown by smaller arrows. After Mercier (1981).

Huchon *et al.* (1982) who report the results of a SEABEAM sonar survey of the detailed morphology and structure of parts of the trench system. Three main NW–SE segments, the Matapan, Gortys and Poseidon trenches, are separated by NE–SW linear troughs, the southern of which offsets the trench sinistrally (Figure 5.19). The trench shallows southeastwards from over 5 km to around 3 km depth. At its eastern end, the Poseidon trench bends sharply into the NE–SW Pliny trench. The Strabo trench lies further to the southeast, parallel to the Pliny trench but not directly connected with it. The Strabo trench is relatively shallow and poorly defined. Both these southeastern trenches have little sedimentary cover. The present convergence vector across the trench is about perpendicular to the main trench but makes an angle of about 35° with the two southeastern trenches (Figure 5.19).

The results of the detailed survey by Huchon *et al.* revealed a set of ridges and troughs on the outer slope of the Matapan trench, generally parallel to the trench axis. These are interpreted as folds. Within the Pliny trench, however, the structure is quite different. A series of en echelon troughs about 10 km long and 2 km wide occur within the main trench, oriented about 15° anticlockwise of the axis. These are interpreted as sinistral strike-slip fault segments, which are consistent with the oblique convergence vector.

The inner wall of the Matapan trench was examined directly by submersible. It seems to be relatively inactive tectonically and is dominated by large normal faults with well-defined slickensides. Huchon *et al.* believe therefore that the inner wall falls within the extensional tectonic province to the north of the subduction zone. In contrast, the outer trench slope and trench floor are characterized by obvious compressional features in the form of folds, small thrusts, and conjugate strike-slip faults. This zone therefore belongs to the compressional domain indicated by the earthquake focal mechanism solutions.

Le Pichon *et al.* (1981) report the results of several SEABEAM sonar traverses across the *Mediterranean ridge*. It had previously been suggested by Ryan *et al.* (1970) that most of the sedimentary cover on the Mediterranean floor is being tectonically thickened along this ridge, rather than being accreted at the trench. The survey by Le Pichon *et al.* showed the presence of an extensive fold system affecting the 3–4 km of uppermost Miocene (Messinian) to Quaternary sedimentary cover. A set of conjugate strike-slip faults cuts the folded sediments. The compressional tectonic regime, which occupies the northern half of the 250 km-wide ridge, extends as far as the trench axis as described above.

The Mediterranean ridge is thus interpreted as an accretionary structure, similar to the Barbados ridge already described, where an upper layer of soft sediments is being deformed above a décollement horizon. It is suggested (Le Pichon *et al.*, 1982) that a thick layer of evaporites may play an important role in the decoupling of the deeper subducting layers from the shallow shortening layers.

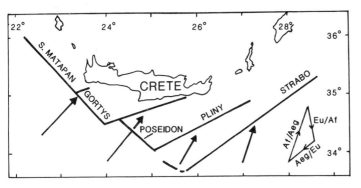

Figure 5.19 Schematic map of the Hellenic trench system, showing the five main segments in relation to the convergence direction between the African and Aegean plates; see also inset vector triangle relating to African (Af), Aegean (Ae) and Eurasian (Eu) relative plate motion. After Huchon *et al.* (1982).

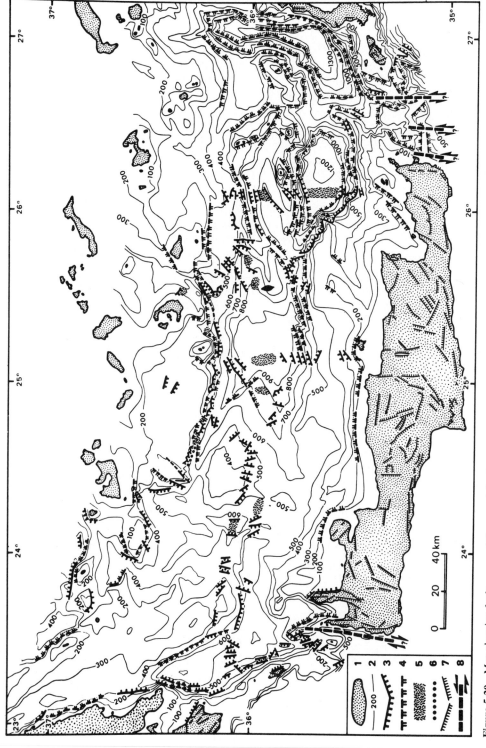

Figure 5.20 Map showing fault scarps of Crete and the Sea of Crete. 1, Land; 2, isobaths in m; 3, fault scarps observed in SEABEAM sonar mapping; 4, fault scarps obtained from previous bathymetric mapping; 7, fault scarps on land; 8, inferred strike-slip faults. From Angelier *et al.* (1982)

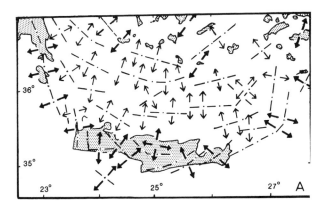

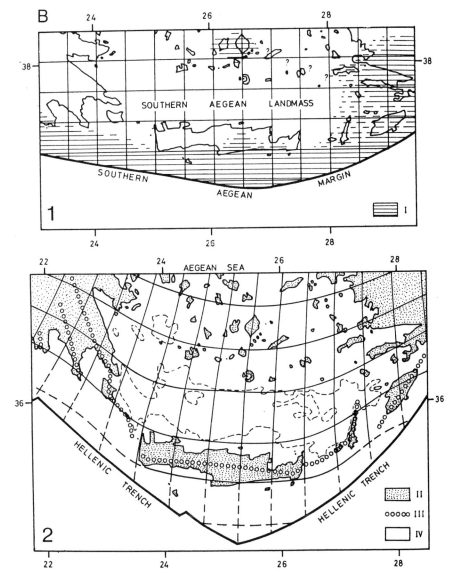

Figure 5.21 Tectonic interpretation of the Aegean Sea. (*A*) Tectonic summary, based on information from Figure 5.20 and adjoining areas, showing inferred extension directions on land (heavy arrows) and sea areas (light arrows). Note the swing in trend (broken lines) from NW–SE in the west to NE–SW in the east. After Angelier *et al.* (1982). (*B*) Computed restoration to its original shape (1) of the present area (2); the method uses estimates of the extension factor, β, which varies in the range 1.1–1.9, restoring to equal squares. From Angelier *et al.* (1982)

McKenzie (1978a) showed that rapid exten-sion is taking place in the Aegean Sea and adjoining mainland areas, and that the crust beneath the Aegean Sea is only about 30 km thick, compared with 50 km in the mainland areas. Focal mechanism solutions of earth-quakes in the region of the Hellenic trench indicate thrust motion with a mean slip vector oriented at 211° on a shallow plane dipping NE. Thus the convergence direction across the Hellenic trench appears to be constrained by southwestwards motion of the Aegean region perpendicular to the central sector of the trench (Figure 5.18). In contrast, the focal mechanism solutions for shallow earthquakes in the Aegean Sea region indicate a prepon-derance of normal dip-slip motion, but with a strike-slip component in some cases. On main-land Greece, earthquakes with similar focal-plane solutions are associated with surface normal faulting. The strike of the dip-slip motion appears to vary from NW–SE to E–W. McKenzie suggests an overall NNE–SSW ex-tension. In-situ stress determinations using the overcoring technique (2.6) indicate a pre-dominantly N–S horizontal extension over the whole area of mainland Greece and the Aegean basin (Paquin et al., 1982).

An analysis of the geometry of recent fault patterns in the *Hellenic arc* and adjoining Cretan Sea basin is reported by Angelier et al. (1982), using data from land surveys, satel-lite photographs and SEABEAM surveys. The authors find that most of the late Miocene to Recent faults of the southern Hellenic arc (principally Crete) are pure normal faults which strike either parallel or oblique to the trend of the arc (Figure 5.20). The *Cretan Sea basin* is also dominated by normal faults, some of which can be traced onshore. Most faults here are parallel to the trend of the arc, E–W in the centre, swinging round to NW–SE in the west and to NE–SW in the east. These results confirm the extensional nature of the back-arc region (Figure 5.21A). It is of interest to observe that this extensional region includes the present volcanic arc, unlike the oceanic examples discussed earlier (see 4.3). Angelier

et al. note that if the arc is restored to its presumed pre-stretching position, the orienta-tion of the faults becomes much more uni-formly E–W (Figure 5.21B).

The uplift of the arc might at first appear to be incompatible with an extensional regime. The authors suggest, however, that the lower 3–6 km of sedimentary cover, which is initially subducted at the trench, becomes underplated onto the upper plate below the Hellenic arc, thus causing the uplift.

The Aegean basin therefore appears to form a separate micro-plate, representing the upper slab of the subduction zone, which is moving rather rapidly southwest over the African plate creating a zone of extension, particularly in its northern part (Figure 5.18). This basin repre-sents a special type of back-arc spreading basin formed on continental crust, but is in some respects analogous to the oceanic spreading basins discussed in 4.4.

Le Pichon and Angelier (1979) suggest that this movement is a result of the withdrawal southwards of the subducting slab (compare the trench rollback mechanism discussed in 4.3). This mechanism is tested using a finite-element model by De Bremaecker et al. (1982) and shown to give a much better match to the observed stress field than two alternative models: the Arabian indenter model and the gravity-spreading model based on the ele-vation difference between Aegea and the Mediterranean floor.

5.3 Collision

The collision of two pieces of continental crust is an inevitable consequence of the continued subduction of oceanic lithosphere. The much greater buoyancy of continental compared with oceanic crust makes the former difficult if not impossible to subduct. The relationship be-tween subduction, collision and orogeny in the new plate tectonic theory was clearly illus-trated by Dewey and Bird (1970). They re-cognize two types of collision, continent–island arc and continent–continent, and demonstrate that the collisional orogenic belts so formed differ fundamentally from the asymmetric subduction orogenic belts of island arc or continent-margin type already described. Dewey and Bird illustrate the simplest such situation (Figure 5.22): the collision of a passive continental margin with an active or subducting continental margin. Clearly subduction must precede collision, so that the effects of the subduction 'orogeny' must be incorporated into the subsequent collisional orogenic belt. The polarity or asymmetry of the subduction structure controls, at least initially, the collisional structure. This principle is illustrated in Figure 5.22, where the left-dipping underthrust structure of the subduction zone continues as rightward-directed overthrusting after collision. Various alternative and more complex scenarios are possible of course: both opposing margins may possess

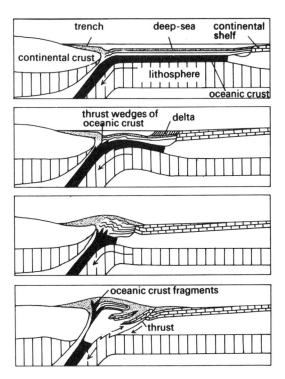

Figure 5.22 Diagrammatic sequence of stages in the transformation of a subduction zone to a continental collision zone by the approach of two continents and the closure of the intervening ocean. After Dewey and Bird (1970).

subduction zones for example, and complex collisional belts may be formed by multiple accretion of island arcs or small continental fragments. The wide Central Asian orogenic belt has been created in this way, as we shall see.

Convergence continues after initial contact of the opposing pieces of buoyant crust. The extent of this continued convergence is the single most important factor in the creation of an orogenic or mountain belt, since continental convergence leads to crustal thickening and consequently to isostatic uplift.

Active collision belts

Recent mountain belts attributable to continental collision are found mainly along a complex zone connecting the Alps, the Himalayas and Indonesia. This zone is the result of the Mesozoic–Cenozoic convergence of Eurasia with Africa, Arabia, India and Australia, with the consequent elimination of the Tethys Ocean that formerly intervened (see Figure 3.5). Other examples are found in the circum-Pacific belt, where the collision of relatively small continental fragments and island arcs has taken place against a background of continued subduction orogeny throughout the Mesozoic and Cenozoic. The occurrence of substantial strike-slip displacements in this belt (for example in North America and Japan, see 6.2) often precludes accurate reconstruction of the events relating to previous collisions.

Looking in more detail at the mountain belts in the Alpine–Himalayan system, we find that the main mountain ranges do not form a continuous belt, but are arranged in a number of separate linear or arcuate chains. In the Mediterranean region, these form a very complex pattern, the northern boundary of which is marked by the tortuous course of the Pyrenees, Alps, Carpathians and Caucasus chains. In the south, the Atlas mountains in northwest Africa are linked to the Taurus mountains of Turkey through the Mediterranean Sea, via the Hellenic arc already described. The mountain ranges bordering the Adriatic Sea strike obliquely across this region connecting the northern and southern branches. The complexity of this arrangement was attributed by Dewey *et al.* (1973) to the movements and interactions of various microplates that existed in the Mediterranean region during the convergence between Europe and Africa. These movements were controlled in part by major changes in the convergence vector between the main plates.

Since the Mediterranean Sea has not completely closed, the full effects of continental collision have yet to be experienced. The Alps and the Pyrenees are examples of the collision of relatively small plates (Iberian and Adriatic) with Europe. The climax of the compressional movements that produced these chains occurred in Oligocene and Miocene times, and the belts exhibit only residual tectonic activity. The Alps are described in Chapter 8 as an example of a Phanerozoic orogenic belt.

Further east, the Zagros mountains mark the site of collision between the Eurasian and Arabian plates (Figure 5.18). Dewey and Bird noted that the Zagros crush zone marks the suture between the two plates. A Mesozoic sequence of ophiolite, chert, flysch and melange marks the site of a subduction zone along the southern margin of the Iranian plateau. The thick Phanerozoic carbonate-shale cover of the Arabian shield to the south, which is underthrusting the Iranian plateau, is deformed in a series of asymmetric folds facing southwest, developed above a basal décollement horizon in late Precambrian to Cambrian salt deposits. This zone passes laterally into the Makran subduction zone already described.

A strike-slip zone connects the eastern end of the Makran complex to the major collision complex of Central Asia, produced by the collision with India. This zone (described below) consists, in addition to the main Himalayan range, a number of related morpho-tectonic units including the Tibetan plateau and the Tien Shan and Altai ranges far to the north.

The eastern end of the collision zone connects via a largely strike-slip boundary with the Indonesian subduction zone. Here the oceanic part of the Indian plate is being destroyed in an arcuate zone from the Andaman islands along the Sunda arc as far as Timor. The eastern continuation of this belt, from Timor to New Guinea, is affected by the collision of the Australian continental crust. This area exhibits a relatively juvenile phase of continental collision, and the Banda arc is an example of a continent-island arc collision orogeny (see 5.5).

Gross structure of collision belts

Active and recent mountain belts are characterized by thickened crust (to between × 1.5 and × 2 normal thickness). Thus thicknesses of 50 km are common, and over 70 km are found locally, for example in Tibet. The topographic elevation of mountain belts (typically 3–7 km) is approximately compensated isostatically, and it has been assumed for many years that the excess mass of the mountain ranges is balanced by a thick root of low-density crustal material. Seismic refraction studies of deep crustal structure confirm this interpretation (Figure 5.23).

We may envisage the process of collision as an overlapping of the lithosphere of the converging plates, resulting in progressive lithosphere thickening. This will produce initially the reverse effects to those described in Chapter 4. The geotherms will be depressed, and surface heat flow will diminish. However, over a longer period of time, as the geotherm is restored towards its normal gradient, the thickened crust heats up and the lower part undergoes prograde metamorphism and possibly melting. These processes are accompanied by progressive uplift as the crust attempts to restore isostatic equilibrium.

Surface heat flow over modern collision belts varies. The mean value of about $72\,\mathrm{mW\,m^{-2}}$ for Mesozoic–Cenozoic orogenic belts given by Vitorello and Pollack (1980) (see Figure 2.7) conceals a wide variation from low values typical of the initial stages of continental accretion, where magmatic effects are absent, to much higher values associated with belts

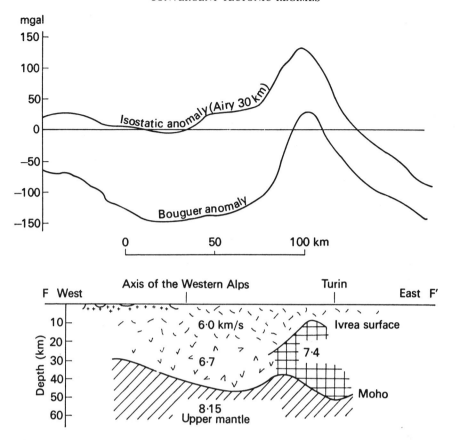

Figure 5.23 Bouguer and isostatic gravity anomaly and crustal structure profiles across the Western Alps. P-wave seismic velocities are given in km s⁻¹. Note that the mountain range is isostatically compensated but that a positive anomaly is associated with the dense Ivrea peridotite, interpreted as upthrust mantle material (see section 8.1). From Bott (1971)

containing young granite batholiths such as the Andes.

Flake tectonics and obduction

It was shown by Oxburgh (1972), based on studies in the Eastern Alps, that collision may involve detachment and overriding of part of the crust on to the opposing continent while the remainder of the crust and lithosphere descended below it (Figure 5.24). This process was termed *flake tectonics*. In the Eastern Alps, a pre-Mesozoic metamorphic basement of the European plate has been overridden from the south by an allochthonous thrust sheet of pre-Upper Palaeozoic crystalline base-

ment derived from the southern plate. Between the two basement sheets lies a highly deformed pelagic and volcanic sequence derived from the intervening oceanic area, now largely subducted. Oxburgh suggests that the initiation of the flaking process is due to the buoyancy and topographic elevation of the opposing continental margin, and that a low-angle crustal split propagated back into the adjoining plate along a convenient zone of weakness. He points out that the separation of the upper third of the continental crust would reduce the buoyancy of the subducted crust to one-half its original value. This would facilitate continued subduction and allow convergence to proceed. The existence of mid-crustal de-

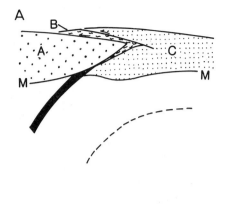

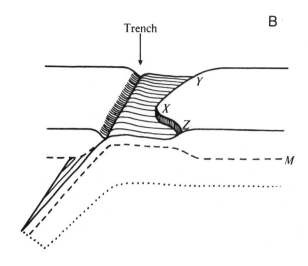

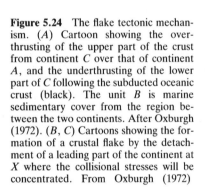

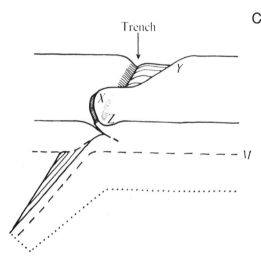

Figure 5.24 The flake tectonic mechanism. (A) Cartoon showing the overthrusting of the upper part of the crust from continent C over that of continent A, and the underthrusting of the lower part of C following the subducted oceanic crust (black). The unit B is marine sedimentary cover from the region between the two continents. After Oxburgh (1972). (B, C) Cartoons showing the formation of a crustal flake by the detachment of a leading part of the continent at X where the collisional stresses will be concentrated. From Oxburgh (1972)

tachment horizons discussed in 2.7 (see Figure 2.29) would assist this process. A similar interpretation is applied to the Himalayas (see below).

Where detachment of the whole crust takes place, the process has been termed *Ampferer subduction* or *A-subduction*, to distinguish it from subduction of the whole lithosphere (*B-subduction*). The basal crustal weak zone (see Figure 2.29) is a particularly favourable site for detachment and explains the occurrence of very high-pressure metamorphic rocks within orogenic belts.

It was realized by Coleman (1971) and

Dewey and Bird (1971) that the ophiolite complexes of orogenic belts could represent fragments of oceanic crust emplaced on to continental crust by a process which was termed *obduction*. Their presence could therefore be used as a valuable indicator of a suture representing a former subduction zone. This idea is now generally accepted. An initial problem with the obduction process was why dense oceanic crust should sometimes be detached and thrust over less dense continental crust rather than be subducted. Dewey and Bird (1971) illustrate three possible ways in which ophiolite obduction could occur (Figure

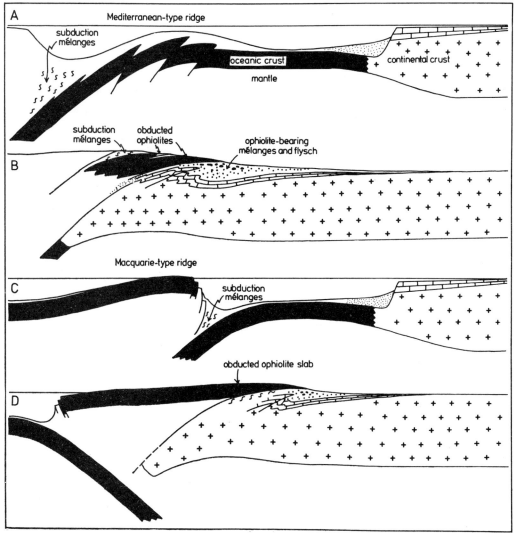

Figure 5.25 Possible mechanisms of obduction. (*A*, *B*) Obduction resulting from the back-thrusting of oceanic lithosphere of the leading plate onto the continental part of the leading plate. (*C*) Uplift of the upper oceanic plate in an intra-oceanic subduction zone results eventually in its emplacement on the approaching continent. (*D*) Obduction onto the upper plate of marginal basin lithosphere belonging to the upper plate. From Dewey and Bird (1971)

5.25): (i) compressional deformation of the descending slab, as found in the Mediterranean ridge, involving thrust wedges of oceanic basement; (ii) overthrusting of an oceanic upper plate during subduction; and (iii) back-thrusting of oceanic crust in the upper plate on to the continent.

Ophiolite sequences form linear belts extending up to several hundreds of km along

strike but never exceed 15 km in thickness. A typical cross-section exhibits the four main oceanic crustal layers: pelagic sediments, pillow lavas, sheeted dykes and layered gabbros, overlying ultrabasic mantle-type material, but there are significant differences between many ophiolites and the standard ocean ridge crustal section. For example, typical ophiolites exhibit a basal crustal layer which is much reduced in thickness, and there is considerable variation among ophiolites in the development of the sheeted dyke layer — in some, it is completely absent. It has been suggested that, for these and other reasons, many ophiolites represent anomalous ocean lithosphere produced in back-arc spreading basins rather than in true oceans (Miyashiro, 1973). Spray (1983) draws attention to the significance of the basal tectono-metamorphic zone or 'sole' found in many ophiolite complexes. These highly deformed zones acquired their fabric at high temperatures, up to granulite facies in some cases, and were formed, according to Spray, while the oceanic lithosphere was still hot, and within 5 Ma of their initial magmatic crystallization. This suggests that the initial decoupling of the ophiolite took place, at or near its origin at a spreading centre, along the lithosphere–asthenosphere boundary. This boundary would be situated at a depth of only about 25 km in lithosphere less than 5 Ma old. The possibility therefore arises that the detachment along which an ophiolite eventually becomes obducted was created as a result of tectonic activity at the spreading centre, perhaps unrelated to the convergent movements which caused the obduction.

Thrust belts

It is already apparent that the geometry of collision zones favours the initiation and development of thrust belts within the continental crust. We have seen that such belts are fundamental to the accretion process in subduction zones, and it is to be expected that these zones should to some extent control subsequent shortening of the crust during the collision process. Thrust belts fall naturally into two classes of polarity: *synthetic* belts dipping towards the continent, parallel to the initial subduction zone, and *antithetic* belts, dipping in the opposite direction, typically found at the outer margin of an orogenic belt, separating it from the undeformed stable craton or *foreland*. Such belts are termed *foreland thrust belts*. Complexities occur if the flake or

A-subduction process superimposes antithetic thrusting on synthetic, or where collision takes place between continental margins with thrust belts of opposed polarity. These problems were first clearly stated by Roeder (1973) in an analysis of the geometric relationships between thrusting and plate movements.

The basic polarity of thrust-driven collisional shortening is determined by the pre-existing subduction zone. Continued convergence after the initial continental contact is ensured by the fact that the negative buoyancy force provided by the sinking slab is still acting on the underthrust plate, as long as it remains attached. Most colliding slabs will have sections along strike that are still subducting. In all the major collision orogenies discussed here, subduction of oceanic lithosphere continues along part of the destructive boundary. For example, the Indian plate is still partly driven by slab-pull in Indonesia, the Arabian plate at the Makran, and the African plate at the Hellenic trench. Moreover the ridge-push force continues to operate as before. These forces are counteracted by a collisional resistance force (see 2.5) which must increase with the extent of crustal overlap and thickening. If we take the Himalayan collision as an example, this process of crustal convergence may last for up to about 40 Ma. The way in which the process seems to operate, by underthrusting of continental crust, forces the orogen to deform internally in an asymmetric manner.

Thrust belts may be divided into *thin-skinned* or *thick-skinned* (Figure 5.26) depending on whether the basal or sole thrust shallows at depth or steepens downwards to meet the base of the crust. Recent geometric and kinematic models of thrust belts have been derived mainly from work in the thin-skinned Rocky Mountains belt (Bally *et al.*, 1966; Price, 1981), in well-bedded sedimentary rocks, and applied to the Moine thrust zone (Elliott and Johnson, 1980; McClay and Coward, 1981), the Scandinavian Caledonides (Hossack, 1978), the Appalachians (Hatcher, 1981; Brewer *et al.*, 1981), and elsewhere.

Useful summaries of the geometry and

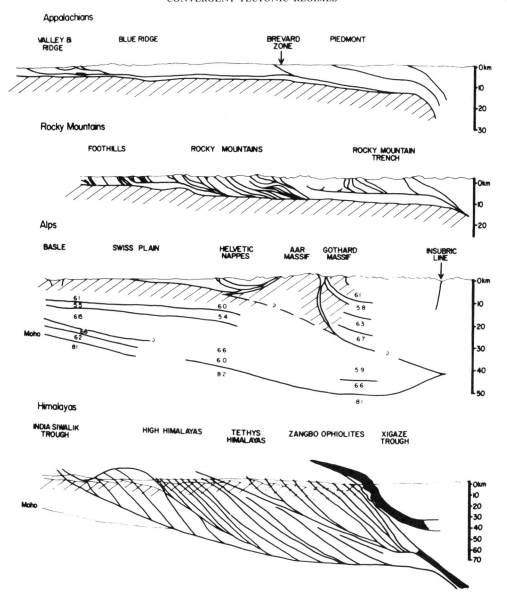

Figure 5.26 Profiles illustrating thin-skinned (upper two) and thick-skinned (lower two) thrust tectonics. All sections are true scale. From Soper and Barber (1982), with permission, after Hatcher (1981), Price (1981), Hsu (1979) and Shackleton (1981), respectively.

mechanism of thrust zones are provided by Dahlstrom (1970), Boyer and Elliott (1982) and Butler (1982). In essence, shortening is achieved by a process of thickening by crustal overlap, whereby older, or structurally lower, material is stacked upon younger, or structurally higher, material. The stacking is achieved by transfer along thrusts which have a *staircase* trajectory of alternating *flats and ramps*. The geometry is similar to that for extensional faulting, described in 4.4 (see Figures 4.25, 4.26). Superimposition of hangingwall ramps upon footwall flats produces geometrically necessary folds in the hanging-

wall, and the lateral movement of a thrust
sheet from flat to ramp to flat produces
continuously migrating zones of internal strain
in the moving sheet. When movement of the
first thrust becomes difficult, due to increasing
resistance, a new thrust propagates forwards,
connects upwards with the old, and transfers
the now inactive upper thrust passively forwards in 'piggyback' manner. A series of
imbricate thrust wedges (*horses*) formed in this
way forms a *duplex* structure. The duplex has
an active *floor* thrust and an inactive *roof*
thrust. Stacked duplexes may form to produce
nappe complexes such as those of all the major
thrust belts. Excellent examples may be seen in
cross-sections of the Himalayas (Figure 5.36),
the Rocky Mountains (Figure 8.11) and the
Moine thrust zone of the Caledonides (Figure
8.23).

The above system achieves the objective of
shortening the cover in an orogen, but avoids
the problem of how the basement is shortened,
and how the displacements are transferred
through the lower crust and mantle lithosphere. This problem is addressed by Coward
(1983) who points out that the evidence from
the inner parts of orogenic belts such as the
Alps and the Himalayas indicates the importance of steep thrusts or shear zones which
transfer deep crustal rocks to the surface
(Figure 5.35). Mattauer (1986) suggests that
shortening in the Himalayas has been achieved
by sub-horizontal displacements along major
décollement horizons at (i) the basement–
cover contact, (ii) the mid-crustal seismic
discontinuity, and (iii) the base of the crust.
These displacements are transferred upwards
along steep ramps connecting the major de-
tachments.

The style of deformation varies considerably
with crustal level. Typical thin-skinned fault–
fold morphology associated with cataclastic
deformation processes in discrete zones gives
way downwards to more pervasive plastic
deformation with the development of slaty
cleavage, and to wide zones of ductile deformation at high metamorphic grades. In the
Himalayan model (Figure 5.35), the thin-

skinned thrusting is a high-level outer expression of displacements of an essentially thick-
skinned nature involving the whole crust. Since
shear zones widen with depth due to rise in
ambient temperature (see e.g. Lockett and
Kusznir, 1982), displacements within middle
and lower crustal rocks are distributed through
wide zones of ductile deformation, which may
amalgamate to involve most of the lower crust.
The origin of such wide belts of deformation,
when they are found in old orogenic belts, may
not be obvious. In an ideal shear zone, a thrust
displacement is transformed into a zone of
simple shear. However, in real shear zones a
component of shortening or extension across
the zone results in the superimposition of a
pure shear component to the simple shear
strain of the ideal zone (see Figure 3.15). The
pure shear component will become increasingly
important with depth due to the combined
effects of gravitational load and elevated temperature, enhancing the ductility of the rocks
(Figure 5.27).

Unless the strain patterns of highly deformed metamorphic belts can be geometrically related to high-level displacements, as
is possible to some extent in certain young
mountain belts, their origin may not be obvi-

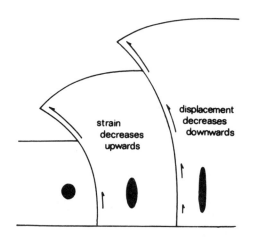

Figure 5.27 Cartoon to illustrate the variation in style of
deformation downwards in the crust, from displacement-
dominated at upper levels to bulk strain-dominated at
lower levels. From Coward (1983)

ous. This difficulty has led to great debate and controversy among structural geologists who have studied old orogenic belts. It is in fact difficult to establish the extent to which thrust-driven collision shortening is fundamental to the deformation of orogenic belts, or whether other mechanisms are equally important.

Indentation

The concept of indentation was developed by Molnar and Tapponnier in their study of the Central Asian collision zone of India and Eurasia (Molnar and Tapponnier, 1975; Tapponnier and Molnar, 1976, 1977; Tapponnier *et al.*, 1982). They observed that the active tectonic areas of Central Asia indicated by current seismicity formed a number of discrete zones affecting a region up to 4000 km wide, northeast of the Himalayan front (Figure 5.28), whereas India, in contrast, is relatively unaffected. They showed on the basis of magnetic stratigraphy and palaeomagnetic evidence that India must have moved at least 2000 km into Asia since the time of initial contact. It is clear, however, from the Asian crustal structure that 3000 km of crustal shortening has not occurred. Tapponnier and Molnar therefore suggest that India has acted as a 'rigid' indenter driven into the more 'plastic' Asian continent (Figure 5.29A) which has reacted by a combination of thrust and strike-slip displacements. The ability of Asia to shorten by lateral displacement is influenced by the boundary conditions of the Asian plate. In the east, the presence of a continuous subduction zone was held to allow lateral 'extrusion' of Asian continental lithosphere over the oceanic Pacific plate. To the west, continuous continental lithosphere extends to Europe and the Atlantic with no comparable possibility of extrusion. Only to the southwest is some lateral movement possible, where the westwards-directed wedge of Afghanistan can move towards oceanic 'space' in the Arabian Sea and ultimately the Mediterranean.

The indentation process commences at the protrusions of India that are presumed to be the first points of contact. These act to concentrate the stress and initiate failure. The authors consider that two wedges of Asian crust, the Indo-China block and the China block, have escaped to the southeast as a result of the northward progress of the indenter (Figure 5.29A). Of the 2500–3500 km of convergence estimated by Molnar and Tapponnier between NE India and Asia, between 1000 and 2500 km is considered to be achieved by strike-slip movements. Tapponnier *et al.* (1982) illustrate the applications of 'extrusion tectonics' to the deformation of Central Asia by means of indentation experiments using plasticine (Figure 5.29B).

The indentation principle has been applied to other orogenic belts. For example Thomas (1983) shows how the irregular margin of the Appalachian–Ouachita orogenic belt of eastern North America could be explained in terms of a series of *recesses* and *salients* of the orogenic front. These are explained as the result of respectively stronger and weaker sectors of the original continental margin, corresponding perhaps to basement domes or rift depressions.

A mathematical model of a collision zone

England and McKenzie (1981) note the limitations imposed by the two-dimensional nature of the indentation model, and report the results of numerical experiments which take account of vertical as well as horizontal strain in a block of material subjected to a constant rate of shortening. They assume that variation of the horizontal component of velocity with depth is negligible, and that the gradients of crustal thickness variation are small. These assumptions imply that the strain-rate of the lithosphere is governed by the strength of its strongest part (see 2.7) and that the effects of heterogeneous brittle fault deformation in the uppermost layers can be ignored. They obtain the most realistic results using viscous material with a non-Newtonian power-law rheology. Their model predicts that, for a wide range of rheological parameters, thickening of the con-

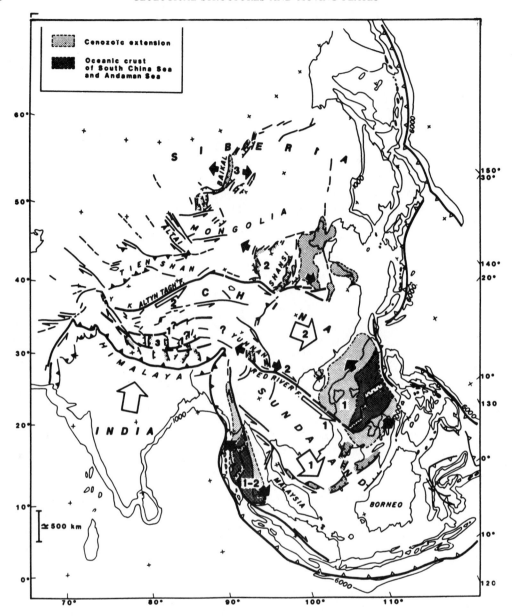

Figure 5.28 Schematic summary map of the tectonic pattern of eastern Asia. Heavy lines, major faults or plate boundaries; open-toothed lines, active subduction zones; closed toothed lines, major intracontinental thrusts; large open arrows, major block movement directions relative to the main Eurasian plate; small black arrows, recent extension; numbers represent phases of extensional movement considered to be related to the continental convergence; (1) 50–17 Ma BP; (2) 170 Ma BP to present; (3) active and projected future extension. From Tapponnier *et al.* (1985)

tinental crust occurs over areas with dimensions at least as large as those of the indenting continent. To give crustal thicknesses approximating to those of the Himalayas after 32 Ma,

a power-law rheology is required where the stress term is raised to about $n = 3$ (see 2.7). The crustal thickness in front of the indenter is limited by the strength of the lithosphere,

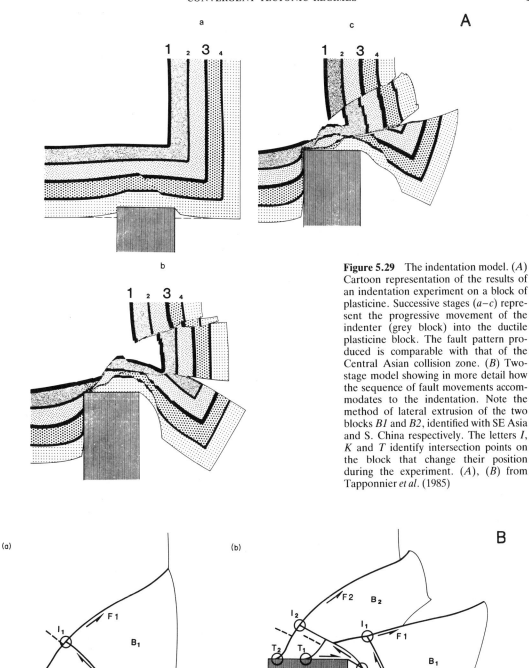

Figure 5.29 The indentation model. (*A*) Cartoon representation of the results of an indentation experiment on a block of plasticine. Successive stages (*a–c*) represent the progressive movement of the indenter (grey block) into the ductile plasticine block. The fault pattern produced is comparable with that of the Central Asian collision zone. (*B*) Two-stage model showing in more detail how the sequence of fault movements accommodates to the indentation. Note the method of lateral extrusion of the two blocks *B1* and *B2*, identified with SE Asia and S. China respectively. The letters *I*, *K* and *T* identify intersection points on the block that change their position during the experiment. (*A*), (*B*) from Tapponnier *et al.* (1985)

and as the maximum permissible thickness is approached, lateral stretching occurs in this region. This is an interesting analogue of the structure north of the Himalayas (see Figure 5.30). An important result of the model is that realistic results are obtained where the forces arising from the crustal thickness contrasts are as important in determining strain as those arising from the original boundary conditions. To maintain crustal thickness contrasts of about 30 km, similar to those of the Himalayas, the lithosphere is required to sustain shear stresses of about 30 MPa at strain rates of about 10^{-15}/s. These stresses and strain rates are consistent with estimates of available stresses from plate boundary forces (see 2.5) and with deformation rates for modern mountain belts.

5.4 The Himalayas and Central Asia

The recent explosion of interest in the Central Asian collision zone is due largely to the work of Molnar and Tapponnier, discussed above. In three influential papers, these authors examine the pattern of recent tectonic activity in the region and attempt to explain it by a series of movements related to plate collision (Molnar and Tapponnier, 1975; Tapponnier and Molnar, 1976, 1977).

Current tectonic activity as indicated by seismic and recent morphotectonic data covers an enormous region extending over 3000 km northeast of the Himalayas (Figure 5.28). This activity is concentrated in a number of active belts of deformation that are separated by comparatively stable blocks. The principal tectonic units are indicated in Figure 5.30. The Himalayan fold-thrust belt is bounded on both sides by major strike-slip belts — the Quetta–Chaman fault system in the west, and the Sittang zone in Burma in the east. These belts define the margins of a large piece of continental lithosphere which, according to Molnar and Tapponnier, has driven in a NNE direction into the Asian crust. The plate boundary lies along the Indus–Zangbo (Tsangpo) suture which lies on the north side of the Himalayas.

This suture connects through complex strike-slip zones of deformation with the Owen fracture zone dividing the Indian plate from the Arabian plate in the west (see Figure 3.6), and with the Andaman trench, at the northern end of the Indonesian subduction zone, in the east. The southern limit of the Himalayan fold-thrust belt is the main Himalayan boundary, or frontal, thrust which lies about 300 km south of the suture with the Indian plate.

North of the Himalayan belt are several other major fold-thrust belts, notably the Pamir, Tien Shan, Altai and Nan Shan ranges, separated by stable blocks such as the Tibetan plateau and the Tarim basin. Focal mechanism data from all these belts yield mostly N–S thrust solutions. The other major component of recent tectonic activity is strike-slip faulting. A number of major strike-slip faults extending for distances of the order of 1000 km account for much of the recent seismic activity. North of the Himalayas these form a conjugate set with NW–SE dextral and NE–SW sinistral displacements. In the southeast, both sinistral and dextral faults appear to be bent into a more N–S orientation. Similarly, in the west, the E–W dextral Herat fault meets the N–S to NE–SW Quetta–Chaman lineament defining the Afghanistan wedge, which is moving southwest in relation to India. These movements are explained by Tapponnier and Molnar as lateral extrusions resulting from northwards indentation of India into Asia (see Figure 5.29).

The third main element in the recent tectonic pattern is extensional. The NE–SW Baikal rift system lies at the northern margin of the active tectonic zone, and the Shansi graben system at the eastern margin (Figure 5.28). Studies of the active faulting of Tibet (Molnar and Tapponnier, 1978; Ni and York, 1978) revealed that the most recent faults are N–S normal faults (Figure 5.30). Focal mechanism solutions of earthquakes in Central Tibet yield approximately E–W slip vectors. These results indicate that the Tibetan plateau has been subjected to E–W extension since the late Cenozoic. Both sets of authors explain the

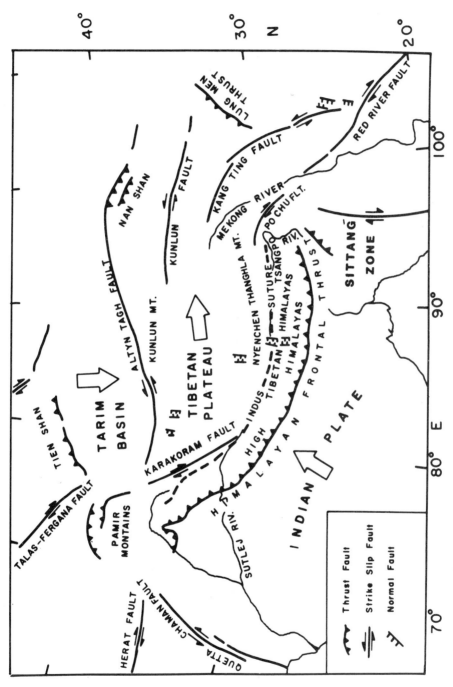

Figure 5.30 Tectonic summary map of the Central Asian collision zone, showing the main fault pattern and the inferred relative movement directions of the Indian, Tibetan and Tarim blocks. From Ni and York (1978), with permission.

extension as a secondary result of the N–S convergence of India and Asia. Ni and York suggest an easterly spreading mechanism resulting from a wedging effect in the west, due to convergence of the E–W thrusts in the south and the NE–SW Altyn Tagh strike-slip fault in the north. Molnar and Tapponnier attribute the extension to E–W lateral flow of lower-crustal material in response to crustal shortening. It is interesting to note that lateral stretching of this type is predicted by the mathematical model of England and McKenzie referred to earlier.

The magnetic stratigraphy record in the Indian Ocean (see 3.6) allows accurate reconstructions to be made of the convergence of India and Eurasia since the late Cretaceous (Figure 5.31). At around 38 Ma BP, at the Eocene–Oligocene boundary, the rate and direction of convergence changed abruptly.

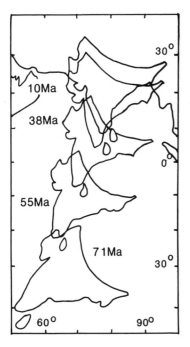

Figure 5.31 Successive positions of India at various times from 71 Ma BP, reconstructed from oceanic magnetic anomaly data. The northern boundary of India, and the position of Asia are arbitrarily fixed for times before the present. Note the anticlockwise rotation of India from 71 to 10 Ma BP. After Molnar and Tapponnier (1975).

Prior to 38 Ma BP, the convergence rate was between 10 and 18 cm/year in a NNE direction. After 30 Ma BP, the rate slowed to about 5 cm/year in a northwards direction. It is likely, but unproved, that this change relates to the initial contact between the two continental masses, which would probably have been in the northwestern 'horn' of the Indian continent, in northern Pakistan. Since that time, approximately 1500 km of convergence between the two continents has taken place.

Stratigraphic evidence from the Indus–Zangbo suture zone (Mitchell, 1984) indicates that early Cretaceous ocean-floor sediments and ophiolites were subjected to thrusting in late Lower Cretaceous times indicating the presence of a subduction zone dipping to the north. The N-facing structures relating to the final collision of India with Asia are of Eocene age.

The manner in which this convergence has been accommodated has been the subject of much debate. It had been thought for many years that India had underthrust Asia and thus effectively doubled the crustal thickness beneath Tibet (see e.g. Holmes, 1978, Figure 29.4). Molnar and Tapponnier suggest that probably only 500–1000 km of horizontal shortening can be taken up in the fold-thrust belts and that the remainder has been accommodated by E–W extension using displacements on the major strike-slip faults. We shall examine structural evidence bearing on this question later.

The Central Asian collage

Many of the active movement zones of Figure 2.8 are re-activated tectonic belts of much older derivation. Central and Eastern Asia is a tectonic collage or composite continent formed by the accretion of separate blocks at various times (Figure 5.32). The main blocks are the Siberian continent in the north, the North China or Sino-Korean block, the South China block, and the Southeast Asia block in the east, the Tarim and Tibet blocks in Central Asia, and the Kazakhstan and Afghanistan

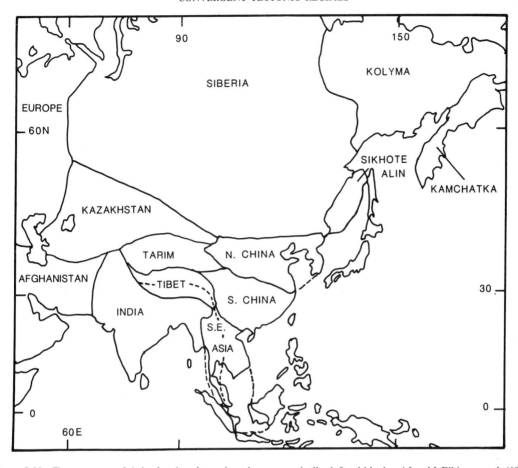

Figure 5.32 Terrane map of Asia showing the main palaeomagnetically defined blocks. After McElhinny *et al.* (1981).

blocks in the west. The palaeomagnetic evidence for the separate existence of these blocks in pre-Mesozoic times is discussed by McElhinny *et al.* (1981), who suggest that the three eastern blocks were situated near the equator in Permian times, and were successively accreted to the Siberian block during the Mesozoic. According to Lin *et al.* (1985), the North China block joined Siberia in the late Permian and the South China block in the early Jurassic, from a position along the northeast margin of Gondwanaland.

Mitchell (1981) discusses the complex area extending from Tibet into SE Asia, and recognizes a Central Tibet/Indo-China block thought to have accreted to Asia in the late Triassic. The southern Tibet block is consi-

dered to be continuous with a N–S block extending through eastern Burma, east of the Sittang suture, and to have collided with Indo–China in the late Triassic also. However, the South and Central Tibet blocks do not appear to have joined until the early Cretaceous. The geology of the Bangang–Nujang suture separating these two Tibetan blocks is discussed by Allegre *et al.* (1984). They demonstrate that highly deformed middle to Upper Jurassic sediments are overthrust by ophiolites and overlain unconformably by Upper Cretaceous volcanic rocks. The authors suggest that an island-arc subduction zone dipping to the south was terminated in Upper Jurassic to Lower Cretaceous times by collision between the two blocks. Molnar and Chen (1978) point out that a palaeomagnetic reconstruction places Central Tibet at about latitude 8°N in the late Cretaceous, and that about 3000 km of subsequent convergence must have taken place before it reached its present position. This implies that Tibet may not have been accreted to Siberia until later than the other accretion events just referred to. The precise history of accumulation of the collage is still subject to considerable uncertainty.

The sutures separating these blocks are zones of subduction and collision which represent major structural weaknesses in the Asian continent. These zones of weakness have been re-activated during the Eocene to Recent convergence, and explain the pattern and extent of recent tectonic activity.

Deep structure of the Himalayas and Tibet

A combination of gravity and deep seismic profiles (Mishra, 1982) indicates a crustal thickness of 58 km below the Himalayas, 71 km under the Karakorum, at the northwest end of the Himalayan range, and 55 km below the Pamir range (Figure 5.33A). A mid-crustal reflector about 14–15 km above the Moho appears to extend across the whole central section, and is interpreted as the basal decollement separating the Asian plate from underthrust Indian plate. Hirn et al. (1984) report a step in the Moho a few tens of km north of Mt. Everest between the 70 km-deep Tibet Moho and the 55 km-deep Himalayan Moho. Above the Moho is another prominent reflector which is interpreted as a probable crust–mantle interface at 35 km depth, reinforcing the underthrust model. However, the arrangement was not thought to support a simple doubling of the crust, but rather a separate decoupling and thrusting of the upper and lower crustal layers. Allegré et al. (1984) report results from a seismic refraction study across the Indus–Tsang Po (Zangbo) suture near Lhasa, that reveal a complex Moho topography involving several steps (Figure 5.33B). These are interpreted as the sites of overthrust mantle wedges, directed southwards.

The Tibetan plateau has an average eleva-tion of about 5 km and an average crustal thickness of about 70 km. The main period of uplift appears to be post-Miocene (Guo, 1980). A study of P_n and S_n seismic wave properties beneath the Himalayas and Tibet, reported by Barazangi and Ni (1982) indicates efficient propagation of S_n waves in the uppermost mantle beneath the Tibetan plateau, together with relatively high velocities of both P_n and S_n waves beneath most of the Tibetan plateau, similar to those found below stable Precambrian shield regions (see 2.2). These results were held to be consistent with a model in which shield-like Indian continental lithosphere underthrusts Tibet at a shallow angle (about 15°) as originally suggested by Argand (1924). An alternative model, in which hot, weak crust and upper mantle is being shortened beneath Tibet (Dewey and Burke, 1973) is not supported by these results. Recent structural work in the Western Himalayas, which we shall now discuss, suggests that crustal structure is much more complex in detail than either of these models envisages.

Structure of the Western Himalaya

A crustal profile across the Karakorum range in North Pakistan is described by Coward et al. (1982). The section extends from the Indus–Zangbo suture in the north to the main Himalayan boundary thrust in the south (Figure 5.34) and crosses a major shear zone termed the Main Mantle Thrust (MMT). On its south side is a 10–20 km-wide intensely deformed zone of blue-schists and amphibolites. These comprise interfolded basement and cover rocks belonging to the Indian plate, and are overthrust by highly deformed and metamorphosed rocks of the Kohistan sequence. This sequence commences with the basic-ultrabasic Chilas complex, more than 8 km thick and 300 km long, which is overlain by pillow lavas and greywackes, intruded by gabbros, diorites and tonalites. These rocks are deformed under granulite-facies conditions and are interpreted as a slice of the lower crust upthrust along the Main Mantle thrust.

The early high-grade fabrics and associated

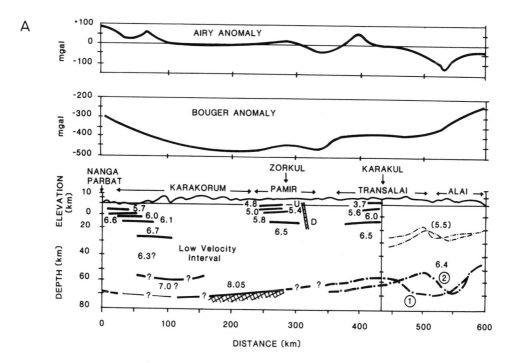

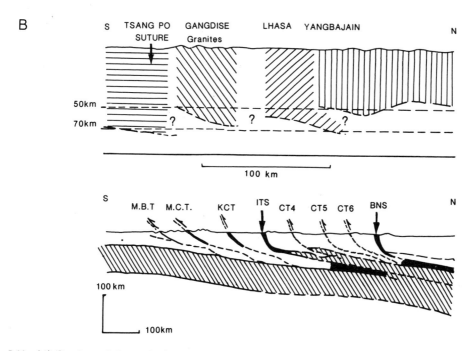

Figure 5.33 (*A*) Gravity and deep seismic structure of a crustal section across the western Himalayan, Pamir and Alai ranges. Note (1) that the main high ranges of the Karakorum and Pamirs appear to be largely compensated isostatically, and (2) the base of the crust at about 70 km beneath the Karakorum. Data west of Karakul are from Pakistan sources and east of this line from the USSR. After Mishra (1982). (*B*) Deep-seismic refraction profile across the Tsang Po (Zangbo) suture near Lhasa, showing the apparently stepped nature of the Moho. The lower diagram is an interpretative cartoon indicating a possible crustal structure. Mantle, hachured; oceanic crust, black; *MBT*, main boundary thrust; *MCT*, main central thrust; *KCT*, Kangmar thrust; *ITS*, Indus–Tsang Po suture; *BNS*, Bangong–Nujiang suture; *CT4–6*, un-named thrusts within the Lhasa block. After Allegré *et al.* (1984).

155

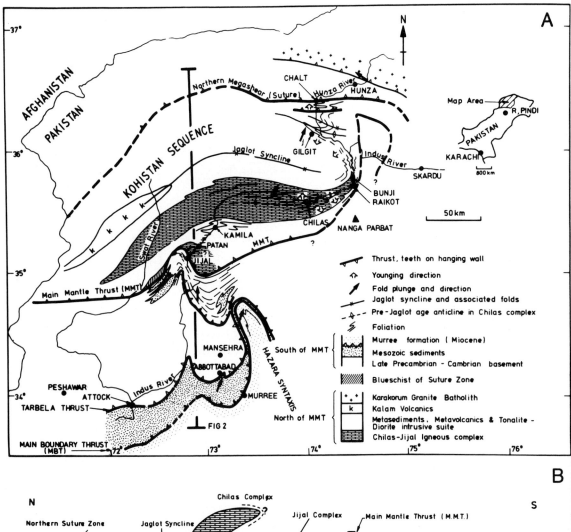

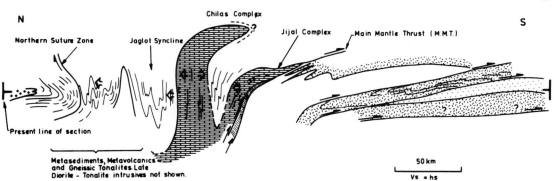

Figure 5.34 (A) Map of the Kohistan region showing the principal structures. (B) Simplified structural profile across the Kohistan region. See (A) for location and key. From Coward *et al.* (1982)

folds are refolded by a major syncline (the Jaglot syncline) that appears to involve the whole upper crust of Kohistan. Most of the Kohistan sequence is steeply dipping and the later folds are tight with steep axial planes. However, south of the MMT the structure is dominated by gentle south-dipping thrust wedges (Figure 5.34B). The steepening of the MMT and the Kohistan sequence may be due to passive back-tilting produced by movements on the younger thrusts as they move up ramps to the south, or possibly to northwards back-thrusting on a southwards-dipping thrust.

Figure 5.35 shows a model (Coward and Butler, 1985) of the post-collisional tectonic evolution of the Karakorum region. Note firstly that the deep earthquakes beneath Tibet are attributed to a north-directed back-thrust involving the whole Indian lithosphere, and secondly that the isostatic response of over-thickening of the lithosphere results in the uplift of the region between the Pamir range and Kohistan. A balanced section from the MMT to the unde-formed Indian plate (Coward and Butler, 1985) leads to a shortening estimate of 64% (Figure 5.36).

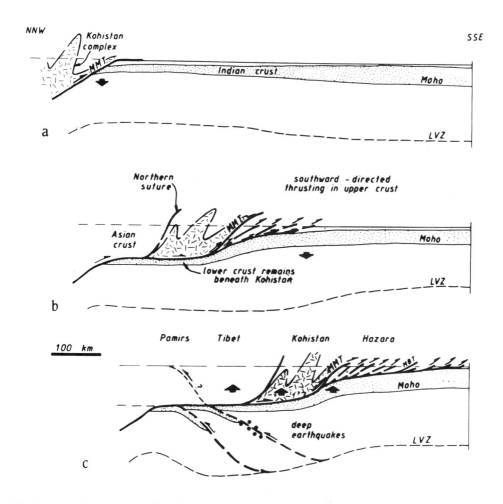

Figure 5.35 Interpretative cartoon profiles illustrating the post-collision tectonic evolution of the northern margin of the Indian plate. Random dashes, mantle rocks of Kohistan complex; stipple, Indian crust in footwall to Himalayan sole thrust; large arrows, predicted vertical movements arising from lithosphere loading; *MMT*, main mantle thrust; *MBT*, main boundary thrust; *LVZ*, base of Indian plate lithosphere (hypothetical). From Coward and Butler (1985)

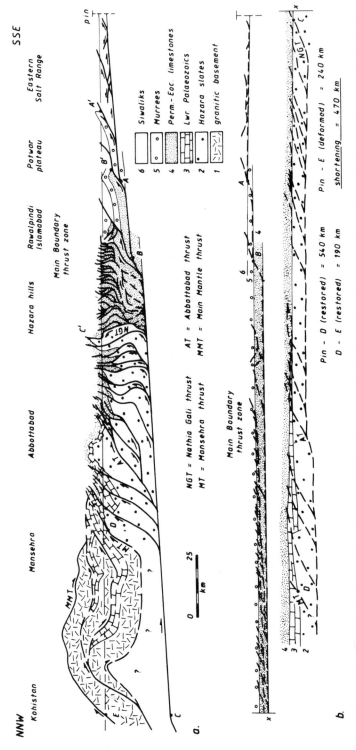

Figure 5.36 Simplified balanced (*a*) and restored (*b*) sections across the Pakistan Himalayas, from the main mantle thrust outcrop to the foreland. Section is constructed to minimize necessary displacements. From Coward and Butler (1985)

The Himalayan range thus represents a foreland thrust belt, currently active at its southern limit in the Salt Range where the basal thrust is still moving, resulting from collision many hundreds of km to the north. The site of the most recent collision has yet to be established; it may lie along the northern margin of the Tarim basin, along the line of the Pamir–Tien Shan ranges much further north than originally thought. Until much more geological field work is undertaken in these remote regions, this question may not be finally resolved.

The whole process of crustal thickening and shortening involved in this collision orogeny appears to have taken about 40 Ma to reach its present state, and is not yet complete. An instructive comparison may be made with the Caledonian orogeny in Britain (see 8.4) where the late Caledonian foreland thrust belt of NW Scotland is linked with closure along a suture 300 km to the south, across an intervening collage of blocks with a much earlier orogenic history.

5.5 Southeast Asia

An incomplete collage

Southeast Asia may be taken as an example of a collision orogenic belt at an early stage in its development. It is instructive to speculate on the extreme complexity of the accretionary terrane that would result from complete continental collision of this region with closure of all the oceanic basins. This notional accretionary terrane may be usefully compared with Central Asia, or indeed with older orogenic belts, as a warning against over-simplistic reconstructions!

The present tectonic framework of the region is summarized in Figure 5.37, and represents the complex interaction of three main plates: the Indian plate to the south with Australian continental crust in its eastern half; the Southeast Asian part of the Eurasian plate to the northwest, and the Pacific plate to the northeast. Subduction of oceanic Indian plate

is taking place at the Java trench below the Sunda arc. This subduction zone extends south of Sumatra and Java eastwards to the edge of the Timor Sea where oceanic crust of the Indian plate gives way to continental Australian crust. The currently active volcanic arc extends from western Sumatra through Java and the smaller islands to the east.

The Australian continental crust extends northwards to include New Guinea (Irian). North of this plate lies a very complicated region consisting of comparatively young back-arc spreading basins and island arcs which lie between the main Pacific plate to the east and the Asian continental margin in the west. The Neogene volcanic arc runs through the island of Sulawesi, east of Borneo, and joins the active arc in the southern Philippine islands.

Charlton (1986) explains some of the complexity of the present pattern by postulating movements along a series of NE–SW sinistral strike-slip faults that result from the geometrical arrangement of continental and oceanic plate during the initial collision (Figure 5.38). Due to the small area of initial continental overlap, and the greater ease of northward travel over the oceanic Pacific plate, fragments of the northwest corner of Australia are progressively sliced off, and attached to Asia. The position of New Guinea, about 1500 km to the northeast of the present Indian/Asian plate boundary at the Sunda arc, is a consequence of this cumulative strike-slip displacement.

At present, continent–island arc collision is taking place along the southern side of the Banda Arc on Timor and the adjacent islands. Charlton believes that the initial collision, the products of which are now to be found in eastern Sulawesi, took place prior to mid-Miocene times when the major Indian plate reorientation referred to earlier took place. The present convergence vector between the Indo-Australian and Eurasian plates is 020° and that between the Asian and Pacific plates is 110° (see Figure 3.1). Prior to the mid-Miocene rearrangement, the Indo-Australian plate was travelling approximately northwards relative to

Figure 5.37 Simplified tectonic map of the eastern Indonesian region. From Norvick (1979) with permission.

Eurasia (see Figure 3.5C–E). However, a complicating factor is the southeastwards movement of the Indo-China block resulting from the India–Asia collision (see Figure 5.28). This movement must have caused an eastwards shift in the Pacific/Asian boundary relative to the Indian plate, so that the convergence direction between the Australian part of the Indo-Australian plate and SE Asia is actually NNW. Figure 5.39 shows a reconstruction of the relative positions of Australia and SE Asia in late Cretaceous to late Pliocene times which may be compared with the simplified model of Figure 5.38.

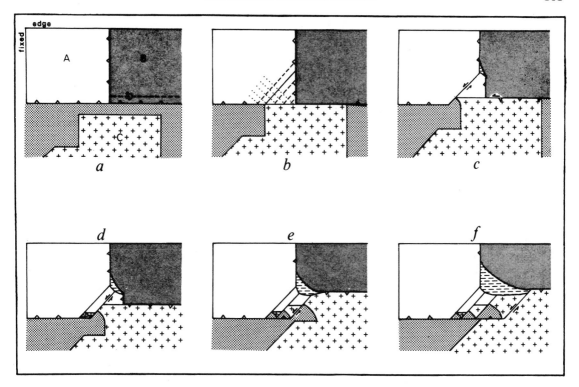

Figure 5.38 Sequence of cartoon maps illustrating the tectonic evolution of the eastern Indonesian region. The three plates are the Eurasian plate (*A*, blank), the Pacific plate (*B*, dark stipple) and the Indo-Australian plate (*C*, oceanic, light stipple; continental, crosses). Subduction zones, toothed lines. Local sea-floor spreading associated with transtensional zones is indicated by dashed ornament. The dotted lines in (*b*) are potential complementary strike-slip faults that do not become active. The effect of the collision is to develop a series of transform faults which transfer pieces of the northwestern corner of the Australian continent to the southeastern portion of the Eurasian plate. From Charlton (1986), with permission.

Timor

The island of Timor forms part of the collision zone between the Australian continental shelf and the Banda island arc. According to Charlton's model, the collision orogeny of Timor is the result of a secondary collision between a piece of the northwest corner of Australia already welded to Asia, and a more south-westerly portion of Australia (see Figure 5.38).

An important feature of the Banda arc is that the Permian and Mesozoic rocks of the para-autochthonous units display Australian affinities, whereas the overlying allochthonous thrust sheets contain strata of the same age range but with tropical facies comparable with the contemporary rocks of the Sunda arc and

Borneo. These allochthonous thrust sheets are directed southwards, away from the Banda Sea and towards the Australian continent. The nearest counterparts to these allochthonous units lie on the north side of the Banda Sea, in Sulawesi (Figure 5.37). The former continuity of these units has therefore been disrupted by the opening of the Banda Sea. The formation of this oceanic spreading basin may relate in part to the regional southeastwards extension of the Indo-China continental margin already discussed, and partly to back-arc extension relating to the Banda subduction zone. A geological cross-section of Timor (Figure 5.40*A*) shows the basic structural framework. A basal low-angle sole thrust (*T*1) carries the allochthon across the para-autochthonous Permian

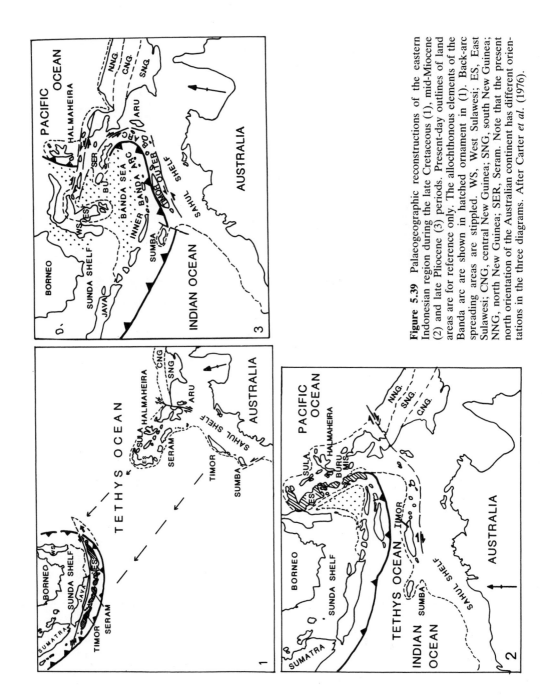

Figure 5.39 Palaeogeographic reconstructions of the eastern Indonesian region during the late Cretaceous (1), mid-Miocene (2) and late Pliocene (3) periods. Present-day outlines of land areas are for reference only. The allochthonous elements of the Banda arc are shown in hatched ornament in (1). Back-arc spreading areas are stippled. WS, West Sulawesi; ES, East Sulawesi; CNG, central New Guinea; SNG, south New Guinea; NNG, north New Guinea; SER, Seram. Note that the present north orientation of the Australian continent has different orientations in the three diagrams. After Carter *et al.* (1976).

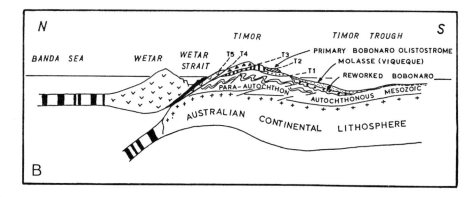

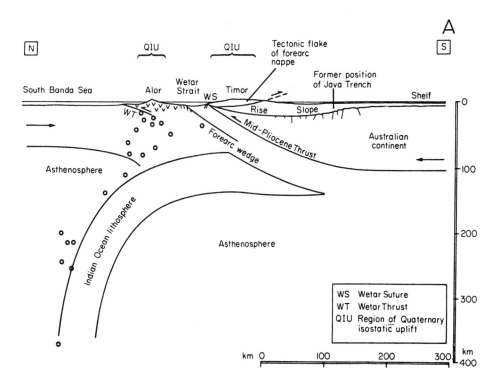

Figure 5.40 (A) Diagrammatic section across Timor, showing the sequence of allochthonous sheets carried on thrusts *T1–T5* overlying the para-autochthonous Australian assemblages. Uneven vertical stripes, oceanic crust; V-symbols, volcanic-arc assemblage; black, Atapupu sheet of amphibolite and serpentine; closed triangles, Ocussi sheet of Permian–Jurassic sediments and volcanics; even vertical stripes, Lolotai metamorphic complex; small regular dots, Kobano–Iliomar sheet of bathyal sediments. Note post-emplacement deposits of primary Bobonaro olistostrome, and reworked Bobonaro sediments. From Carter *et al.* (1976) (B) Tectonic cartoon illustrating an interpretation of the geophysical and geological structure of Timor. Note the postulated rupture of the Australian continental lithosphere along a thrust with a surface outcrop in the Wetar strait; this interpretation differs significantly from that of (A). Earthquake foci represented by open circles. From Milsom and Audley-Charles (1986), with permission.

Age BP	Stratigraphic and structural events in Timor	Tectonic events

0 Ma

Uplift of Pleistocene coral reefs and
alluvial terraces

2 Ma

Deposition of coral reefs and alluvium during
Pleistocene (N. 22 – 23)

2 Ma

— — — —Local angular unconformity — — — —

Widespread subaerial erosion
Gentle folding of Viqueque turbidites

Deposition of Viqueque turbidites (Noele
Marl Formation – Late Pliocene – Early
Pleistocene N. 21 – 22) with subaerial erosion
in Northern Timor
— — — — — — — — — — — — — — — — — — —

Deposition of Sabaoe Limestone Formation
(Late Pliocene N. 21) in shallow water
— — — — — — — — — — — — — — — — — — —

Deposition of upper part of Batu Putih
Limestone in deep water
(Late Pliocene N. 21)

~ 3 Ma

— — — —Local angular unconformity — — — —

3 Ma

(Cessation of volcanic activity in Atauro and
Wetar of inner Banda Arc)

~ 3·5 Ma

Emplacement of thrust sheets with overlying
Bobonaro Scaly Clay olistostrome and lower
part of Batu Putih Limestone in the Mid-
Pliocene (N. 20)

~ 4 Ma

Deposition of lower part of Batu Putih
Limestone on the Bobonaro Scaly Clay in the
Early Pliocene (N. 18 – 19)

Erosion of para – autochthon
Folding, faulting and local imbrication of
para – autochton

~ 5 Ma

Deposition of youngest member of para-
autochthonous Australian continental margin
facies (Early Pliocene N. 18)

~ 7 Ma

Sedimentation in the allochthonous elements
changes from shallow water Cablac Limestone
(Early Miocene N. 8) to deep water Miomaffu
tuffs (Late Miocene N. 17)

Vertical labels:
- Viqueque Group sequence
- ~ 3 km
- Uplift of collision zone
- Compression of collision zone
- Subsidence of allochthon

to Cretaceous sequence. Above $T1$ is the Kalbano thrust sheet ($T2$), which is an imbricate stack or duplex containing early Cretaceous sediments. This thrust sheet is overlain by three higher thrust sheets with ophiolitic material on the topmost and most northerly. The thrust sequence is unconformably overlain by late Pliocene to late Pleistocene turbidites.

The allochthonous units reveal a different deformation history from that of the underlying para-autochthonous Australian units. There is evidence in the allochthon for deformation and igneous activity before emplacement on the Australian margin. The youngest unit in the allochthon is of late Miocene age. The para-autochthonous sequence reflects the deep-water continental slope and rise environment of the northern Australian continental shelf. The youngest unit in this sequence is early Pliocene in age. There is no evidence for any deformation in these rocks prior to this date. The deformation accompanying the emplacement of the allochthon therefore took place in mid-Pliocene times, which is taken as the date of collision. After the collisional deformation, the Timor region experienced vertical uplift of at least 3 km in the last 3 Ma. This uplift is well documented from the shallowing and erosion of marine sequences, and the formation and uplift of coral reefs and alluvial terraces (Table 5.1).

The gravity field in the Banda arc region is discussed by Milsom and Audley-Charles (1986). A prominent negative Bouguer anomaly extends through Timor and Jamdena, following the arc round to Seram. A smaller positive anomaly follows the active volcanic arc to the north and also extends over the Banda Sea. The northern limit of the negative anomaly runs through the island of Timor. Recalculating the gravity field assuming isostatic compensation still leaves a substantial negative anomaly, although the positive anomaly is virtually eliminated. These results suggest that the southern part of Timor is depressed below its isostatic level, whereas the north coast, which has been subjected to rapid recent uplift, is now near equilibrium. Figure 5.40B shows an interpretative model profile of the area. The negative anomaly is attributed to the subducted oceanic slab which must still exist below the inner arc from seismic evidence. Complications arise from horizontal shortening, firstly on the south-directed thrusts on Timor, and secondly by a postulated north-directed thrust in the straits between Timor and the volcanic arc. The rise of northern Timor is attributed to isostatic adjustment resulting from its detachment from the sinking slab.

Table 5.1 Summary of late Cenozoic stratigraphic and tectonic history of Timor. From Milsom and Audley–Charles (1985)

6 Strike-slip and oblique-slip regimes

6.1 Characteristics of strike-slip regimes

In the early days of the plate tectonic theory, attention was concentrated on the tectonic effects of destructive and constructive plate boundaries. However, the detailed study of oceanic transform faults and of major continental transform faults, particularly the San Andreas fault system, has led to the recognition that strike-slip or transform regimes are also of fundamental tectonic importance.

If we examine again the plate boundary network in Figure 3.1, we see that conservative or strike-slip boundaries make up a significant proportion of the total boundary length. In addition to the large number of minor transform offsets of the ocean ridges, there are long sections of boundary made up entirely of transform fault. The most striking example forms the northeastern boundary of the Pacific plate. Here the San Andreas fault joins the end of the East Pacific ridge in the Gulf of California to the Juan de Fuca ridge west of Washington and Oregon in the northwestern USA. North of this short section of ridge, another major strike-slip fault system connecting the Chugach–Fairweather–Queen Charlotte islands faults extends off the coast of Western Canada and Alaska. Another major continental strike-slip boundary is the Alpine fault of New Zealand, which connects subduction zones to the north and south marking the boundary between the Pacific and Indo-Australian plates.

There are a number of major oceanic transform faults. Among the more important are the Owen fracture zone between the Arabian and Indian plates (see Figure 3.6B), the Azores fracture zone connecting the mid-Atlantic ridge and the Mediterranean subduction zone, and the major fracture zone connecting the Scotia arc with the Chile trench. The morphology and structure of oceanic transform faults is discussed below (6.4).

Major strike-slip movement never takes place along a single fault plane, but is distributed through a zone. In the case of the San Andreas fault system, this zone is about 100 km in width. Some of the great oceanic fracture zones are over 50 km wide. A simple model of a strike-slip boundary consists of two plates sliding past each other, with complete conservation of plate area, and no convergence or divergence across the boundary. This model must be replaced by a model involving a boundary zone of finite width within which complex tectonic effects take place.

The general kinematic relationships were described in 3.3, where the significance of oblique relative movements across plate boundaries was stressed. It was concluded that, in general, plate movements at boundaries were transpressional or transtensional, with components of compression or extension across the boundary, and that the boundary should be considered as a deformable sheet rather than a plane.

The importance of the continental strike-slip regime was highlighted in an influential paper by Reading (1980). He pointed out that a strike-slip tectonic regime created a special type of orogenic belt characterized by intense seismic activity and deformation, by important differential vertical movements, by rapid and varied sedimentation, but by comparatively feeble magmatic and metamorphic activity.

Major strike-slip zones are common on the continents, and it is frequently unclear whether or not these are plate boundaries. For example, the major strike-slip zones resulting from the India–Asia collision (see Figure 5.28) are mostly the result of internal deformation of the Eurasian plate, and it is not practicable to use the rigid plate model in this area. Nevertheless, individual fault zones may constitute major belts of strike-slip deformation similar in their effects to the plate boundary type. Some may penetrate the whole thickness of the

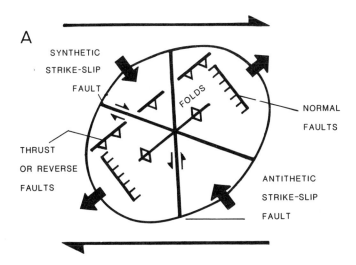

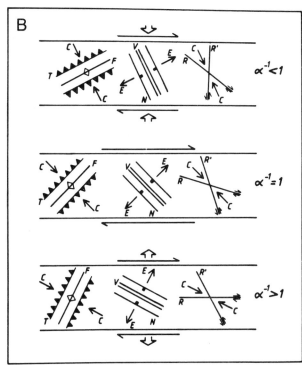

Figure 6.1 (*A*) Diagrammatic representation of the structural pattern produced by a dextral simple-shear couple, after Harding (1974), and Reading (1980). (*B*) Diagrams showing the orientation patterns of faults and fold axes during dextral simple shear (middle diagram), under transpression (top diagram) and transtension (bottom diagram). *C*, compression axis; *E*, extension axis; *N*, normal faults; *T*, thrust faults; *R*, *R'*, Riedel shears or strike-slip faults; *V*, veins, dykes or extension fractures; *F*, fold axes. Note that transpression results in clockwise rotation of compression and extension axes, and transtension in anticlockwise rotation of stress axes. The opposite would of course hold for sinistral shear. From Sanderson and Marchini (1984), with permission.

lithosphere, while others may detach on low-angle décollement planes within or at the base of the crust.

Studies of deeply eroded Precambrian orogenic belts demonstrate the importance of major strike-slip shear zones at deeper crustal levels. For example the Precambrian of South Greenland exhibits several major orogenic belts that represent middle- and lower-crustal ductile counterparts of the high-level strike-slip fault zone (see Figures 9.17, 9.18).

Causes of geometrical complexity

If we assume the strike-slip boundary to be a deformable sheet, the bulk strain can be

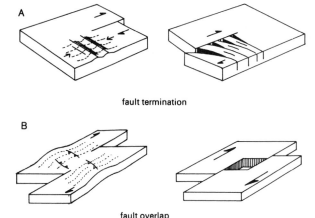

fault termination

fault overlap

Figure 6.2 Local compressional and extensional structures produced by fault terminations (A) and fault overlaps (B) in strike-slip faulting. After Reading (1980).

considered as a combination of pure and simple shear. The pure shear component arises from the extensional or compressional component across the sheet, and the simple shear component from the strike-slip displacement (see Figure 3.14). Transtension and transpression involve a change in surface area but not necessarily in volume. Compression across the sheet at constant volume will be compensated by extension, most probably in the vertical dimension, leading to crustal thickening and uplift. Extension across the sheet may be compensated by vertical shortening leading to crustal thinning and depression. The geometric rules governing transpression and transtension are described by Sanderson and Marchini (1984) and their effects are summarized in Figure 6.1B.

The importance of volume changes should not be overlooked: extensional movements, particularly in the oceanic lithosphere, will normally be accompanied by emplacement of new mantle material, thus adding to lithosphere volume, and volumetrically less important changes also accompany metamorphic effects in both compression and extension.

The effect of simple shear strain is summarized in Figure 6.1A. In a block of rock deforming heterogeneously, the directions of extension and compression are given by the orientation of the bulk simple-shear strain ellipsoid. Various types of structure may form,

and different combinations of structure will be appropriate in different materials. Fold axes will parallel the long axis of the strain ellipse. Conjugate sets of strike-slip faults may form, one synthetic, making a small angle with the main strike-slip direction, and the other antithetic, making a large angle with this direction. Inclined faults will have normal dip-slip components parallel to the long axis of the ellipse, and thrust components parallel to the short axis. Thus where the sense of movement of the major strike-slip displacement is unknown, it can be deduced from the relationship of any of these subsidiary structures to the boundaries of the deformation zone.

Other geometrical effects arise out of the nature of the fault movements. Only one small sector of a fault is active at any given time, and the displacement must therefore be taken up elsewhere by heterogeneous strain. This fault termination effect is illustrated in Figure 6.2A. Local zones of compression or extension are produced at the ends of displaced segments. Simultaneous movements on en-echelon faults also produce local zones of compression or extension (Figure 6.2B).

The most important geometrical effects are produced by changes in direction of strike-slip faults (Figure 6.3A). As the two opposed blocks move past each other, local zones of convergence or divergence occur, which produce compressional and extensional effects respectively. In a complex fault network, this process will lead to alternate zones of raised and depressed fault blocks (Figure 6.3B). These effects are analogous to the geometrical effects created by the ramp-flat geometry in dip-slip fault displacements (see e.g. Figure 4.26). The combination of folds and faults produced by these local zones of compression and extension have been termed *flower structures* by Harding and Lowell (1979). Positive

flowers are uplifted zones with a compressional component across the strike-slip belt, and negative flowers are depressed zones with an extensional component (Figure 6.4).

Strike-slip duplex structures may be created in an analogous manner to thrust and extensional duplexes (Figure 6.5). Pieces from one side of the main fault may be sliced off and transferred to the other side as the active fault takes a new course. On a large scale, this

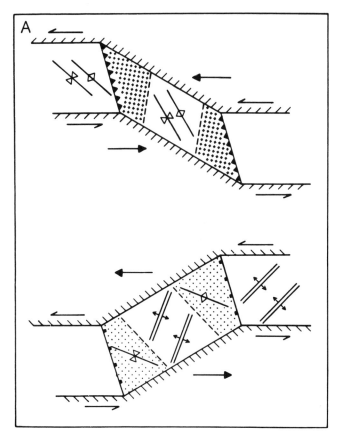

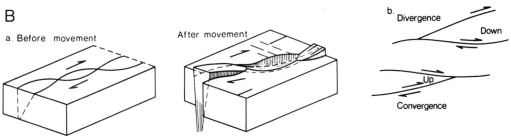

Figure 6.3 The effect of changes in fault orientation. (*A*) Diagrams showing the generation of transpressive (upper diagram) and transtensional (lower) regimes within a region of offset in an otherwise pure strike-slip zone. Transpressional and transtensional regions are stippled; fold axes, single lines, and extensional fissures, double lines. Boundaries of transpressional and transtensional regions are shown as faults (toothed lines) to indicate accommodation of the strain discontinuity. From Sanderson and Marchini (1984). (*B*) Diagrams illustrating the formation of raised and depressed blocks by convergence and divergence along curved fault segments, during strike-slip motion. From Reading (1980)

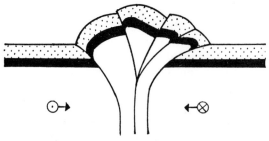

POSITIVE FLOWER STRUCTURE

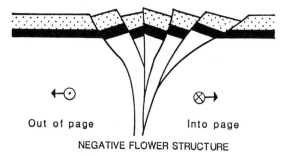

Out of page   Into page

NEGATIVE FLOWER STRUCTURE

Figure 6.4 Positive and negative *flower structures* produced by convergence and divergence respectively in strike-slip motion. Dot and cross symbols within circles indicate out-of-page and into-page components of motion, respectively. After an unpublished diagram of N. Woodcock.

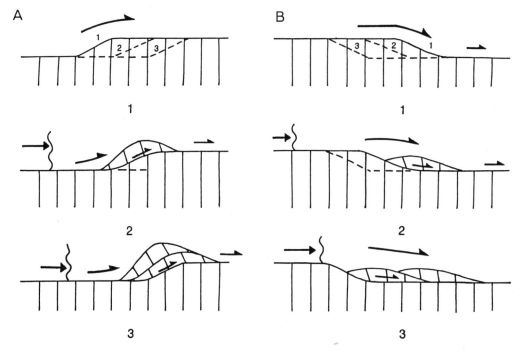

Figure 6.5 Diagrams illustrating the formation of strike-slip duplex structure in transpression (*A*) and transtension (*B*). Note that the structures are analogous morphologically to compressional and extensional dip-slip fault duplexes, respectively.

process may produce far-travelled blocks that are allochthonous or *exotic* in relation to the block with which they are now associated. Large displaced blocks of this type are termed displaced or exotic *terranes* and are discussed below. (Note that the American form of the word 'terrain' is now universally used in this context.)

Sedimentation in strike-slip zones

Despite the fact that the main displacements in strike-slip fault zones are horizontal, the most obvious displacements locally are usually dip-slip (see Figures 6.1–6.3). It is the vertical movements of fault blocks relative to each other which produce the most important stratigraphic effects of a strike-slip zone. There are obvious differences between transtension and transpression in this respect, since transtension leads predominantly to depression and transpression to uplift. Thus transpressional zones will be eroded and the derived sediments will be carried outside the strike-slip zone, possibly to distant areas. For example the major transpressive Alpine fault zone of New Zealand is not an area of major sedimentation. In transtensional zones on the other hand, major sedimentary basins form, which on land are normally within reach of abundant sediment supply.

Transtensional basins are typically lacustrine, and bordered by alluvial fans. An excellent example of such a zone is the Dead Sea rift, which forms part of the western margin of the Arabian plate. Freund *et al.* (1968) discuss the evidence for a major sinistral displacement along this zone, and relate it to the sedimentary record. Several rhomb-shaped graben occur in the zone (Figure 6.6A, B), some of which contain lakes, of which the Dead Sea and the Sea of Galilee are well-known examples. These graben are pull-apart features formed in regions of extensional fault overlap (see Figure 6.2B). The total sinistral movement is 110 km as measured by the displacement of various igneous and sedimentary markers of Precambrian to late Cretaceous

age. Quennell (1959) attributes 67 km of this movement to the early Miocene, after which an inactive period allowed the rift to fill up with red beds until rivers from the east were able to flow across the rift towards the Mediterranean. Figure 6.6C demonstrates the evidence for sinistral displacement of these Miocene rivers by 43 km in the Pliocene to Pleistocene period. Some movements are very recent. Zak and Freund (1966) demonstrated displacements of 150 m in the Lisan marl of the Jordan valley. This formation has been dated at 23 000 year BP by the radiocarbon method.

A well-documented example of a Palaeozoic strike-slip zone is the Midland Valley of Scotland in the Devonian period, described by Bluck (1980) — see 8.4. The best active example is the intensively studied San Andreas fault zone which we shall examine in 6.3.

6.2 Displaced or exotic terranes

The concept of *displaced* or *exotic terranes* arose from observations in the North American Cordilleran belt, over 70% of which is regarded as a collage of *suspect terranes* of probably allochthonous origin (Wilson, 1968; Monger *et al.*, 1972; Jones *et al.*, 1972). Although displaced terranes may of course be found at any convergent plate boundary, and are prominent for example in the orthogonal convergent regime of the Central Asian collision zone, they are particularly associated with oblique convergence or strike-slip regimes, and tend to accumulate at geometrically favourable locations along such boundaries.

The distribution and nature of more than 50 suspect terranes in western N. America (see Figure 8.9) is summarized by Coney *et al.* (1980) who lay down certain principles in their recognition. A terrane exhibits internal homogeneity and continuity of stratigraphy, and of tectonic style and sequence, and is distinguishable from adjoining terranes by discontinuities of structure or stratigraphy that cannot be explained on the basis of normal facies or tectonic changes. Most terrane boundaries separate totally distinct rock sequences and/or

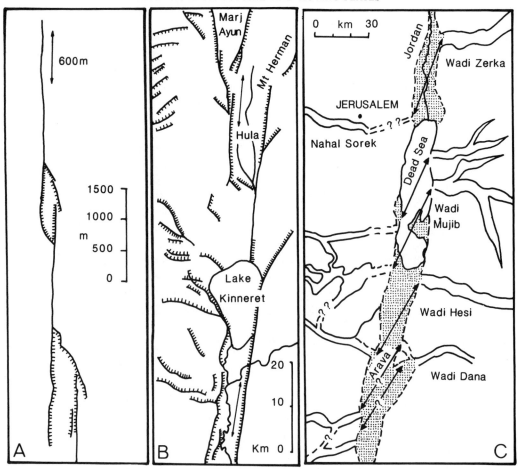

Figure 6.6 (*A*, *B*) Rhomb-shaped graben formed in en-echelon segments, resulting from strike-slip movement on the Dead Sea transform fault zone. Note difference in scale between (*A*) and (*B*). (*C*) Displacement of Miocene river system across the Dead Sea fault zone. (*A*)–(*C*) after Freund *et al.* (1968).

faunas, and many contain palaeomagnetic records that differ strongly from those of the stable craton or of adjacent terranes. A suspect terrane may be proved to be allochthonous or exotic if its faunal or palaeomagnetic signature proves it to originate a considerable distance from its present position. Most of the suspect terranes of N. America appear to have collided and accreted to the craton margin during Mesozoic and early Cenozoic time. Many show evidence of an origin far distant from their present position, and may also have undergone translations of hundreds of km after collision. Palaeomagnetic evidence also indicates significant rotations about the vertical in many cases.

During much of the 120 Ma period during which the terrane displacement process occurred, the continental margin was a subduction zone, so that displacement and accretion took place by a process of oblique convergence, combining underthrusting and strike-slip movements. It appears that the strike-slip component was dextral throughout, and the terranes seem to have originated far to the south of their present position, in some cases possibly on the other side of the Pacific Ocean.

Silver and Smith (1983) discuss the western Pacific Ocean as an active example of the terrane displacement process. As is clear from Figure 6.7, oblique convergence is taking place between the main Pacific plate and Australia. The authors point out that this motion appears to have sliced off pieces of oceanic plateau and island arc, together with fragments of continental Australia, and carried them northwards. Large ophiolite masses have been emplaced in New Guinea and New Caledonia as a result of this oblique convergence.

The boundary between the Indo-Australian and Eurasian plates is also the scene of oblique convergence (see Figure 5.37, 5.38), the effects of which have already been discussed. Major strike-slip displacements along NE–SW faults have transferred portions of the Australian continental plate to the Eurasian plate, creating a collage zone on the southeastern margin of the Eurasian plate. Large rotations can be proved in this region. For example Haile (1978) has shown from palaeomagnetic evidence that Seram, in the northern arm of the

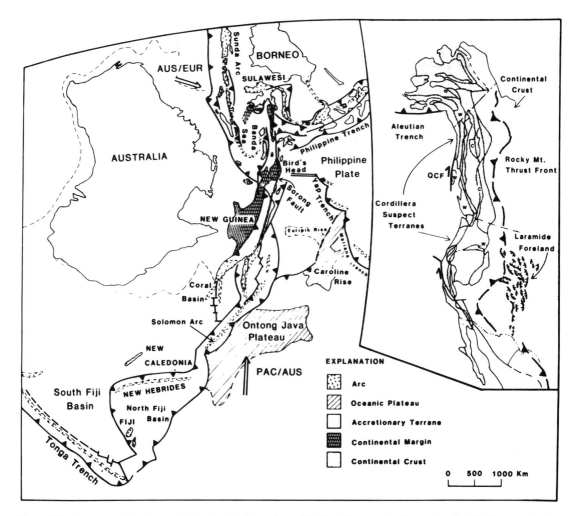

Figure 6.7 Summary tectonic map illustrating the formation of displaced terranes in the western Pacific Ocean region due to the oblique convergence of the Indo-Australian and Pacific plates, and of the Indo-Australian and Eurasian plates (see movement vectors). The region between the New Hebrides and western New Guinea (Irian Jaya) consists of a tectonic collage of terranes detached from both main plates on either side, many of which have suffered sinistral strike-slip displacement relative to their place of origin. Inset for comparison is a map of the Cordilleran strike-slip tectonic collage (see 8.2 and Figure 8.9). From Silver and Smith (1983)

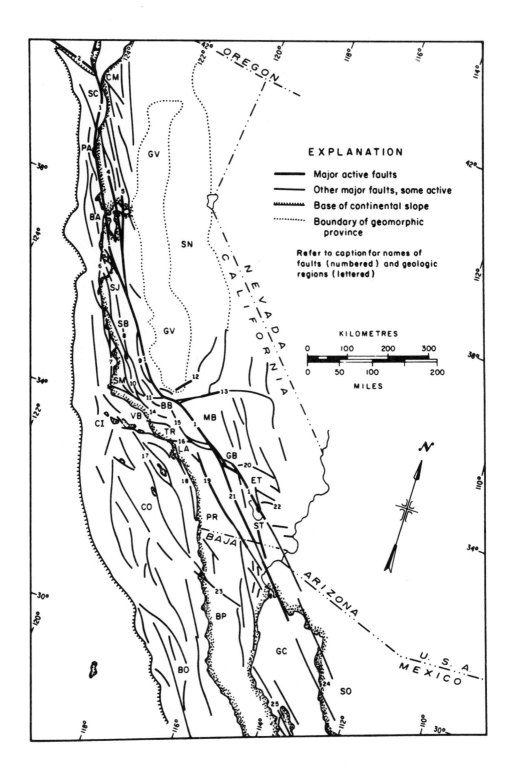

Banda arc, has undergone a large clockwise rotation due to the opening of the Banda Sea. Once accretion is complete, the reason for this rotation will no longer be obvious. The large rotations noted in the terrane collage of the North American Cordillera may have originated in the same way.

The oblique convergence between the oceanic part of the Indo-Australian plate and Southeast Asia has resulted in large dextral strike-slip displacements along the continental margin in Burma, Thailand and western Indonesia (Figure 5.28). In Sumatra, the Asian plate has responded to the oblique subduction by forming a dextral strike-slip fault that traverses the island, and displaces the leading edge of the volcanic arc northwards in relation to the remainder of the overriding plate. In this case, the combination of steep strike-slip faulting and subduction has clearly been favoured, rather than straightforward oblique subduction, as a means of achieving the oblique convergence (Beck, 1983). Beck discusses the possible conditions controlling the mechanism of oblique convergence, and concludes that three factors favour the combined strike-slip/ subduction mechanism: (i) a large angle of convergence (i.e. the convergence direction makes a small angle with the trench); (ii) a shallow angle of subduction; and (iii) the thermal 'softening' effect of the volcanic arc.

6.3 The San Andreas fault zone

The San Andreas fault system is probably the most widely known and intensively studied in the world. Attention has been focused particularly by two major earthquakes; the San Francisco earthquake of 1906 with a magnitude of 8.25 (Richter scale), and the San Fernando earthquake of 1971 with a magnitude of 6.6. The devastation caused by the 1906 earthquake, and the potential for a major catastrophe if a shock of comparable magnitude affects one of the more heavily populated parts of California, have stimulated a major programme of study in this fault zone.

The kinematic history of the region is summarized in 3.1. Before about 38 Ma BP, the region was a subduction zone with oblique convergence taking place between the oceanic Farallon plate and the overriding North American continent (see Figure 3.4A). As we have seen, oblique convergence during the Mesozoic and early Cenozoic was responsible for the displacement of a large number of blocks northwards along the continental margin to form the collage of exotic or suspect terranes making up the Cordilleran orogenic belt (see Figure 8.9). However, when the East Pacific ridge met the American plate boundary at c.38 Ma BP, the movement pattern at the boundary changed abruptly from subduction to strike-slip. This change occurred because the direction of relative motion between the Pacific and American plates was now approximately parallel to the boundary (see Figure 3.4A). The northern end of the fault zone is determined by the intersection of the Murray transform fault with the North American plate boundary, forming a trench–fault–fault triple junction. This junction migrates north with the Pacific plate. Thus the San Andreas fault zone is a consequence of a step in the East Pacific ridge along the Murray transform. This step has resulted in the division of the ridge into two segments, the Juan de Fuca in the north and the main East Pacific ridge in the south,

Figure 6.8 Simplified map of the principal faults and other structural elements in the San Andreas fault zone of California and northern Mexico. CM, Cape Mendocino; SC, Shelter Cove; PA, Point Arena; GV, Great Valley; BA, San Francisco Bay; SN, Sierra Nevada; SJ, San Juan Bautista bend; SB, Salinian block; SN, Santa Maria basin; BB, Big Bend (of the San Andreas fault); SB, Ventura basin; MB, Mojave block; CI, Channel Islands; TR, Transverse Ranges; LA, Los Angeles basin; GB, San Gorgione bend; CO, south California offshore borderland; ET, eastern Transverse Ranges; PR, Peninsular Ranges; ST, Salton trough; BP, Baja California peninsula; GC, Gulf of California; BO, Baja California offshore borderland; SO, Sonora. Numbered faults: 1, San Andreas; 2, Mendocino fracture zone; 3, Oregon subduction zone; 11, Big Pine; 12, White Wolf-Kern; 13, Garlock; 15, San Gabriel; 19, Elsinore; 21, San Jacinto; for names of other numbered faults, see source. From Crowell (1979)

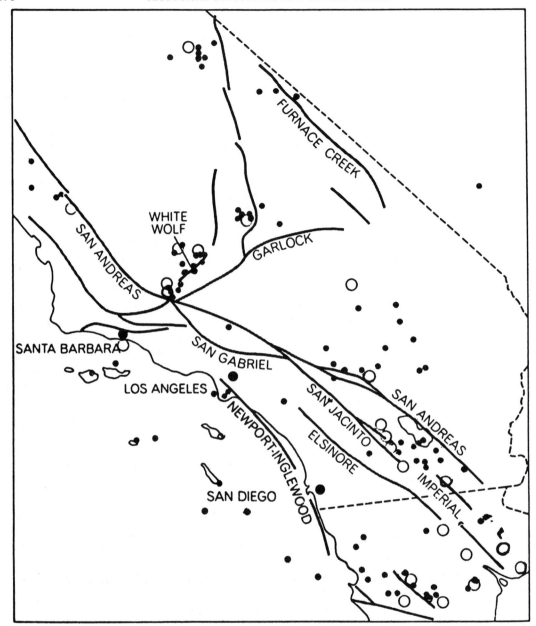

Figure 6.9 Distribution of earthquakes of magnitude 5 or greater in the period 1934–69 in southern California, in relation to the principal faults. Dots indicate magnitudes of 5.0–5.9; circles, magnitudes of 6 or greater. From Anderson (1971)

separated by the San Andreas fault zone. The growth of the zone from the initial point of contact northwards is documented by the northward progression of the ending of activity on the Neogene volcanic arc (see Figure 4.18).

The fault zone is about 1200 km long and 100 km wide, and consists of a complex network of faults (Figure 6.8; see Anderson, 1971). Most of these trend NW–SE sub-parallel to the main San Andreas fault (see for example the San Jacinto and Elsinore faults), but many trend in a more NNW–SSE direction, making an acute angle with the main faults and branching from them. Many of these faults are also provably dextral strike-slip. A significant number of faults are NE–SW in trend, approximately perpendicular to the main faults. Some of them (e.g. the Garlock fault) are antithetic sinistral strike-slip faults, while others are steep reverse faults.

Seismicity

The region is very active seismically: 7300 earthquakes were recorded with a magnitude of 4 or greater in southern California in the period 1934–69. The location of the larger recent earthquakes is shown in Figure 6.9. Most are unconnected with the San Andreas fault itself, but are associated with parallel strike-slip faults such as the San Jacinto and Imperial faults, and with the White Wolf fault, which intersects the San Andreas fault at 90° in the region of the 'Big Bend'. Several of such faults with trends making a large angle with the main fault show overthrust displacements. For example, the large earthquakes in Kern County and San Fernando were associated with overthrust movements. Many of the faults of Figure 6.8 appear to be currently inactive.

Stress and heat flow

In-situ stress determinations in the vicinity of the San Andreas fault using the hydraulic fracture technique are reported by Zoback *et al.* (1980). They investigated the variation of stress with distance from the fault and also with depth. In shallow wells located along profiles across two sections of the fault where slow creep is taking place, the direction of maximum horizontal compressive stress was found to be N–S, approximately 45° to the trend of the fault. The shear stress was found to increase with distance from the fault to a value of about 5 MPa 34 km from the fault. At 4 km from the fault, the shear stress increases from about 2.5 MPa at 150–300 m depth to about

8 MPa at 750–850 m depth. This increase suggests that the mean stress at seismogenic depths (10–15 km) must be several tens of MPa in magnitude.

Heat flow measurements reported by Lachenbruch and Sass (1980) indicate that there is no evidence for local frictional heating along any part of the San Andreas fault. A broad region of high heat flow, around 80 mW m^{-2} exists throughout the Coast Ranges, but the heat flow decreases eastwards across the San Andreas fault towards the Great Valley where it is much lower. The source of this heat-flow high is thought perhaps to be associated with the presence of warmer asthenospheric mantle below the Coast Ranges, due to the absence of a subducting slab south of the Mendocino triple junction. Alternatively, if the heat is frictional in origin, the lack of direct association with the main fault suggests that the high-level fault network is partly decoupled on a broad low-angle detachment horizon extending below the Coast Ranges at or below seismogenic depths. Shear resistance, and consequently shear heating, at these levels could be sufficient to produce the thermal anomaly.

Displacement geometry

The rate of displacement along the transform boundary can be calculated from the magnetic sea-floor stratigraphy. This varies from 1.3 cm/year in the early Miocene to 5.5 cm/year since the Pliocene (Atwater and Molnar, 1973). Seismic techniques yield a value of around 4 cm/year for recent movements, and geodetic measurements give an estimate of 5–7 cm/year (Anderson, 1971). According to Crowell (1979) the total displacement along the fault system is around 1000 km. However geological evidence indicates only about 300 km of dextral displacement on the San Andreas fault itself since the mid-Miocene (Crowell, 1979), and the remainder of the movement is probably distributed among many smaller faults. A significant proportion may be taken up by a major fault along the continental margin offshore. According to Hein (1973), for example, a submarine fan lying on the ocean floor has been displaced from its source by a distance of 300–550 km.

The total observed dextral displacement on the main San Andreas fault is 600 km as measured by the offset of the Sierras basement. Thus the fault is interpreted as having a two-stage history of movement: the first, of late Cretaceous to Palaeocene age, produced about 300 km of displacement; the second is the current phase commencing in the Miocene. The early movement is tentatively linked with a possible transform boundary between the Kula and North American plates and ended when the Kula–Farallon ridge swept northwards up the American continental margin (see Figure 3.7A). These early movements were associated with the major terrane displacement process referred to earlier.

The straight sections of the main San Andreas fault appear to be precisely parallel

with small circles about the pole of rotation for the Pacific–American plate motion, as determined by Minster et al. (1974). Slip along these straight sections is exhibited in the form of slow 'creep' and by many small-magnitude earthquakes. The curved sections, in contrast, appear to be locked, and such sections are obvious sites for large earthquakes such as those already mentioned. The most prominent curved section is known as the *Big Bend* region (see below), where activity is presently concentrated on the White Wolf and other faults (see Figures 6.9, 6.12). In other places, bends appear to be in the process of being by-passed. For example, at the San Gorgione Bend (Figure 6.8), major movement appears now to be taken up along the San Jacinto fault, thus 'straightening out' the main fault line, transferring the major movement to the west and isolating an inactive slice on the eastern side of the new fault line.

At the southeastern end of the fault zone, the San Jacinto and Elsinore faults end at the margins of pull-apart basins. These form part of a system of side-stepping spreading centres within the Gulf of California, which has opened by a process of oblique rifting (Figure 6.10). The pull-apart basins are bounded by normal faults with trends oblique or perpendicular to the strike-slip faults (see Figure 6.2B). This process has resulted in the northwestwards movement of the Baja California peninsula by a combination of E–W spreading and NW–SE strike-slip faulting. Rhomboid depressions are also forming in the offshore regions of California by the same mechanism.

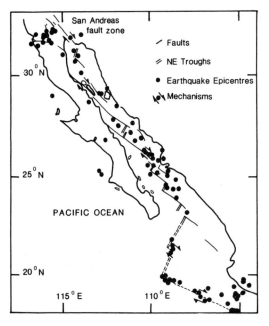

Figure 6.10 Combination of spreading ridges and transform faults in the Gulf of California by which northwestwards displacement of the Baja California peninsula has been achieved. Note the pattern of earthquake epicentres, many of which are clearly associated with strike-slip movements. After Isacks et al. (1968).

Structure of the Santa Maria district

In northern and central California, the San Andreas fault lies relatively close to the continental margin, both on and offshore, for about 450 km from its northern end until it cuts inland towards the Big Bend region (Figure 6.8). The Coast Ranges, which lie on its western side, contain a number of elongate Neogene basins that trend slightly oblique to, and anticlockwise of, the line of the San

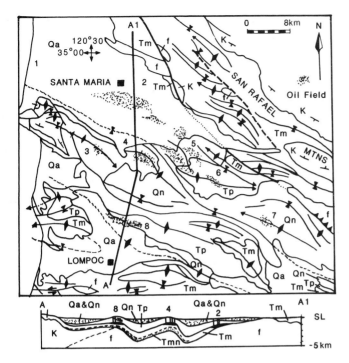

Figure 6.11 Simplified structural map and section of the Santa Maria oil district, California. Note the oblique relationship between fold axes (conventional symbols) and strike-slip faults (heavy lines) characteristic of strike-slip regimes. Qa, alluvium; Qn, non-marine Quaternary; Tp, Pliocene marine; Tm, Miocene marine; Tmn, Miocene or Oligocene non-marine; Cretaceous marine; f, Franciscan deposits. Numbers refer to oil fields (see source reference); oil fields are stippled on the map and shown in black in the section. After Blake *et al.* (1978).

Andreas fault. These basins are separated by parallel elongate uplifts. The structure of these basins, which is related to movements on the fault zone, is of interest because of the oil fields occupying many of the anticlinal areas. The Santa Maria district lies at the southeast end of the Coast Ranges, northwest of Santa Barbara. The structure is described by Blake *et al.* (1978) and summarized in Figure 6.11. The Neogene Santa Maria basin contains a marine sequence, up to 4500 m thick, of mainly Pliocene–Pleistocene age, lying unconformably on Franciscan basement. The sedimentary sequence was folded and locally thrust in late Quaternary time, forming broad, open, periclinal folds with WNW–ESE axes. The oil is contained in fractured shale of Upper Miocene age and occurs mainly in anticlinal traps (Figure 6.11, section).

The oblique relationship between the folds and both the San Andreas fault and the parallel fault to the northeast, is consistent with the geometric model of Figure 6.1 for a dextral strike-slip regime, and suggests that the whole sedimentary cover of the Coast Ranges block has been shortened in a NNW–SSE direction and extended in a WNW–ESE direction due to progressive simple shear. This model applies generally to the San Andreas fault zone, which exhibits a complex arrangement of uplifted blocks undergoing compression and depressed basins undergoing extension, whose sedimentary infill is being folded as a result of shortening in a direction anticlockwise of the strike-slip orientation, in accordance with the simple-shear model.

Structure of the Big Bend Region

The complex region of the intersection of the San Andreas and Garlock faults (Figure 6.12) is discussed in detail by Bohannon and Howell (1982). They point out the geometric incompatibility of simultaneous movement on the dextral San Andreas fault and the sinistral Garlock and Big Pine faults which intersect it. The Garlock fault has a displacement of 60 km, much of which is probably late Cenozoic to Quaternary in age, and it was active historically. The Big Pine fault displacement is

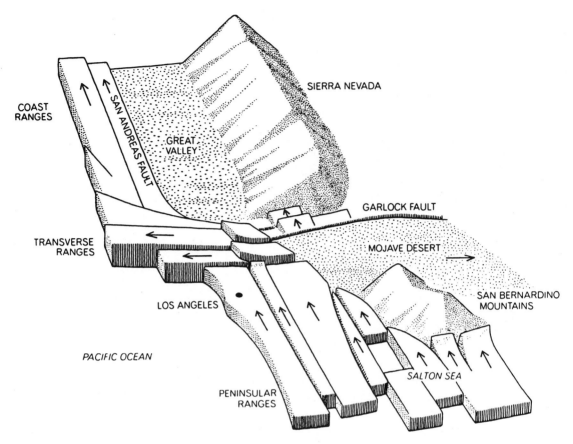

Figure 6.12 Block model to illustrate the relationship between the various fault blocks in the region of the 'Big Bend'. Note the necessary upthrust structures along the Garlock fault (cf. Figure 6.13). From Anderson (1971)

probably mostly of Pliocene to Pleistocene age. The authors suggest that the San Andreas fault probably originated as a straight linear feature, but was deformed by extension in the Basin-and-Range Province to the north (see 4.3).

Figure 6.13 shows how the structure might have evolved. Continuous sinistral movement on the Garlock fault produced a bend in the San Andreas fault, since neither fault could cut the other while both were active. The effect of the bend was to produce a local compressional region in the northeastern quadrant, where E–W overthrusts are found. The northward-moving western block may either have moved laterally westwards or deformed internally as it slid past the bend. A similar but less intense zone of compression exists in the southwestern

quadrant where the sinistral Big Pine fault meets the San Andreas fault. These sinistral faults are unconnected but have moved closer together as a result of the dextral displacement on the main fault. The bending of the San Andreas fault would cause internal deformation within the Mojave block in the south-western quadrant, which would suffer N–S shortening and E–W extension (compare Figure 6.13A,D). This deformation is a consequence of the change in angle of the boundaries of the block. This model is compatible with the late Cenozoic deformation pattern in the Mojave desert.

If the whole thickness of the crust were involved in these movements, it is difficult to visualize how the excess volume represented

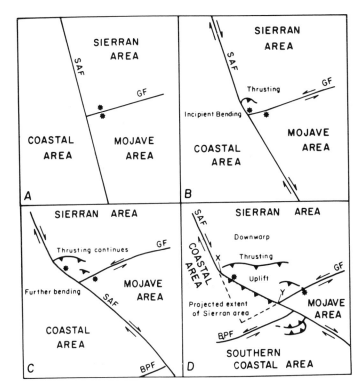

Figure 6.13 Model to illustrate the structural evolution of the intersection of the Garlock (GF) and San Andreas (SAF) faults (*A*) initial situation with a uniform NNW–SSE orientation of the SAF. (*B*) Early slip on the GF is accompanied by bending and anticlockwise rotation of the southern SAF relative to the fault north of GF, and by thrusting in the corner of the Sierran area. (*C*) Further development of the features referred to in (*B*); the Big Pine fault (BPF) has now moved into the field of view, on the west side of the SAF. (*D*) Continuation of development up to the present; note the area of crustal overlap (stippled) produced by the bending, delimited by an overthrust segment of the SAF. *X*, original junction with GF; *Y*, junction at time of stages (*D*). From Bohannon and Howell (1982), with permission.

by the triangle of overlap (shown with stippled ornament in Figure 6.13*D*) could be accommodated. The problem is overcome if, as seems likely, the faults detach on a low-angle décollement surface at depth, or are otherwise transferred at mid-crustal levels to a broader zone of ductile deformation.

6.4 Oceanic transform faults

Transform faults make up a significant proportion of the oceanic plate boundary network (see Figure 3.1). They were originally defined and their significance explained by Wilson (1965), and their orientations were used in the early attempt by McKenzie and Parker (1969) to define and describe the mechanism of plate motion.

Oceanic transform faults may be divided into major plate boundary faults and the relatively minor, often transient offset faults associated with the spreading process at ridges. The latter are discussed in 4.2. Some major transform

faults can be directly attributed to irregularities in the shape of the initial continental rift system (see Figure 4.9). As the passive continental margins spread apart to form an ocean, the transform faults propagate continuously inwards to form permanent boundaries to the spreading segments. The major fracture zones of the equatorial Central Atlantic (e.g. the St. Paul, Romanche and Chain zones) originated in this way. Other transform faults originate as a geometrical response to kinematic changes, as illustrated in Figure 4.5.

The morphology and general structure of the major oceanic fracture zones are described by Menard and Chase (1970). Most possess a prominent scarp, often with an associated ridge and trough, although any of these features may be absent. There is often considerable topographic variation along the length of a given fracture zone. The width varies from a few km to over 50 km in the case of the largest zones. Linear gravity and magnetic anomalies are often associated with the zones. The origin of

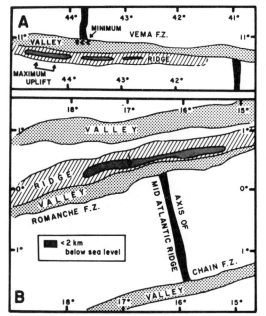

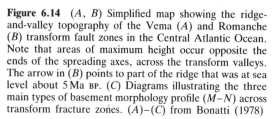

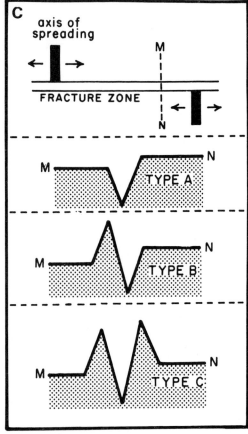

Figure 6.14 (*A*, *B*) Simplified map showing the ridge-and-valley topography of the Vema (*A*) and Romanche (*B*) transform fault zones in the Central Atlantic Ocean. Note that areas of maximum height occur opposite the ends of the spreading axes, across the transform valleys. The arrow in (*B*) points to part of the ridge that was at sea level about 5 Ma BP. (*C*) Diagrams illustrating the three main types of basement morphology profile (*M–N*) across transform fracture zones. (*A*)–(*C*) from Bonatti (1978)

these anomalies has caused some debate; they have been attributed both to magmatic intrusion and to hydrothermal alteration, and both processes probably contribute.

The morphology of fracture zones is discussed in detail by Bonatti (1978) who attributes the topographic variation primarily to the effects of differential cooling and subsidence. Figure 6.14*C* shows the different types of topographic profile found at fracture zones. The trough itself is considered to be a thermal contraction effect related to the shape of the magma chamber at spreading axes (see Figure 4.5). Type *A* would be the expected normal profile if this were the only control on the topography. The elevated side is closer to the spreading axis and has therefore cooled and subsided less than the other. However type *A* fracture zones are comparably rare, and only

found in certain Pacific zones. Profiles of type *B* or *C*, or of an intermediate type, are typical of most fracture zones. In the Atlantic Ocean in particular, major topographic ridges are associated with the great transform faults marking the sinistral offset of the mid-Atlantic ridge between the Central and South Atlantic. The Vema fracture zone, which offsets the mid-Atlantic ridge at 11°N, shows a topographic profile approaching type *B* (Figure 6.14*A*). A prominent ridge occurs on the south side of the valley marking the transform fault zone. The ridge rises to a height of over 5 km from the adjacent ocean floor, and is presently about 600 m below sea-level. There is even evidence of recent emergence. The Romanche fracture zone (Figure 6.14*B*) appears to approach type *C* in topography, since major transverse ridges border the seismically active transform valley, and extend from the ridge intersection to the African and South American continental margins. Similar topographic profiles with one or more ridges have been observed in the Owen fracture zone in the Indian Ocean, in the Alula fracture zone in the Gulf of Aden, and in the Charlie Gibbs fracture zone in the Atlantic (see below).

Bonatti notes that large fracture zones are generally characterized by ridges, running parallel to the main transform fault-zone valley, whose summits may reach to 1 km or more above the expected level for 'normal' oceanic crust of that age. The nature of the rocks forming these ridges (mostly serpentinized peridotite) indicates that they are mostly formed by the uplift of normal oceanic lithosphere material, rather than by magmatic intrusion. The fracture zones appear to be affected by intense vertical motion requiring subsidence rates much higher than expected from normal ocean crust. Bonatti argues that these vertical motions cannot be explained on the basis of standard ocean-spreading theory, and that some additional mechanism is required.

The zone of maximum elevation of the transverse ridges occurs near the spreading axis on the opposite side of the fracture zone valley, suggesting that the ridge might be due to the thermal anomaly generated by the spreading axis. It has been calculated that temperature changes of 100°C extending to 20 km from the fracture zone and to a depth of 70 km may be produced (Louden and Forsyth, 1976). However, this effect would only produce an uplift of a few hundred m at most,

whereas topographic relief of several km needs to be explained.

Bonatti concludes that uplifts of this magnitude can only be caused by compressional forces, and cites several lines of evidence for the existence of both compression and extension across fracture zones, as originally suggested by Dewey (1975). (i) Focal mechanism solutions of oceanic intraplate earthquakes indicate a state of general horizontal compression (see 2.6). In particular, two earthquakes in the region of the Chaim and Charcot fracture zones in the Central Atlantic exhibit compressional thrust solutions with σ_1 perpendicular to the strike of the fault (Sykes and Sbar, 1974). (ii) Seismic reflection profiles across sediment-filled fracture valleys in several zones show folding and other compressional features. (iii) Mylonitized rocks dredged from many of the large fracture zone ridges suggest that their uplift was tectonic. Extension is suggested in several zones by the graben–horst morphology, both in a gross sense and within the sediment fill, and by the occurrence of minor volcanism.

There are several possible causes of horizontal compressional and extensional forces. As pointed out in 3.3, relative plate motion vectors are normally oblique to plate boundaries. Even although a transform boundary normally originates parallel to the relative plate movement vector, changes in this vector occur during the course of the plate motion, resulting in a component of convergence or divergence across the boundary.

On a more local scale, the presence of sinuous or en-echelon faults within the fracture

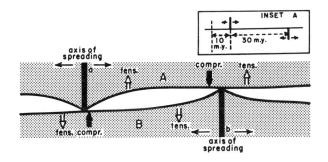

Figure 6.15 Diagrammatic plan illustrating the effects of differential horizontal thermal contraction of oceanic lithosphere along a transform fracture zone. Maximum compression occurs at the ends of ridge segments, and extension occurs between these. From Bonatti (1978)

zone will produce zones of compression or extension as described earlier (see Figure 6.2, 6.3). Thermal contraction of the cooling plate as it moves away from the spreading axis is an important source of minor extensional stress, as shown in 2.5. Collette (1974) has estimated that a horizontal contraction of 0.3 km would result from the cooling through 200°C of a 150 km segment of axial ridge. As shown in Figure 6.15, a zone of maximum compression must exist across a fracture zone at the end of each spreading sector, and a zone of maximum extension between the offset spreading sectors. Of these sources of stress, changes in relative plate motion are probably the most effective in producing large stresses across fracture zones, and are the most obvious reason for uplift on only one side of zones such as the Vema and Owen fracture zones.

The Charlie Gibbs and Gloria fracture zones

Searle (1979, 1986) describes the results of a side-scan sonar study, using the GLORIA system, of several Atlantic fracture zones. He notes a marked asymmetry of fracture valleys on the short-offset sections of the mid-Atlantic ridge, with a steep scarp on the older side, facing the younger lithosphere, and a gentle slope on the younger side. Seismic reflection and gravity profiles across the Kurchatov fracture zone (Figure 6.16) indicate that the asymmetry of the topography is a reflection of the crustal structure. The asymmetry is attributed to differential uplift of the original ridge crest compared with the transform valley floor. The topographic effect is preserved on the walls of the transform valley as the ocean floor cools and moves away from the spreading axis.

The combination of differential uplift and simple shear produces a set of oblique normal fault scarps as shown in Figure 6.17. These have been found in a number of fracture zones, and are particularly clearly demonstrated in the *Charlie Gibbs fracture zone*, which displaces the mid-Atlantic ridge around 50°N. Two separate faults were found, at 52°06′N and at 52°36′N, separated by a short N–S spreading sector. The southern fracture valley (Figure 6.18) is almost sediment-free, and displays the nature of the basement structure clearly. The valley is V-shaped, about 2 km deep, and 15–20 km across at the top, which lies at around 2000 m depth. A steeper inner region can be recognized, which probably corresponds to the 'active transform domain' recognized by Francheteau *et al.* (1976). This inner valley is 5–10 km wide and lies below

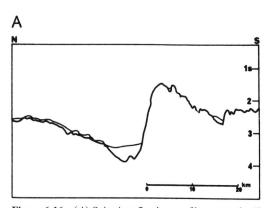
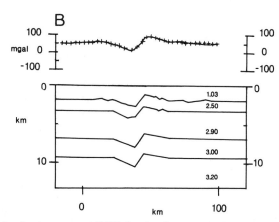

Figure 6.16 (*A*) Seismic reflection profile across the Kurchatov fracture zone at 29°W showing the asymmetric nature of the valley. Heavy line is basement, fine line is sediment surface. From Searle (1979), with permission. (*B*) Free-air gravity anomaly profile and crustal model along the line of (*A*). Note that the asymmetry in the surface topography expresses a deep-seated asymmetry in crustal structure. The figures in the crustal model are specific gravities of the layers used for the model. The crosses in the gravity profile are the computed values. After Searle (1979).

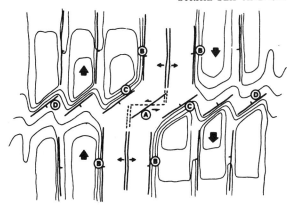

Figure 6.17 Diagram illustrating a method of formation of oblique scarps in transform fracture zones. Light lines represent schematic contours; heavy lines are faults; large arrows show regional dips; small arrows local stress fields. At *A*, tension gashes and normal faults form in newly created lithosphere near spreading axis–transform fault intersections. Uplift from median valley floor to crestal mountains occurs at *B*. Thus sea floor on the older side of fracture zone is uplifted, while that on the younger side remains temporarily on the valley floor, causing major vertical movements on oblique faults and producing large scarps (*C*). By position *D*, the faults are essentially locked. With a large enough offset (greater than about 20 km), a true transform fault (dashed line) may develop. From Searle (1979)

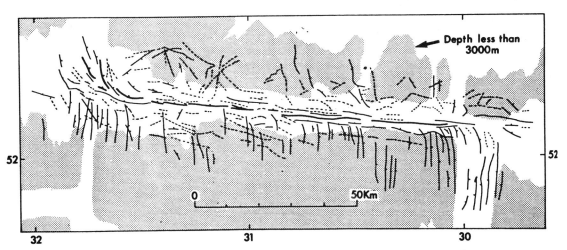

Figure 6.18 Side-scan sonar photograph mosaic, and tectonic interpretation, of the Charlie Gibbs transform fracture zone. Inferred major strike-slip fault, heavy lines; other faults, light lines, with ticks indicating inferred downthrow; other lineaments, dashed. From Searle (1979)

3 km depth. The walls are moderately steep, with slopes of up to 30°, and contain many E–W scarps attributed to normal faulting. The N–S mid-Atlantic-ridge fault structures curve sinistrally into the top of the transform valley but do not extend into the inner 'active' portion. The bottom of the valley exhibits long straight E–W scarps that are thought to

represent the active strike-slip fault traces. A linear E–W basement ridge, a few km across, borders the inactive portion of the northern transform. Major oblique scarps, trending NW–SE at 45° to the spreading direction, occur on inactive fracture zone walls, especially on the south side of the northern transform valley. These are about 15–20 km long and spaced about 15–20 km apart. Searle attributes these to oblique normal faults formed under a dextral simple-shear regime.

Another fracture zone described in some detail by Searle is the Gloria fracture zone which runs from the mid-Atlantic ridge near the Azores to Gibraltar, and forms the boundary between the Eurasian and African plates

(see Figure 3.1). According to McKenzie (1972), this section of the plate boundary is extensional at the Azores end, but becomes compressional through the Straits of Gibraltar, with a pure strike-slip section between. Along this central section, seismic reflection profiles indicated the presence of a scarp, 100–500 m high, on the south side of the valley marking the active fault. This valley is V-shaped and between 5 and 16 km wide. Minor WNW–ESE to NW–SE weak linear features on both sides of the main fault are interpreted as oblique normal faults originating in the same way as those in the Charlie Gibbs zone. The NNE–SSW linear ocean-floor fabric continues to the edge of the main valley from the north, and is

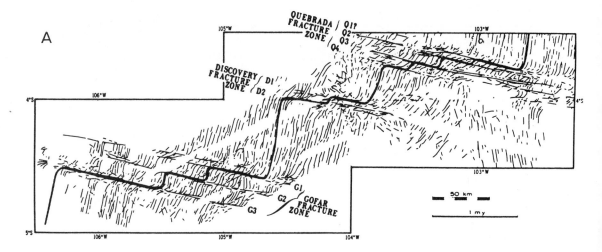

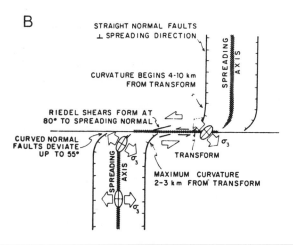

Figure 6.19 (A) Simplified sketch of tectonic lineament pattern, inferred from sonographs, of an area of the East Pacific ridge between 3° and 5°S. The three main fracture zones each consist of several smaller-scale fracture zones. Note the bending of the ocean-floor lineament fabric into the fracture zones. The heavy line is the inferred plate boundary. (B) Summary model illustrating the proposed tectonic pattern at a ridge–transform intersection for a fast-spreading ridge. Full arrows indicate inferred deviatoric stress direction; half arrows, shear stress. (A) and (B) from Searle (1983)

truncated by it. A broad ridge on the north side of the valley, between 23° and 24°W, carries the older structures on it. A less prominent and narrower ridge occurs on the south side between 20° and 22°W.

These detailed observations on fracture zones confirm the transform fault model described by Bonatti (1978). Transverse ridges carrying basement structures are clearly the product of differential uplift rather than intrusion. An important refinement to the Bonatti model is suggested by the oblique scarps which imply the operation of a narrow zone of simple shear, about 10 km wide, within which extensional normal faults form at 45° to the spreading direction. With long transform offsets (probably > 20 km), a true strike-slip fault develops in a narrow valley in the centre of this zone.

Fast-slipping fracture zones on the East Pacific ridge

Searle (1983) describes a rather different morpho-tectonic pattern in the multiple offset zone of the East Pacific Ridge between 3° and 5°S using the GLORIA sidescan sonar system (Figure 6.19). Nine separate transform faults were identified in three groups corresponding to the Quebrado and Gofar fracture zones and to the previously unidentified Discovery fracture zone. The spacings between the individual transform faults range from 5 to 16 km, and the offsets range from 24 to 93 km. As in the Atlantic fracture zones, the individual transform faults occupy narrow valleys a few km in width, but in the case of the closely spaced zones, the valleys merge into a single broad valley. Several fault scarps occur in the valleys. The active fault is not confined to the bottom of the valley, but occupies various positions on the flanks or even the top of the slope.

The pervasive spreading fabric found on the Pacific ocean floor is modified near the transform faults, beginning about 4–10 km from the fault with a gentle curve. This brings the fabric to a trend of about 55° from the normal direction within 2–3 km of the transform fault. In some cases the fabric becomes nearly asymptotic to the fault. The sigmoidal nature of the curvature, and the degree of obliquity, contrasts with the Atlantic examples described above. The direction of curvature is similar to that of the Atlantic fracture zones and is in the opposite sense to the strike-slip displacement (Figure 6.19). Searle explains this curvature by a gradual change in orientation within a narrow zone of simple shear along the transform fault. Linear features formed at low angles to the transform fault are interpreted as Riedel shears (synthetic shears with the same sense of strike-slip displacement as the main fault — see Figure 6.1).

The pattern of multiple, closely-spaced transform faults is thought by Searle to be typical of fast-spreading ridge offsets. He also suggests that this particular zone may have developed in response to a small (10°) clockwise change in spreading direction, which would encourage the development of transform faults with a small extensional component. The net effect would be to produce an overall clockwise change in divergence direction.

7 Intraplate tectonic regimes

7.1 Types and characteristics of intraplate structure

According to classical plate tectonic theory, plates are essentially stable internally, and tectonic effects are concentrated at their boundaries. However it has always been recognized that this was only true to a first approximation, and that all regions of the Earth's surface experience tectonic effects to some degree. The most common type of intraplate (within-plate) tectonic activity is undoubtedly vertical movements. Accurate geodetic measurements involving precise levelling and other techniques have shown that most parts of the crust are undergoing slow uplift or depression. The frequency and distribution of intraplate earthquakes is also an indication of widespread tectonic activity, albeit to a notably lesser degree than at plate boundaries.

The vertical movements, however, are typically at least an order of magnitude slower than the movements associated with plate boundary activity. Lateral variations in these intraplate vertical movements create a system of basins and intervening uplifts that is the characteristic structure of all continental intraplate regions.

The origin of these vertical intraplate movements is still the cause of considerable uncertainty and debate. It is convenient to divide the possible modes of origin into those relating to horizontal plate movements and those that are not directly related to such movements. Movements in the former category take place as a result of horizontal distortions of the plate interior, either by extension or by shortening, and thus violate the principle of the laterally 'rigid' plate. The widely used model of the extensional basin developed originally by McKenzie (1978) is dependent on an initial assumption of lateral intraplate extension.

Plate boundary processes can produce vertical movements at considerable distances from the boundary, as we have seen in the case of Central Asia (5.4). Lithosphere loading and the resulting crustal thickening will produce isostatic uplift in the region of the load, and flexural depression in the region beyond the load, producing *foreland basins*. These cannot be considered as genuinely intraplate phenomena. However, many structures situated far from plate boundaries may have no obvious connection with boundary structures. For example the Baikal rift, north of the Himalayan plate boundary (see Figure 5.28) and the Cenozoic compressional fold belt of southern England, situated 700 km north of the Alpine front, are both examples of enigmatic structures that appear to be intraplate, but for which a connection with plate boundary tectonic effects may be postulated.

Processes that are unrelated or only indirectly related to plate movements include isostatic responses to changes in the density structure of the lithosphere or asthenosphere, such as intraplate hot spots, and unloading due to the removal of ice sheets.

A special type of intraplate basin is formed at passive continental margins. Such basins typically show a lower faulted section corresponding to an early rifting stage in the evolution of the basin. At this stage, the basin is the product of a divergent plate boundary regime (see 4.3). However, during the later stages of its evolution, the basin becomes part of the stable plate interior, as the boundary moves away from its initial position with continued sea-floor spreading. The Atlantic coastal basins provide excellent examples of such structures. Continued depression in these basins is produced by cooling and sinking of the oceanic lithosphere aided by the gravitational load of the sediment pile.

Vertical movements in the oceanic lithosphere are relatively simply explained on the basis of the cooling model described in 2.3. Ocean floor is progressively depressed with increasing age and distance from the spreading

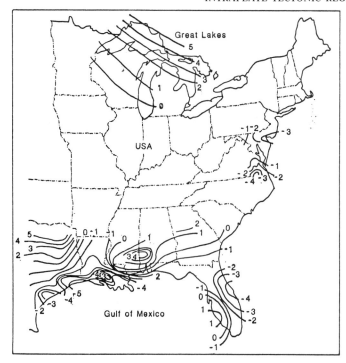

Figure 7.1 Recent vertical movements in parts of the USA. Note particularly the uplift of the Great Lakes region and depression in the Gulf of Mexico. Contours, in mm/year, from re-levelling and sea or lake level changes. After Brown and Reilinger (1980)

centre. Departures from this general rule occur in areas of hot-spot activity, where the ocean crust becomes more buoyant due to the thermal effects on the density structure. Similar variations associated with fracture zones were discussed in 6.4.

Cratonic basins vary widely in morphology and structure. Many appear to be underlain by rift systems, in which case it is possible to use the lithosphere stretching model to explain their generation. The North Sea is an excellent example of this type of basin, and is described in detail below.

Determination of recent vertical movements in plate interiors

The most widely used method of gaining accurate information concerning present-day movements is by the geodetic technique of repeated precise levelling and triangulation. Information about rather longer-term movements can also be obtained from measuring old lake and sea shore levels. The distribution of recent vertical movements in the stable American continental plate is shown in Figure 7.1. Note that the rates of movement are low — in the range 0–5 mm/year — an order of magnitude smaller than those associated with horizontal plate movement. Several depressed regions are associated with subsidence along the passive continental margin, particularly in the Gulf of Mexico. Equally prominent is the uplifted region of the Great Lakes, attributed to post-glacial rebound. Recent movements in the Russian platform (Figure 7.4) display a similar pattern.

The average rates calculated for recent vertical movements are much higher (by more than $\times 10$) than those calculated from the study of sedimentary basins measured over millions of years. As we shall see, these average only about 0.2 mm/year. This suggests that cratonic movements are episodic or oscillatory over periods of the order of 10^4 years or more. Perhaps periods of relatively rapid depression or uplift are separated by much longer periods of comparative quiescence.

There is some evidence pointing to a correlation between present movement and topography. Areas currently undergoing depression are already low-lying, and those undergoing uplift are already elevated, confirming that the current movements are part of a longer-term process.

7.2 The Russian platform: a typical intraplate region?

The stable interior of the European part of the Eurasian plate consists of the Fennoscandian or Baltic shield and the *Russian platform* (Figure 7.2). This large region has existed as a craton since mid-Proterozoic times and is surrounded by orogenic belts of Phanerozoic age: the Caledonian belt in the northwest, the Hercynian belts of western Europe and the Urals on the southwestern and eastern sides respectively, and the Alpine belt in the south.

In a review of the Russian platform by Aleinikov *et al.* (1980), eleven major sedimentary basins and six uplifts are recognized. In the central part of the platform, deposition

of sediments commenced in late Proterozoic time, around 1400 Ma ago, on highly deformed and metamorphosed early Proterozoic basement. In the northeastern part of the platform however, deposition did not commence until the Ordovician.

The authors subdivide the structures found within the platform into three main types: (i) large circular depressions (*syneclises*) or uplifts (*anteclises*) caused by flexure of the Earth's crust; (ii) large elongate graben (*aulacogens*) due mainly to faulting; and (iii) embayments at the margins of the platform associated with subsiding basins of plate boundary type.

Most of the large structures are at least 1000 km across and represent crustal flexures with amplitudes of around 1500 m on average. Aulacogens are the dominant type of basin: of the eleven major basins recognized, seven may be classified as aulacogens. The formation of aulacogens was virtually confined to two periods, the late Proterozoic and the mid-Palaeozoic (Figure 7.3A,B). Most aulacogens appear to be replaced by much wider basins with time: thus the late Precambrian aulaco-

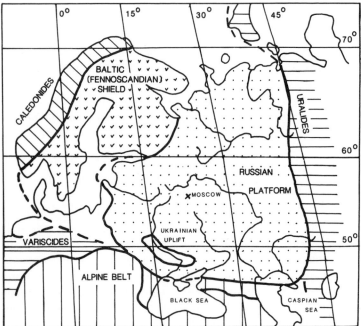

Figure 7.2 Tectonic setting of the Russian platform. The platform and Baltic shield together comprise a Phanerozoic craton surrounded by orogenic belts.

gens are replaced by early Palaeozoic basins (compare Figure 7.3*A,B*). For example, the mid-Russian aulacogen is the site of the much larger Moscow syneclise during the latest Precambrian and early Palaeozoic. By Ordovician and Silurian times, only two major basins existed within the platform: the Baltic syneclise, which represents an extension westwards of the Moscow syneclise, and the Pechura syneclise, which formed on the site of the late Precambrian Timan geosyncline (Figure 7.3*B*). In the mid-Palaeozoic, however, a new major aulacogen formed, the Dneiper–Donetsk aulacogen, across the centre of a large domal uplift consisting of the Voronezh anteclise to the northeast, and the Ukrainian anteclise to the southwest (Figure 7.3*C*). The Moscow syneclise became re-activated, and the Pre-Caspian syneclise formed in the south.

During Mesozoic and Cenozoic time, the Pre-Caspian syneclise continued to subside, but elsewhere an almost completely different pattern had been established (Figure 7.3*D*). The other major basins formed during this period were the Dneiper–Donetz syneclise, formed on the site of the earlier aulacogen, a much reduced Moscow basin, and the new Ul'yanovsk–Saratov depression formed on the site of the Volga–Ural anteclise.

Several important conclusions can be drawn from the analysis of these intraplate structures. Almost all the major basins originated as aulacogens (i.e. rifts in our previous terminology) that formed in only two main periods. Many parts of the platform appear to have alternated between active basin subsidence and either neutral or uplift behaviour. Such areas exhibit the property of *inversion* (see Beloussov, 1962) where a part of the crust reverses a long-continued direction of vertical movement. Inversion is characteristic of intraplate regions as well as of orogenic belts and several good examples may be noted in Figure 7.3. Thus parts of the Upper Palaeozoic Voronezh-Ukraine uplift formed late Precambrian or early Palaeozoic basins, and other parts formed a Mesozoic basin.

Overall movement of the platform was downwards, in contrast to the neighbouring Baltic shield, which was generally uplifted during the same period. The major basins existed for periods ranging from 160 to 770 Ma with an average life of 350 Ma. The aulacogens, on the other hand, had a shorter life-span of 180 Ma on average. The average rate of depression from 1400 Ma BP to the present is 7 m/Ma, or 0.007 mm/year. In the major aulacogens, the rate of subsidence was much higher, about 60–199 m/Ma (0.06–0.2 mm/year) in the case of the Devonian examples.

The aulacogens are bounded by major faults several hundred km in length, with a throw of up to 1 km. Many of these faults have been intermittently active over long periods of time. Smaller faults with throws of the order of tens of metres form step-like systems in the basement, and are associated with the earlier stages of depression of the major basins. However, faulting, though abundant in the Precambrian basement, is uncommon in the Mesozoic strata, which are normally affected only by flexures.

The formation of the major platform structures (i.e. those larger than 100 km across) is considered to be related to processes occurring at depths of several tens of km. The thickness of the crust of the Russian platform varies from 37 to 53 km, with an average value of 46 km. The greater the thickness of the sedimentary cover, the smaller is the thickness of the basement: that is, deep basins correspond to areas of thin basement crust. For example, in basins with more than 5 km of sediment fill, the basement thickness averages only 38 km. Areas with around 1 km of sediment cover have average basement thicknesses of 46 km. Thus regions of below-average basement thickness are only partly compensated by thicker sediment cover.

The pattern of recent vertical movements on the Russian platform is illustrated in Figure 7.4. These show rates of movement ranging up to over 5 mm/year. Major zones of recent uplift occur in the region of the Baltic shield, and in the Ukrainian and Voronezh uplifts, and depression is still occurring in the Moscow basin

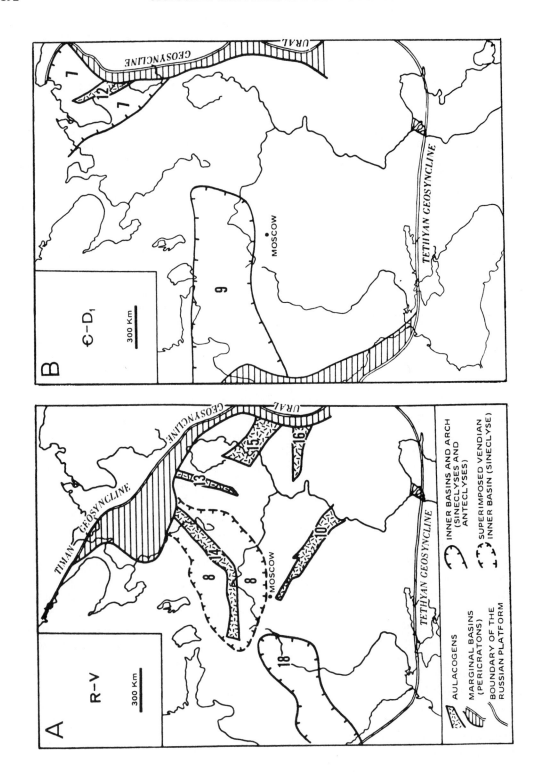

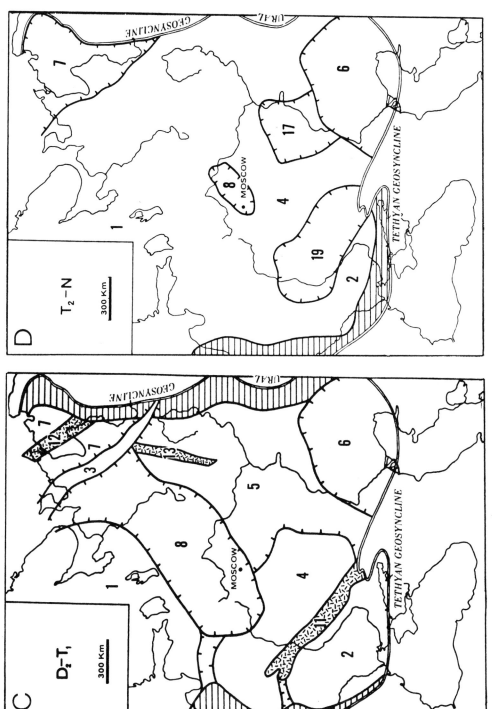

Figure 7.3 Four stages in the tectonic evolution of the Russian platform, showing the location of the major intraplate structures: (*A*) Upper (Late) Proterozoic; (*B*) Early Palaeozoic; (*C*) Mid–Late Palaeozoic; (*D*) Mesozoic and Cenozoic. 1, Baltic shield; 2, Ukrainian anteclise; 3, Timan ridge; 4, Voronezh anteclise; 5, Volga–Ural anteclise; 6, Pre-Caspian syneclise; 7, Pechora syneclise; 8, Moscovian syneclise; 9, Baltic syneclise; 10, Pachelma aulacogen; 11, Dneiper–Donetsk aulacogen; 12, Pechora–Kolvin aulacogen; 13, Vyatka aulacogen; 14, Mid-Russian aulacogen; 15, Kaltasa aulacogen; 16, Sernovodsk-Abdulino aulacogen; 17, Ul'yanovsk-Saratov depression; 18, Orshansk syneclise; 19, Dneiper-Donetsk syneclise. From Aleinikov *et al.* (1980), with permission.

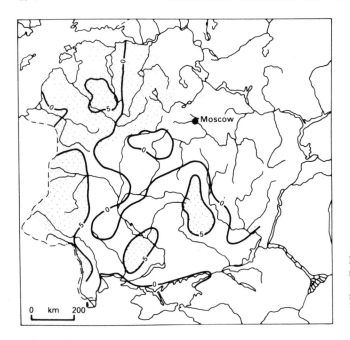

Figure 7.4 Pattern of recent vertical movements in the Russian platform. Contours in mm/year. Based on re-levelling data. The stippled area is undergoing subsidence. From Vita-Finzi (1986), after Mattskova (1967).

and in the large depression extending southeast of Moscow towards the Caspian Sea. This pattern indicates that the major platform structures of the Mesozoic (Figure 7.3D) are still active, but that the rates of movement are around two orders of magnitude greater than those measured over periods of 100–1000 Ma.

The Russian platform is a particularly well-studied example of a major intraplate region. Other continents, however, display a similar pattern of long-continued equidimensional or non-linear basins separated by broad uplifts. This pattern may be seen on tectonic maps of all the continents and is particularly clearly displayed on the tectonic maps of North America and Africa.

7.3 Intraplate basins

We have seen from our study of the Russian platform that major tectonic basins can be recognized there, characterized by continued depression over periods of 100–1000 Ma. These basins are large features with dimensions approximately 500–1000 km across, and represent depressions in the basement of, in some cases, over 5 km depth. They are there-

fore major crustal structures and form, with the associated uplifts, the most important and widespread type of intraplate regime. Over a long time-scale, the vertical movements responsible for these structures constitute a significant departure from intraplate stability.

We shall now look at several classical and well-studied examples of intraplate sedimentary basins from different continents: the Paris basin in Western Europe, the Michigan basin in the northern USA, and the Taoudeni basin in West Africa. These are examples of former marine basins that are now inactive or which form shallow continental depressions. In a later section (7.4) we shall examine in more detail the North Sea basin and the Atlantic continental margin of the northern USA, which are examples of currently active marine basins whose structure is comparatively well known because of oil and gas exploration.

The Paris basin

This basin, described by Megnien and Pomerol (1980), about 600 km in diameter, is situated in northern France, and is one of the best-defined basin structures in Europe. It rests on Palaeozoic basement to the west (Massif Armoricain), south (Massif Central) and east (Ardennes massif), but extends to the north into the English Channel, and northeastwards towards the North Sea.

The basin probably originated during Upper Triassic time, but existed as a well-defined basin structure only during the Jurassic and Cretaceous periods. During this time, a total maximum thickness of 2900 m of sediments accumulated. However, the area of maximum subsidence migrated southwards for a distance of 60 km during this period, so that the total thickness at any one point along this line is rather less than this. The authors demonstrate, from a study of the variation in cumulative thickness with time, a rather steady rate of accumulation of sediments with an average rate of 21.5 m/Ma (0.2 mm/year). The authors attribute the growth of the basin over this period to gradual flexing of the crust induced by the sediment loading, but do not comment on the initial mode of origin of the basin.

The Michigan basin

This basin has been intensively studied by means of numerous boreholes, and gravity and magnetic surveys. The basin was formed in mid-Ordovician time and lasted at least until the late Carboniferous. The basal mid-Ordovician sediments rest unconformably on Precambrian basement. Structural contours on the basement surface (Figure 7.5) indicate an unusually regular, almost perfectly circular, shape with a gradual increase in depth towards a well-defined central point. Most Palaeozoic units thicken towards the centre of the basin, and facies variations indicate consistently deeper-water conditions there. Subsidence therefore has continued throughout most of the lifetime of the basin. The total thickness in the centre is about 3 km, from which an average rate of subsidence of 24 m/Ma (0.24 mm/year) is obtained — very similar to that of the Paris basin.

A small positive free-air gravity anomaly over the basin is attributed by Walcott (1970)

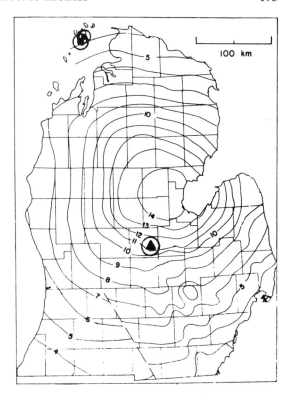

Figure 7.5 The shape of the Michigan basin, shown by structural contours (in ft × 10³) on the Precambrian basement. Circled triangles indicate boreholes. Note gradual and steepening dip of the basement surface towards the centre of the basin. From Sleep (1980), with permission.

to regional subsidence caused by loading. There seems to be no general agreement, however, over the origin of the structure.

The Taoudeni basin

This basin, described by Bronner *et al.* (1980), is one of the most prominent structural features of the African craton. It lies in West Africa, mainly in Mauritania and Mali, in the western Sahara region (Figure 7.6) and is about 1300 km across, with an area of 2 × 10⁶ km². The sedimentary thickness varies from 1000–1500 m. The basin rests on Archaean and Early Proterozoic basement which forms shield areas to the north and south. It is bounded on the west by the Palaeozoic Mauritanide belt, and

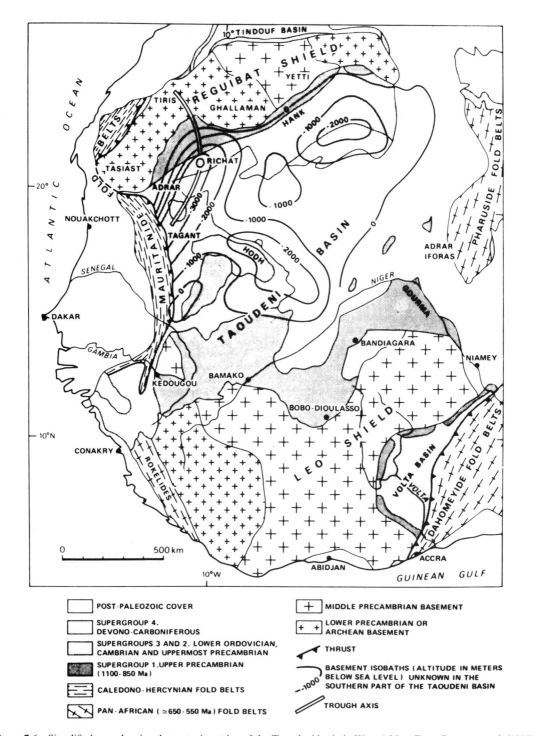

Figure 7.6   Simplified map showing the tectonic setting of the Taoudeni basin in West Africa. From Bronner *et al.* (1980)

to the east by the late Precambrian to early Cambrian Pharuside belt.

The sedimentary fill ranges in age from mid-Proterozoic (1100–1000 Ma) to Carboniferous. The central part of the basin is covered by a thin veneer of Mesozoic and Cenozoic sediments. The sequence within the basin is sub-divided into four supergroups separated by unconformities or disconformities. These show quite different subsidence rates, varying from 4 m/Ma for supergroup 1 (late Precambrian) to 16 m/Ma for supergroup 4 (Devonian). The mean subsidence rate for the whole period of activity is about 5 m/Ma (= 0.05 mm/year).

The main basin contains two smaller basins with depths of about 2500 m, and a curved linear trough with a maximum depth of 5–6 km. This trough appears to have existed as a major rift zone in the late Precambrian, and does not extend into the Phanerozoic, since the second supergroup rests unconformably on the

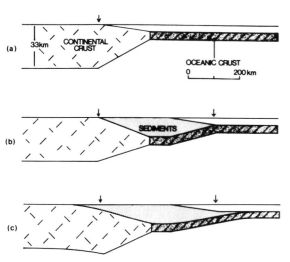

Figure 7.7 Schematic model illustrating the tectonic effect of sediment loading at a continental margin. In (*a*) a 200 km-wide transition between continental and oceanic crust is assumed; (*b*) shows calculated result of loading, using the Airy model, assuming a sediment density of 2450 kg/m³, and mantle density of 3300 kg/m³; (*c*) shows calculated result using a flexural loading model, assuming a flexural rigidity of 2×10^{22} Nm, and densities as in (*b*). Note that the inward edge of the load in (*c*) migrates inwards over the continent. From Bott (1980)

rift-fill sediments of supergroup 1, onlapping the basement at the sides of the rift. This curved trough, which controlled sedimentation in the late Precambrian basin, coincides with a belt of iron-rich gneisses within the early Precambrian basement, and with a strong, local, positive gravity anomaly. The authors attribute the formation of the rift to the gravitational response of the crust to this excess weight.

Origin of intraplate basins

The origin of sedimentary basins is discussed by Bally (1980, 1982), who classifies basins into 19 types based on their tectonic location. His fundamental division is between basins associated with 'megasutures' (i.e. plate boundaries) and those located on 'rigid' lithosphere (i.e. intraplate). Of the intraplate basins, he recognizes the following categories:

(1) Related to formation of oceanic crust (oceanic basins)
   (a) Rifts
   (b) Oceanic transform fault-associated basins
   (c) Oceanic abyssal plains
   (d) Atlantic-type passive margins (shelf, slope and rise) which straddle continental and oceanic crust
      (i) Overlying earlier rift systems
      (ii) Overlying earlier transform systems
      (iii) Overlying earlier back-arc basins

(2) Located on pre-Mesozoic continental lithosphere (cratonic basins)
   (a) Located on earlier rifted graben
   (b) Located on former back-arc basins.

Oceanic basins of types 1(a) and 1(b) have already been discussed (see 4.2 and 6.4). Basins of type 1(c) are controlled by the age and cooling-related depression of oceanic lithosphere as it moves away from spreading ridges. Such basins are important at present in terms of their areal extent, since they constitute most of the ocean floor, but are not of such great geological interest because they are not preserved in the pre-Mesozoic geological record. Passive-margin basins, in contrast, are of great importance because of the large volumes of terrigenous sediment that accumulate in them, and because they become welded to the continental crust during collisional orogenesis. Bally distinguishes three types according to whether the basins were initiated over rifts, transform faults, or back-arc spreading basins. Types d(i) and d(iii) are extensional in origin, whereas type d(ii) is strike-slip. However, in each case the initial depression results from the creation of oceanic lithosphere that becomes isostatically depressed in relation to surrounding areas.

Continental or cratonic basins of intraplate type are divided into two categories: those located on earlier rifts, and those located on former back-arc basins. In both cases, the origin of the basins is regarded as extensional, causing crustal thinning and depression of the surface.

The implication of Bally's classification is that the origin of intraplate basins is related to an inherited crustal structure, that is, one resulting from a former period of tectonic activity that caused the surface depression. There is, however, considerable controversy over the origin of the depressions required to create intraplate basins, and the establishment of a generally agreed model has been hampered by the lack of evidence as to the deep structure of many basins. A number of proposed mechanisms are cited by Bally: (i) sediment loading; (ii) isostatic response to cooling and density increase; (iii) lithosphere stretching and thinning; (iv) emplacement of dense, mantle-derived, igneous material; (v) density increase of lower-crustal rocks due to gabbro-eclogite phase change; (vi) complete basification or *oceanization* of the continental crust by the supply of ultramafic material from the mantle; and (vii) creep of ductile middle- to lower-crustal material towards the ocean at a passive margin.

These mechanisms are by no means mutually exclusive, and all are probably operative to some extent, with the exception of (vi). Complete oceanization, proposed by Beloussov (1968), is implausible since a mixture of crustal and mantle densities will still be lighter than mantle material and will be unable to sink into it (see Artyushkov *et al.*, 1980). However, partial oceanization of the crust is essentially the process envisaged in mechanisms (iv) and (v).

There appears to be widespread agreement that most basins result from an isostatic response to a portion of lithosphere that has become, for some reason, more dense than its surroundings. The increase in density may arise in several different ways: in the oceans, and in continental rift zones, it results from straightforward cooling of an initially heated region; elsewhere, it is a consequence of the emplacement of dense mantle-derived material within the crust. The main area of disagreement appears to be between those who regard lithosphere stretching and rifting as the main, or indeed the only, method for generating this crustal transformation, and those who believe that such a process proceeds independently of the normal plate tectonic processes, and without significant lithospheric stretching.

The mechanism which is now widely referred to as the McKenzie stretching model (McKenzie, 1978) assumes that basin formation is the result of an 'instantaneous' lithosphere extension. This process thins the lithosphere, producing a surface depression and a corresponding upwards bulge in the base of the lithosphere. The thinned lithosphere generates a thermal anomaly, since the warm asthenosphere is now closer to the surface in the stretched portion. However, as the initial basin fills with sediment, the thermal anomaly gradually decays, and the base of the lithosphere returns to its original level. This slow cooling results in gradual isostatic subsidence over a period of about 60 Ma.

The initial McKenzie model envisaged instantaneous stretching for simplicity. However, Jarvis and McKenzie (1980) have shown that for most basins the instantaneous model predicts the subsidence geometry accurately, provided that the duration of the initial stretching is less than $60/\beta$ Ma, where the stretching factor β is less than 2, and $60/[1-(1/\beta)]$ if $\beta \geq 2$. Thus for a value of $\beta = 1.5$ corresponding to the mean value for 11 basins (see Table 2.6) the duration of initial stretching should be less than 27 Ma for the model to apply. The model appears to account successfully for the stratigraphy and subsidence history in the Pannonian basin of Hungary and the North Sea basin (Christie and Sclater, 1980).

Subsidence due to sediment loading is the simple gravitational response to imposing an extra mass on the surface (Dietz, 1963; Walcott, 1972) and requires an initial depression

for sufficient sediment to accumulate. Bott (1980), in an adaptation of the flexure model developed by Walcott (1972), shows that, in a 200 km wide transition zone at a passive continental margin (Figure 7.8), a thick pile of sediment at a major delta, such as that of the Niger, will produce a downwarping extending for about 150 km beyond the area of the initial load. The sediment loading mechanism can produce a sedimentary pile of twice the initial water depth for sediment with a mean density of 250 kg/m^3, and of nearly three times the initial water depth for sediment with a mean density of 255 kg/m^3. Thick shelf successions formed in water with a depth of less than about 200 m cannot form by this mechanism, but total

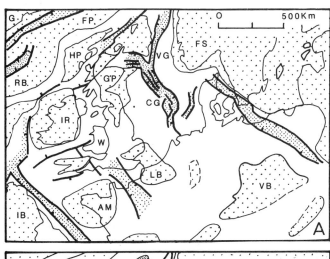

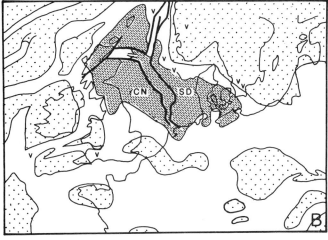

Figure 7.8 Simplified maps illustrating two stages: (A) Early Jurassic, (B) Middle Jurassic, in the tectonic evolution of the North Sea basin. Open stipple, land areas; close stipple, major rifts with boundary faults indicated by heavy lines with tick on downthrown side. The dense dashed area in (B) represents the uplift comprising the Central North Sea dome, the Shetland platform and adjoining land areas of northern Britain. 'V' marks areas of contemporaneous vulcanicity. G, Greenland, RB, Rockall Bank; FP, Faroes Plateau; HP, Hebridean platform; SP, Shetlands platform; FS, Fennoscandia; IR, Irish massif; W, Welsh massif; GP, Grampian–Pennine massif; AM, Armorican massif; LB, London–Brabant massif; IB, Iberian massif; VB, Vindelician–Bohemian massif. After Ziegler (1982).

thicknesses of about 14 km could form near the base of the continental slope if the initial depth is greater than this.

The sediment loading mechanism provides a means of explaining the continued evolution of basins long after their initial causal mechanisms have ceased to operate. The process thus explains the very long life (ranging from 100–700 Ma) of the major inactive basins discussed earlier, despite the relatively short time periods required by the initial extensional or thermal cooling events (of the order of tens of Ma only).

It is significant that many of the major basins discussed above appear to be situated over early rifts. It appears likely therefore that most, if not all, major basins that are genuinely intraplate (i.e. excluding passive-margin types) originated by a process of extensional rifting that allowed dense material to be emplaced locally within the crust, or at the base of thinned crust, giving rise to isostatic depression. The amount of extension need not be large: a narrow zone of severe localized thinning with accompanying basification, or a wide zone of less intense thinning, may each produce large basins due to the subsequent isostatic responses, although the detailed geometry will differ (see Figure 2.29).

Basins originating by crustal flexure require an initial load such as that provided by a thrust sheet. Basins of this type are therefore not genuinely intraplate in origin, but are associated with destructive plate boundaries. They differ from extensional basins in not exhibiting crustal thinning, and can therefore be distinguished from them by geophysical means.

7.4 Examples of active marine basins: the North Sea and the Atlantic continental margin of the USA

The North Sea basin

This large marine basin is situated on continental crust of the northwest Eurasian plate between Britain and Scandinavia, opening into the Atlantic Ocean in the north. It is lozenge-shaped, measuring 1000 km along its longer side in a N–S direction, and about 500 km across from east to west. With the exception of a narrow strip along the coast of Norway, the basin is everywhere less than 200 m in depth. It receives terrigenous sediment from a number of major rivers in eastern Britain, the Netherlands, West Germany and southern Norway.

The history of the basin is summarized by Ziegler (1982). The basin probably originated in the Devonian period as one of a number of continental extensional rift basins developed in the region of the Caledonian orogenic belt in Britain and Norway (see 8.3). During the Permian, two separate basins existed in the northern and southern North Sea respectively, separated by the mid-North Sea high. The present North Sea basin was initiated in the Triassic period as part of an extensive rift system connecting with the major Arctic–Atlantic rift. The two major rifts in the North Sea are the Viking and Central graben, which lie midway between Norway and Shetland, and between Norway and mainland Britain respectively (Figure 7.8). These two rifts meet in a triple junction with a graben that extends westwards into the Moray Firth.

In the early Jurassic, a rise in sea-level produced a widespread marine transgression across the North Sea basin linking the Arctic and Tethyan Oceans via the rift systems of Western Europe. Subsidence appears to have continued throughout the early Jurassic, but was interrupted in mid-Jurassic times by the creation of a large rifted dome, 2–3 km high, in the central North Sea (Figure 7.8*B*). A large volcanic centre was established in the area of the triple junction and several smaller centres along the Viking Graben and to the east. This uplift phase was associated with the major regional kinematic change that resulted in the opening of the Central Atlantic (see Figure 3.5*B*).

By late Jurassic time, the dome had subsided, and deep-water conditions had become re-established throughout the basin. Con-

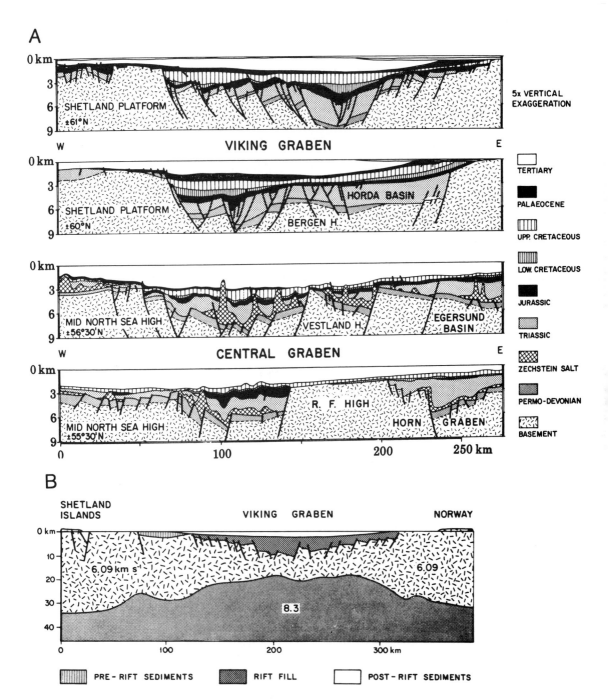

Figure 7.9 (*A*) Simplified structural cross-sections, (*B*), seismic-refraction crustal profile, across the Viking Graben. Figures in (*B*) are mean P-wave seismic velocities. From Ziegler (1982)

tinued extension took place by the westwards movement of Britain in relation to both Scandinavia and mainland Europe. The extension appears to have continued into the early Cretaceous, with renewed movements on faults in the main rift zones producing sea-floor relief of up to 1 km.

In the late Cretaceous, rifting in NW Europe appears to have become concentrated along the abortive Labrador Sea spreading axis between Greenland and North America. Although subsidence continued in the North Sea basin, there is no evidence for significant continued extension. During this period, up to 2 km of Upper Cretaceous chalks and marls infilled the topographic depressions of the Viking and Central graben (Figure 7.9A,C). Further subsidence of the basin took place during the Cenozoic, when a maximum thickness of 3.5 km of sediments was deposited in the central areas of the basin. The region of maximum Cenozoic sedimentation probably corresponds to the zone of maximum crustal thinning produced during the earlier extensional phase (Donato and Tully, 1981).

Gravity profiles indicate a mass excess beneath the Viking and Central graben, which is consistent with seismic refraction data showing a much shallower Moho beneath the graben. A typical crustal profile across the rift (Figure 7.9B) indicates normal crustal thicknesses of 30–35 km beneath Norway and Shetland, decreasing to about 20 km in the Viking graben where there is an 8–10 km sediment fill. The upper mantle displays normal P-wave velocities throughout. Sclater and Christie (1980) propose that the Mesozoic crustal thinning was accomplished by an extension of $\beta = 1.8–2.0$ (80–100%).

Beach (1986) describes a deep seismic reflection profile across the Viking graben from north of the Shetlands to Bergen (Figure 7.11). The profile illustrates the upper crustal structure clearly and shows the asymmetric nature of the rift structure, now known to be characteristic of many extensional provinces (see 4.4). The Viking graben is seen as an asymmetric *half-graben* formed by tilted fault blocks of Jurassic and older strata, with a consistent westerly tilt. The Jurassic block-faulted structure is buried by flat-lying Cretaceous sediments that had completely buried the highest block by end-Cretaceous time (Figure 7.9A).

The blocks are bounded by eastwards-dipping faults, with a weak listric geometry, that appear to detach on a broad zone of lower-crustal reflectors interpreted by Beach as a major, low-angle shear zone (cf. Figure 2.30).

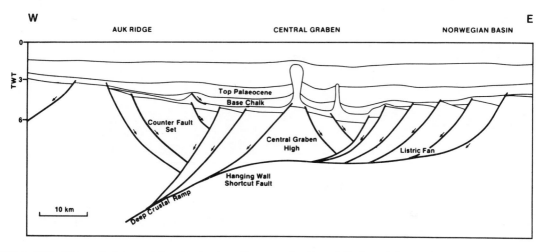

Figure 7.10 Composite interpretative structural profile across the Central Graben of the southern North Sea, based on seismic and well control. Vertical scale is two-way time. From Gibbs (1985), with permission.

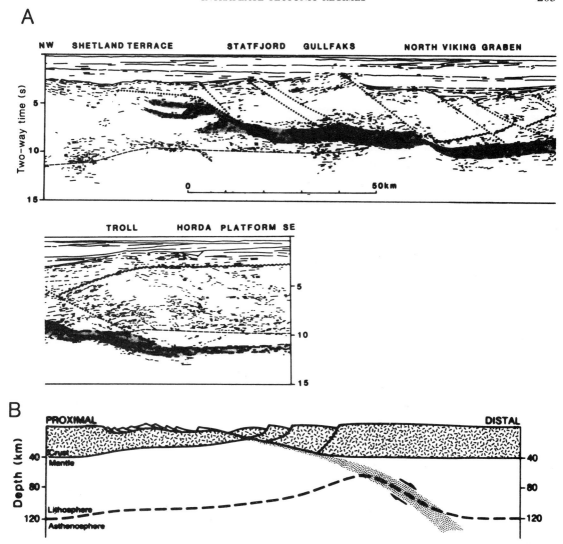

Figure 7.11 Interpretation of deep seismic reflection profile across the Viking graben. The shaded zone in (*A*) is interpreted as a major extensional shear zone passing from the middle crust in the west, into the mantle in the east. East-dipping normal faults detaching on this zone are shown dotted. Note the major crustal-scale rollover anticline on the eastern side of the section. The section shows a fundamental asymmetry that contrasts with the interpretation of Figure 7.9. The schematic model (B) suggests how the lithosphere may have accommodated the extension. From Beach (1986), with permission.

This detachment appears to rise westwards to meet the top of the basement at the western side of the graben. The elevated Gullfaks block, in the centre of the graben, is shown to be a tilted fault block like the others, but situated over the area of greatest extension, and bounded by the fault with the largest throw. The eastern side of the graben is shown

to be structurally quite different. A very large, crustal-scale, roll-over anticline (see Figure 4.25) accommodates the slip along a deep detachment that is believed to lie within the upper mantle at a depth of about 40 km — 15–20 km below the base of the crust. As seen earlier (Figure 7.9*B*), the crust thins to 20–25 km over the graben region. Interestingly, the area of maximum thinning is displaced eastwards of the graben axis, which supports the asymmetric extension model proposed by Wernicke (1985). This model (see Figure 4.24*B*) explains large extensions by block-faulting at high levels, detaching on a low-angle fault/shear zone that extends through the whole thickness of the lithosphere. In this case, displacements at the western margin of the graben are transferred to the base of the lithosphere below the Horda platform, east of the eastern graben margin (Figure 7.11*A*). Note that there is no seismic evidence for the shear zone penetrating to the *base* of the lithosphere in the manner of Figure 7.11*B*: how the middle and lower part of the lithosphere deform is still very much an open question.

Beach suggests that the extensional displacements on the North Sea graben system may be transferred along the major strike-slip Tornquist zone, which runs from Denmark through Poland to the Tethyan margin (Ziegler, 1982). By this means, the North Sea zone of intraplate extension could be explained by displacements at the plate boundary 1800 km to the south.

The long and complicated history of the North Sea basin might perhaps limit its usefulness as a test of general models of intraplate basin, was an important element in its subsequent development. This extensional phase, while not responsible for the initiation of the basin, was an important element in its subsequent development. This extensional phase, lasting approximately 60 Ma, involved a major crustal extension of around 1.5 (50%), calculated by comparing the pre- and post-Mesozoic crustal thicknesses. Total extension, assuming an original crustal thickness of 31 km, is as much as 1.8 (Wood and Barton, 1983). The difference may be explained by an earlier phase of extension for which no detailed evidence is available. The Jurassic extensional phase requires a strain rate of about 2×10^{-16} and corresponds to the slower type of extensional rifting, producing a wide zone of extensional deformation (see 2.7, Figure 2.29).

Sclater and Christie (1980) view the evolu-

tion of the basin in Cretaceous and Cenozoic times as the consequence of isostatic depression, that resulted from cooling of the lithosphere as the thermal anomaly produced by the initial stretching disappeared.

The Atlantic continental margin of the northern USA

The continental shelf and slope bordering the Atlantic coast of the northern USA is an active example of the passive-margin type of intraplate basin. The region has been intensively studied using borehole, seismic reflection, and gravity data.

A model profile across the basin (Figure 7.12) is described by Sawyer *et al.* (1982). The profile crosses the Baltimore trough, located east of the New Jersey coastline, where up to 18 km of sediment has been deposited since the late Triassic. The period of maximum subsidence occurred during the Jurassic, when the initial extension in the Central Atlantic took place. The Jurassic sediments rest on a comparatively thin wedge of Triassic red beds, volcanics and evaporites, which themselves rest directly on the continental basement. Above the Jurassic strata, a much thinner sequence of Cretaceous sediments is draped over the edge of the continental shelf without showing marked thickness variation. The basin is not recognizable as a separate structure in the Cenozoic sedimentary sequence, which is clearly separated by the edge of the continental shelf into shelf and ocean-basin assemblages.

Sawyer *et al.* use this example to test the McKenzie stretching model of basin subsidence. They assume that the main extensional event commenced at about 200 Ma BP with the formation of the late Triassic rift sedimentation, and ceased about 175 Ma BP with the formation of the first oceanic crust. Further extension would be taken up by ocean-floor spreading.

Theoretical subsidence curves based on the McKenzie model show the amount of subsidence at a given time after initiation for

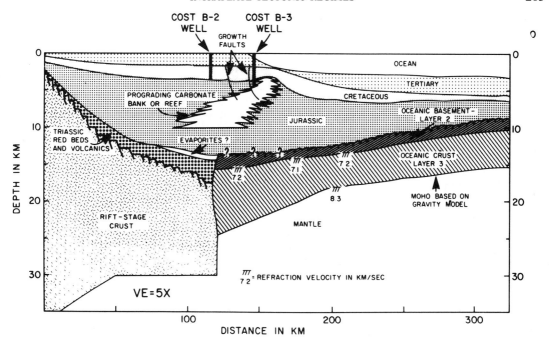

Figure 7.12 Interpretative profile across the continental margin of the eastern USA, in the region of the Baltimore trough, east of New Jersey. The positions of two wells, COST B2 and B3, discussed in the text, are shown. From Sawyer *et al.* (1982)

different values of the stretching factor β. Subsidence is initially linear (i.e. for the first 25 Ma) during the continental stretching phase, and is followed by thermal subsidence beginning at 175 Ma and continuing to the present. The model predicts that the initial subsidence is larger, about 40% of the total, whereas the subsequent thermal subsidence, accounting for the remaining 60%, lasts for over 100 Ma.

Actual subsidence curves (Figure 7.13A) were calculated from sediment thicknesses in two wells (COST B2 and B3), after making corrections for sediment compaction, isostatic responses to sediment loading, palaeo-depth of water, and eustatic sea-level changes. The curve for the COST B3 well shows a reasonable fit to the model extension curves over the latter part of the time range for a stretching factor of between 5 and infinity. However, since these two wells are situated over the edge of the oceanic crust, they cannot be used to give an accurate estimate of continental extension, since the extension at the ocean margin is

effectively infinite. A better guide to the continental extension in the basin is provided by two wells situated in a traverse further north where the basin is rather shallower (about 11 km). The curve for the more westerly well (COST G1) on the northern profile (Figure 7.13B) corresponds to a stretching factor of between 1.66 and 2.5, and the more easterly to a factor of between 2.5 and infinity. These results suggest that the McKenzie model gives a reasonable approximation to the subsidence history of a passive margin basin, at least over the greater part of the cooling stage. Moreover, such a basin would be expected to show varying subsidence rates corresponding to model extensions ranging from a minimum (continental) value to infinity over the oceanic part of the basin.

Passive-margin basins are a product of tectonic processes related to divergent plate boundaries (see Chapter 4). However, since their effects are retained within the lithosphere long after the plate boundary has migrated

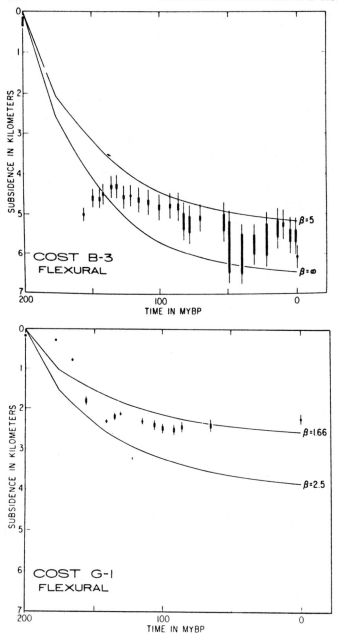

Figure 7.13 Plots of tectonic subsidence *v*. time, compared with theoretical subsidence curves with specified β values derived from the McKenzie extensional model. Plot *A* is derived from the data of the COST B3 well (Figure 7.12) on the margin of the oceanic crust. The data fit between theoretical curves for extension of between 5 and infinity. Plot *B* is derived from well COST G1, situated 450 km NE of B3, on continental crust. Note that the data here fit a lower extension of 1.66–2.5. The curves are calculated assuming a flexural elastic plate model. Bold error bars refer to palaeo-water-depth estimates, and are extended to approximate other sources of error. From Sawyer *et al*. (1982)

away from them, they can properly be regarded as intraplate structures.

7.5 Intraplate uplifts

Intraplate uplifts are as important as basins in the tectonic history of the plate interiors. They occupy a similar surface area, and over long periods of time must approximately balance depressions volumetrically in order to maintain their sediment supply. Uplifts are more difficult to study than basins because, in many cases, the stratigraphic record is either incomplete or totally missing. The major Precambrian shield regions are exposed because they are uplifts, but the detailed history of their uplift has not, in general, been recorded. Those regions of the Northern Hemisphere covered by the Quaternary ice sheets show high recent uplift rates that have been attributed to post-glacial isostatic response to the removal of the load. However, these same areas have acted as uplifts over much longer periods, and the Quaternary movements merely accentuate a long-continued tectonic trend.

Many continental uplifts are associated with rift zones, and are thought to be genetically related to them (see 4.3). Among uplifts of this type are the East African and Ethiopian plateaux, the Rhenish shield (see Figure 4.16A) and the Voronezh–Ukraine uplift of the Russian platform (Figure 7.3C). The origin of such structures is generally presumed to be thermal in view of their association with vulcanicity and, in the recent examples, with high heat flow. Other uplifts are attributed to present plate-boundary processes (e.g. the Tibetan plateau — see 5.4) or are associated with past plate boundaries. The Deccan plateau, for example, marks the site of a Cretaceous constructive boundary. Many uplifts, however, are situated far from plate boundaries or rift zones and their origin is not so obvious.

We shall now consider two examples of active uplifts, the Fennoscandian uplift and the Colorado plateau.

The Fennoscandian uplift

The 'post-glacial' uplifts of Laurentia (northern N. America) and Fennoscandia (Finland and Scandinavia) are familiar examples of recent deformation. These uplifts involve vertical movements ranging up to 100 mm/year over periods of the order of 10 000 years (Table 7.1). The formerly glaciated areas of N. America, Fennoscandia and Scotland yield uplift rates very similar to those of the present-day glaciated regions of Greenland and Franz Josef Land.

The pattern of uplift of Fennoscandia obtained from raised post-glacial shorelines is shown in Figure 7.14. The contours show an oval uplift, 1800 km long in a NE–SW direction, and about 1000 km across. The central portion has been elevated over 250 m since 6800 BC, corresponding to an average uplift rate of 28 mm/year. A very similar recent uplift pattern has been determined by precise re-levelling since 1835. The maximum uplift of 9 mm/year occurs in the Gulf of Bothnia.

The uplift corresponds with a marked, negative, free-air gravity anomaly that has been interpreted as the residual mass deficiency produced by the removal of the ice sheet. The existence of zones of active seismicity, together with evidence of recent fault movements, complicates the tectonic interpretation of the uplift, and suggests that at least part of the isostatic 'rebound' of the region has taken place by fault movements (Vita-Finzi, 1986).

The Colorado Plateau

This structure, summarized by McGetchin *et al.* (1982) is frequently cited as an example of currently active uplift. It lies immediately east of the Basin-and-Range province of the western USA described in 4.4 (see Figure 4.17A). The

Table 7.1 Uplift (in mm/year) in currently and recently glaciated areas. From Vita-Finzi (1986), after Nikonov (1980).

Period (10³ years BC)	Ancient ice sheets						Present-day ice sheets and ice caps			
	Laurentide		European		Barents Sea		Greenland		Franz Josef Land	Severnaya Zemlja
	Centre	East Periphery	Fennoscandian centre	Scottish centre	Spitzbergen	Kolgver Is.	East periphery	West periphery	Alexandra Land	Bolshevik Is.
10–8	100–70	70	30	10	13	18	60–70	30	8–10	12
4–3	30	3–5	10–15	4–5	3	1–2		5	1.5–2.5	

Figures (mm/year) corrected for eustatic sea level. The values for the Barents Sea are very tentative.

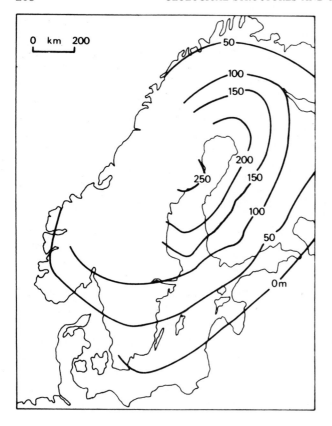

Figure 7.14 Map showing contours, in m, of uplift of the Baltic area since 6800 BC. From Vita-Finzi (1986), after Zeuner (1958).

plateau has a mean elevation of 1800 m and is about 700 km across. Undeformed Mesozoic strata are exposed at the surface.

The crust beneath the Colorado plateau is unusually thick (45 km) and exhibits shield-like properties from surface-wave behaviour. The upper mantle, in contrast, is characterized by an anomalously low P_n wave velocity of 7.8 km/s. Heat flow measurements indicate a surface flux of about 70 mW m^{-2}. The plateau lies within the same region of Cenozoic igneous activity as the Basin-and-Range province, but has not suffered the extensional effects experienced there and in the Rio Grande rift to the east.

The plateau is interpreted as a remnant of a large uplifted region of thickened crust in the western USA produced in the late Cretaceous Laramide orogeny. The bulk of the present uplift of 1.5–2 km, however, occurred in the Miocene, between about 18–10 Ma BP, giving

an average uplift rate of about 200 m/Ma during the period of extension in the neighbouring Basin-and-Range province.

There is no general consensus regarding the uplift mechanism for the Colorado plateau, but the most likely cause appears to be the isostatic response to the Miocene regional thermal event. This event may have produced a region of anomalously warm mantle beneath the plateau crust which, unlike the adjoining regions, was too strong or not warm enough to produce the extensional failure seen there.

Origin of intraplate uplifts

Of the many mechanisms proposed to explain uplifts, we shall consider the two most plausible: isostatic and flexural. Isostatic uplift may be the response to a change in either lithosphere mass, or mean lithosphere density. Thus an uplift may be the response to a reduction in density caused by local heating of the lithosphere, for example above a mantle plume. This mechanism is the same as that proposed for oceanic ridges and plateaux, and is the reverse of the thermal process responsible for basin formation. Uplifts produced in this way exhibit high heat flow, are associated with vulcanicity, and are likely to be related to rifts. Examples of such structures are the East African and Rhenish uplifts referred to earlier. Plateaux on the ocean floor have a similar origin. The removal of material by erosion from an uplift will also produce an isostatic response.

The second important mechanism is lithosphere flexure. Because of the short-term mechanical rigidity of the lithosphere, downward bending to form a basin in one place is inevitably accompanied by an upward bending around the margins of the depression. In other words, the lithosphere is tilted towards the area of the down-bend. The uplift will be compensated by the flow of asthenospheric material into the bulge at the base of the lithosphere created by the uplift.

The processes of uplift and depression are complementary, and probably balance out, broadly, over long periods of time. Both processes must be essentially self-limiting in order to conserve crustal thickness. Erosion of uplifts results in crustal thinning, and overcompensation by dense asthenosphere will produce a counterbalancing downward force. The reverse is true of basins: over-thickening of the crust produces a reduction in overall lithosphere density, and results in a counterbalancing upwards force. This may explain why the intraplate tectonic pattern over time periods of the order of hundreds of Ma is characterized by periodic inversions of the kind seen in the Russian platform.

8 Phanerozoic orogenic belts: some examples

We have established in the foregoing chapters a set of principles and methods, both on theoretical grounds and by looking at 'actualistic' or active examples, linking geological structure with plate movements and plate behaviour. In this and in the following chapter, we shall apply these principles and methods in order to understand the structure of certain orogenic belts. Obviously, very few belts can be discussed here, and much of the essential detail must be omitted for reasons of space. However, it is hoped, by discussing a few selected well-studied examples, to give a general impression of how the various tectonic patterns expressed in these belts have been linked with plate tectonic models.

Plate tectonic interpretation has been achieved with varying degrees of success: in general, the interpretation of young orogenic belts formed during the Mesozoic–Cenozoic period is much more tightly constrained than in the case of Palaeozoic and, more especially, Precambrian examples. The reason for this difference lies in the comparative certainty with which the successive positions of the various plates and plate fragments can be tracked, by ocean-floor magnetic stratigraphy and other means, since the break-up of Pangaea about 200 Ma BP. Before this time, the oceanic record is missing, and we have to rely on much less accurate palaeomagnetic restorations in plotting the former relative positions of the continental fragments.

We shall look at four examples: the Alpine orogenic belt in the Western Mediterranean, the Cordilleran Mesozoic–Cenozoic belt in western North America, the Hercynian system of Western Europe and its North American counterpart, the Alleghenian, and finally the Caledonian belt of the North Atlantic region. These four examples provide a complete range in tectonic environment: they encompass divergent, convergent and strike-slip regimes, subduction, and collision, of both continent–arc and continent–continent type. Most oro-genic belts will find analogues somewhere in these examples.

8.1 The Alpine orogenic belt of the Western Mediterranean

An active destructive plate boundary extends from the Gibraltar fracture zone in the Atlantic Ocean (see Figure 3.1) to the western boundary of the Pacific plate in eastern Indonesia. This boundary includes subduction zones such as the Aegean arc, the Makran and the Sunda arc (see 5.2) and the major collision zones between Eurasia in the north and Arabia and India to the south. The Western Mediterranean sector of this active boundary lies mostly below the Mediterranean Sea, continuing the line of the Azores–Gibraltar fracture zone to Sicily. Here a short subduction arc links it to the Eastern Mediterranean trench system (Figure 8.1). This active boundary lies within a wide zone of Mesozoic–Cenozoic activity comprising several separate but related belts. South of the suture lies the Atlas belt stretching from Morocco to Tunisia. Across the Straits of Gibraltar is the Betic Cordillera, along the south coast of Spain, which is separated by a large stable craton, making up most of the Iberian peninsula, from the Pyrenean belt between Spain and France. The latter belt meets the main Alpine belt east of the Gulf of Lions. The Alps are joined to the south by the Apennine belt, running along the Italian peninsula, to the east by the Dinaride chain of Yugoslavia, and to the northeast by the Carpathian chain.

The explanation for this complex pattern lies in the plate tectonic history of the region, which, according to Dewey *et al.* (1973) and subsequent workers, has involved separate but interrelated movements of a number of microplates or minor continental fragments in addition to the main European and African plates. The microplates include Iberia, the Carnics, Moesia, Apulia and Rhodope (Figure 8.2).

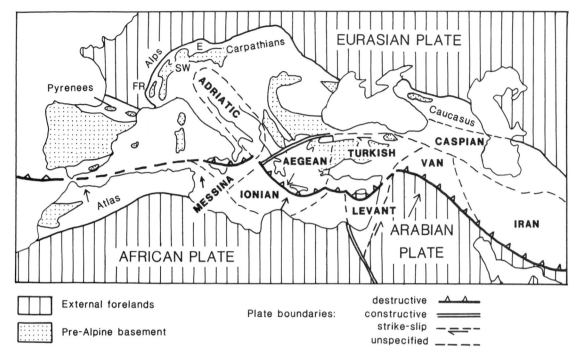

Figure 8.1 Map showing the tectonic setting of the Alpine orogenic belt in the Mediterranean region. Microplates or major crustal blocks within the belt are shown in bold letters. The main plates bordering the belt (ruled ornament) are the Eurasian, African and Arabian plates. The plate boundary (mostly destructive) between these is shown as a heavy toothed line. Arrows indicate directions of movement relative to the Eurasian plate. Note the main sectors of the Alps chain, discussed in the text; FR, French; SW, Swiss; and E, Eastern. After Windley (1977) and Dewey *et al.* (1973).

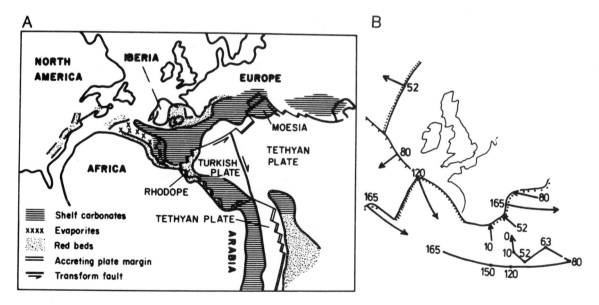

Figure 8.2 (*A*) Late Triassic palaeogeographic and plate tectonic reconstruction of the Alpine orogenic belt system. From Windley (1977), after Dewey *et al.* (1973). (*B*) Schematic map showing timing (in Ma BP), direction and amount of Mesozoic–Cenozoic motion of plates and blocks relative to NW Europe. From Dewey (1982)

The Carnic and Apulian fragments amalgamated to form the present-day Adriatic microplate (Figure 8.1) that forms the southern hinterland of the Alps.

Regional tectonic context

A useful summary of the tectonic evolution is contained in Windley (1977). Mesozoic–Cenozoic history of the Western Mediterranean has been dominated by a major plate reconstruction involving the opening of the Atlantic Ocean to the west and the closure of the Tethys Ocean (see Figure 3.5). In the late Triassic period, Western Europe formed a landmass containing continental red-bed basins bordered by a wide carbonate platform along the northern side of the Tethys Ocean (Figure 8.2A). A similar carbonate platform formed the northern and eastern flanks of the African–Arabian continent. New oceanic crust was formed in extensional basins now recognizable in several obducted ophiolite complexes. The Triassic carbonate platforms began to disintegrate in the early Jurassic, when widespread extensional faulting and rifting were experienced, associated with volcanic activity (Figure 8.3A). Subsidence led to deeper-water, muddy and pelagic facies throughout the region.

The next major change marked the commencement of convergence and subduction (Figure 8.3B). The former existence of trenches is often deduced from the widespread appearance of the so-called 'flysch' facies in the Cretaceous and Eocene periods. This sedimentary assemblage consists of marine shales with intercalated coarse clastic material, typically turbiditic, and associated with olistostromes or massive slump deposits. Assemblages of this type are characteristic of modern trenches but also of extensional basins such as those of the present-day Western Mediterranean, and of major submarine delta fans. They have therefore been regarded traditionally as *synorogenic* deposits created as part of the orogenic process. In plate tectonic terms, they represent both the erosional products of the uplifted zones of convergent boundaries (e.g. volcanic arcs) and also the associated depositional environments such as trenches, and fore- and back-arc basins.

The main flysch deposits, which are of Cretaceous age, are found in the external zone of the Alps, extending eastwards into the Carpathians. These deposits have long been regarded as marking the Alpine 'foredeep basin' receiving the erosional products of the uplifted mountain range to the south and east (cf. Argand, 1916). This foredeep basin has been interpreted in plate tectonic reconstructions of the region as a trench bordering an uplifted arc to the south. However, Hsu (1972) has pointed out that flysch sedimentation in the eastern Alps continued for about 50 Ma, and that in a normal subduction zone, such material would have been removed much more quickly. The explanation he suggests is that the relative motion of the trench was largely strike-slip. This is in agreement with the relative plate movement vector, as we shall see.

Associated with this stage in the Alpine orogeny was the development of blue-schist metamorphism in the oceanic material, interpreted as the effect of subduction; the appearance of acid to intermediate vulcanicity indicating the presence of volcanic arcs; and the development of thrusting and obduction marking the closure of small oceanic basins.

Continued convergence led ultimately to continent–continent collision in several parts of the Alpine belt, and particularly in the main Alpine chain itself, where the Adriatic continental plate came into contact with the European continent. This event led to progressive movement of thrust sheets and to general uplift. The process took place during Eocene to Oligocene time, and was accompanied by high-grade metamorphism. The climax of regional orogenic uplift in the Alps occurred in the Miocene period (Figure 8.3C).

During the final stages of convergence and continental collision, an important change took place on the foreland, where largely continental clastic deposits were formed in foredeep basins situated along the craton margin. Non-marine clastic sequences of the type formed in

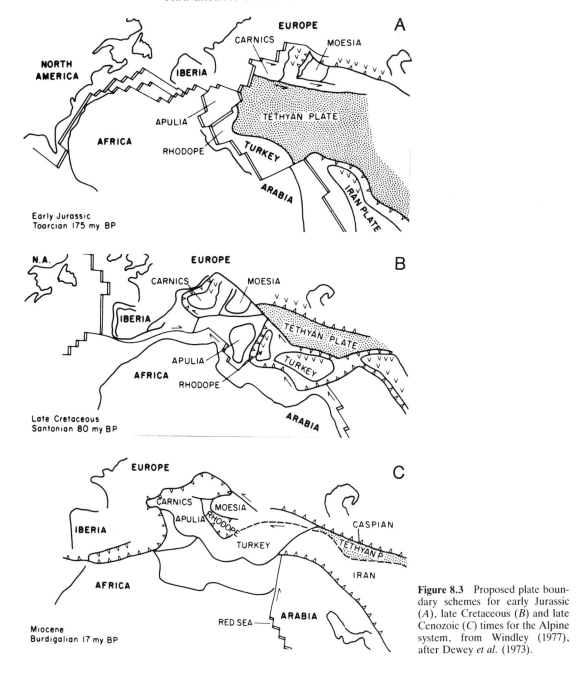

Figure 8.3 Proposed plate boundary schemes for early Jurassic (*A*), late Cretaceous (*B*) and late Cenozoic (*C*) times for the Alpine system, from Windley (1977), after Dewey *et al.* (1973).

the later stages of orogenesis are termed '*molasse*' in classical geosynclinal or orogenic theory. These deposits contain a high proportion of coarse, clastic material, along with intercalated coals and evaporites. Molasse deposition in the Alpine belt overlapped the period of flysch deposition in general, although within the Alps proper, molasse deposition replaced flysch during the Oligocene period and continued into the Pliocene.

History of plate movements

The record of plate movements within the Western Mediterranean region was first documented by Dewey *et al.* (1973) whose account of the plate tectonic evolution of the area forms the basis for most subsequent models. The late Triassic to early Jurassic rifting phase was expressed not only along the line of subsequent ocean opening of the Central Atlantic, but also along the boundaries of several 'microplates': the Carnics and Moesia attached to Europe, and Apulia, Rhodope and the Turkish plate attached to Africa (Figure 8.3*A*). These continental blocks subsequently moved eastwards and northeastwards respectively, creating new ocean basins behind them.

The commencement of opening in the Central Atlantic is generally dated at 180 Ma BP in the early Jurassic. This date corresponds with the collapse of the carbonate platforms noted above. As a result of this opening, the African plate moved in an ESE to E direction parallel to the Azores–Gibraltar fracture zone. At the same time, further opening of the new ocean basins was accompanied by subduction of the main Tethyan oceanic plate along the Black Sea–Caucasus line (Figure 8.3*A*).

This phase of movement continued until about 80 Ma BP in late Cretaceous time, when an important change in relative plate vectors took place. The change resulted from the opening of the North Atlantic Ocean west of Spain, and extending to the Davis Strait between Greenland and N. America, together with the linked opening of the Bay of Biscay and the consequential rotation of Iberia. Because this part of the Atlantic was opening faster than the Central Atlantic, the overall movement vector of Africa reversed to become NW to W relative to Europe (Figure 8.2*B*). The rotation of the Iberian peninsula during this phase has been determined palaeomagnetically (Van der Voo, 1969). During this second phase of essentially sinistral relative movement between Africa and Europe, convergent motion took place in the Alps, marked by flysch sedimentation in the Piemont trough.

As can be seen from Figure 8.2*B*, the convergence vector at this time is approximately perpendicular to the N–S-trending Western Alps, but almost parallel to the E–W-trending Eastern Alps.

At about 52 Ma BP another important change in relative motion occurred, caused by the opening of the main N. Atlantic–Arctic Ocean between Greenland and Scandinavia. This change was marked by a NW-directed movement of Africa towards Europe. During this third phase, from Upper Eocene to Lower Oligocene time, the main deformation and metamorphism of the Alps took place, although flysch sedimentation continued in the Helvetic zone to the north. Note that the convergence vector is now oblique to both the Eastern and Western segments of the Alpine chain, but with a strong convergent component in each case. Thus movement is dextral transpressive in the Eastern Alps but sinistral transpressive along the N–S to NW–SE French Alps, and only truly convergent along the short NE–SW segment of the Central or Swiss Alps. The movement uplifted the earlier fold belt, and created the great gravity-sliding nappes of the Pennine and Ultrahelvetic systems (see Figure 8.5). This stage also corresponds to the commencement of molasse deposition in the foredeeps to the north and in the Po plain to the south.

By Oligocene times, all Mesozoic ocean crust in the Western Mediterranean had disappeared. The present oceanic areas result from the opening of new marginal basins created by the anticlockwise rotation of Italy (and the Dinarides) and of Sardinia–Corsica away from the Iberian peninsula in the Miocene. Again these movements are reliably determined from palaeomagnetic data (Zijderveld *et al.*, 1970*a,b*).

The final change in convergence direction took place about 10 Ma BP during the Miocene (Figure 8.2*B*). From then until the present, Africa has been moving northwards relative to Europe. This change is reflected in the intraplate structures such as the Rhine–Ruhr rift system (see 4.3). During this phase, the main deformation occurred in the Helvetic zone and in the Southern Alps, on the margins of the African plate, and finally, in the early Pliocene, in the Jura.

Structural framework of the Alps

The Alps proper (as distinct from the Alpides or Alpine chains, which are much more widely distributed) comprise three main sectors: the French, Swiss, and Eastern, or Austrian, Alps (Figure 8.1). The French (or French-Italian) Alps extend from the Mediterranean coast near Nice, to the Swiss border in an arc that varies in trend from NW–SE in the south to NE–SW in the north. The change in trend is rather sudden and occurs in the region of Grenoble. The Swiss Alps continue in a NE–SW direction but gradually bend into the E–W trend of the Eastern Alps in Austria and N.

Italy. The latter belt is truncated on its eastern side by the undeformed deposits of the Hungarian plain.

There is considerable variation along the strike of the Alps of which a useful summary is provided by Debelmas *et al.* (1983) in the form of four characteristic profiles across the belt. We shall concentrate on the interesting central region where the NW–SE French sector bends around into the NE–SW Swiss sector (Figures 8.4, 8.5). There are eleven main tectonic zones, only some of which can be recognized in any given profile. (1) the northern *foreland* consists of undeformed Mesozoic–Cenozoic cover on a Hercynian crystalline basement; in the central Franco-Swiss sector, this zone is known as the *plateau Jura*. (2) The *folded Jura* zone, seen only in this central sector, consists of relatively simple fold-thrust cover resting on a shallow basal detachment. The synclines

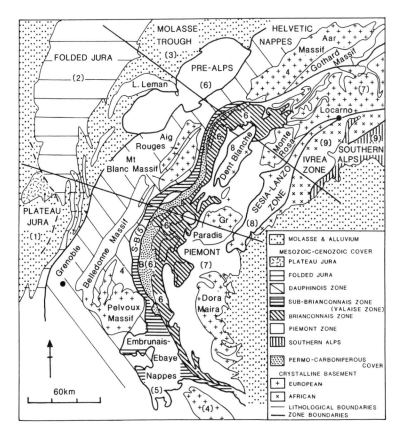

Figure 8.4 Map showing the main tectonic subdivisions of the Swiss and northern French Alps. The zone numbers are referred to in the text. After Ramsay (1963).

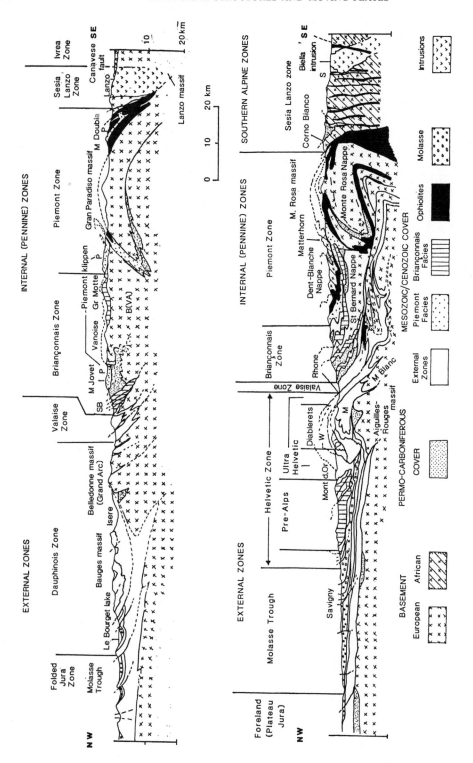

Figure 8.5 Representative structural profiles across the northern French (upper diagram) and Swiss (lower) Alps (see lines on Figure 8.4), showing the outcrop of the major structural units and tectonic zones. *SB*, Sub-Briançonnais zone; *P*, Piemont zone; *W*, Wildhorn nappe; *M*, Morcles nappe; *Di*, Diablerets nappe. After Debelmas *et al.* (1983).

of Mesozoic folded cover contain weakly-deformed Miocene molasse deposits. The amount of deformation increases south-eastwards. (3) The *Molasse trough* is a flexural foreland basin developed during the Oligocene to Pliocene period in response to thrust loading to the southeast. The molasse is undeformed in the central and western parts of the basin but is involved in thrusting on its southeast side. The sole thrust passes beneath the undeformed molasse to link with the next zone. (4) The *Dauphinois zone* (or *sub-Alpine chains*) contains the most highly deformed of the external zones of the Alps. In this zone, Hercynian basement and a thick platform Mesozoic cover has been involved in major thrust sheets which are para-autochthonous in that they have travelled only a short distance from their origin. Gravity sliding in the Tinée nappes of the Alpes Maritimes north of Nice (Figure 8.6*A*) is described by Graham (1981). He attributes the 26 km of shortening seen in the Triassic cover to gliding on weak décollement planes in Triassic evaporite deposits. The gravity gliding is attributed to uplift of the Argentera basement massif to the north (Figure 8.6*B*).

This zone is replaced along-strike in Switzerland by the *Helvetic nappes*. These consist of basement blocks of the Aiguilles Rouge and Mont Blanc massifs together with their para-autochthonous Mesozoic cover. In a study of the Helvetic nappes, Ramsay (1981) and Ramsay *et al.* (1983) integrate the major and minor structure and fabrics developed in the progressive strain history of the nappes. He shows that the folding and internal strain are related to movement along sub-horizontal shear zones that are the deeper-seated equivalent of thrusts. Figure 8.7 is a profile across the Morcles, Diableret and Wildhorn nappes (see Figure 8.5), which consist of detached Mesozoic cover. The profile illustrates the general form and stratigraphy of the Helvetic nappes. By studying the strain history of the various parts of the nappe complex, Ramsay shows that the earliest strains result from NNW elongation arising from sub-vertical shortening. These early strains are only shown in the uppermost nappes which have experienced a longer deformation history than the lower. Moreover, the higher nappes were displaced under conditions of lower confining restraint than the lower and show more variable strain patterns. Later strains are due to marked extension parallel to the fold axes. These observations are consistent with early ductile translations along low-angle shear zones that steepen downwards into the intensely deformed shear zones seen in the basement.

The next three zones constitute the *internal Alpine zones*. (5) The *Embrunais–Ebaye nappes* of the *Valais zone* occur in the southern French sector, where they are thrust over the rocks of the sub-Alpine chains. They contain allochthonous material derived from the internal Pennine nappe zones, and comprise a lower unit of Mesozoic cover slices and an upper unit of Cretaceous flysch. The lower unit continues northwards as the *sub-Briançonnais zone*. (6) The *Briançonnais zone* consists of numerous superimposed units exhibiting a fan arrangement, with more westerly structures verging west and more easterly verging east (Figure 8.5). The stratigraphic sequence is characterized by very thick Triassic shelf deposits on a Permo-Carboniferous basement, overlain by a very thin Jurassic and Cretaceous cover, with many stratigraphic gaps, and is interpreted as a pelagic geanticlinal zone. Deformation is intense, and shows two main phases; an earlier, characterized by north-westwards thrusting, and a later, related to southeastwards back-thrusting or *retrocharriage* (see e.g. Platt and Lister, 1985). The Briançonnais zone in the Swiss Alps is represented by the *Saint Bernard Nappe*. The later back-thrusting phase is dated by Hunziker (1986) from mica cooling ages and apatite fission-track ages, and is attributed to the N–S Miocene collision movement.

(7) The *Piémont*, or *Schistes Lustrés zone*, is the most easterly of the internal or *Pennine nappe* zones. It contains a number of complexly deformed nappes containing a stratigraphic sequence that changes from west to east. The external units possess a thick Triassic

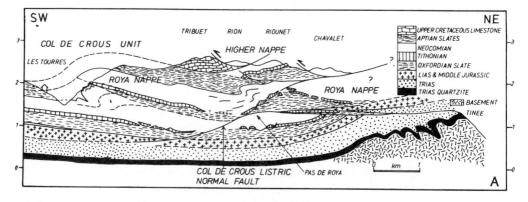

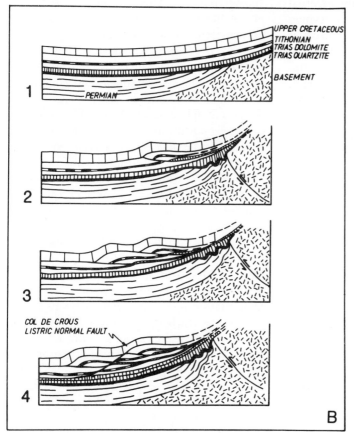

Figure 8.6 (*A*) Representative down-plunge structural profile across the Tinée nappe complex. (*B*) Sequence of diagrammatic profiles illustrating a gravity sliding model for the formation of the Tinée nappes. (*A*), (*B*) from Graham (1981)

shelf sequence on a Carboniferous basement, similar to that of the Briançonnais zone. However, the internal units contain typical ophiolitic ocean-crust assemblages with Jurassic to mid-Cretaceous pelagic sediments. These nappes show a similar change in vergence across the zone to the Brianconnais nappes. It is this zone that contains the evidence for Upper Cretaceous obduction linked with high-pressure metamorphism. In the northern French sector, this zone is reduced to several klippen resting on the Briançonnais nappes (Figure 8.5). In the eastern part of the zone lie the Lanzo peridotites, interpreted as the topmost part of

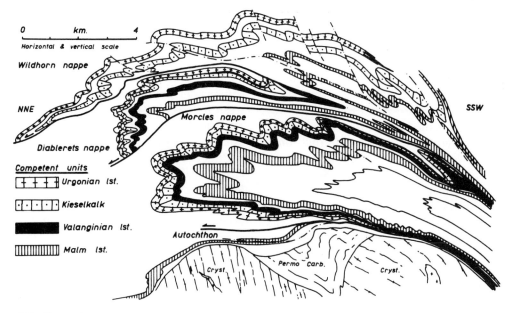

Figure 8.7 Composite structural profile across the western Helvetic Alps, showing the allochthonous Morcles and Wildhorn nappes overlying the autochthon of the Dauphinois zone (see Figures 8.4, 8.5). From Ramsay (1981)

the upper mantle, bounded on its south side by African crystalline basement of the Ivrea zone. In Switzerland, the lateral equivalents of the Piémont nappes are found in the complex Monte Rosa nappe, with its associated ophiolites. The Pre-Alps of Switzerland represent a large klippe of Piémont-zone material resting on the Molasse basin, at least 50 km from the nearest Pennine rocks, having travelled across the intervening Helvetic zone.

(8) In the Swiss Alps and Eastern Alps of Austria, the Pennine nappes are overlain by the next zone, the *Austro-Alpine nappes*. These contain crystalline basement of the Adriatic plate with its Triassic to Jurassic cover. The Dent Blanche nappe in the central Swiss Alps forms an Austro-Alpine klippe resting on the Piémont nappes. These nappes root in the Sesia Lanzo zone on the SE side of the Piémont zone. (9) The *Southern Alps zone* consists of a simple south-verging fold-thrust belt that is separated from the zones to the north by the Insubric and Tonale faults. This zone is only recognized in the eastern Alpine sector. The crystalline basement of the south-

ern Alps zone is known as the *Ivrea zone*. In the southern sector, the *Po basin* (zone 10) with its thick molasse deposits conceals the southern Alpine margin. Zone 11 is the undeformed Adriatic plate or *African foreland*. It is concealed by the Po basin in the south, but is represented by the *Ivrea zone* in the central sector.

In summary, then, the three main tectonic units of the central or Swiss Alps are the Helvetic, Pennine and Austro-Alpine nappe assemblages. The Austro-Alpine sheets are the topmost unit, and represent the relatively thin basement and cover from the African (Adriatic) plate, which have been 'flaked' off the top of that plate as first suggested by Oxburgh 1972 (see Figure 5.24). The Pennine nappes, with their ophiolitic sequence, represent the thinned continental margin and oceanic crust of the subducted margin of the European plate. The Helvetic nappes represent the platform sedimentary cover from the European plate, stripped off and transported towards the foreland.

The metamorphic history of the Alps reflects

the above changes in tectonic environment. The early high-pressure, low-temperature metamorphism, giving rise to blue-schists and eclogites, is associated with subduction and obduction during the early stages of convergence in the late Cretaceous. The later, higher-temperature, lower-pressure phase was superimposed on the former to give greenschist-facies conditions throughout the internal zones coinciding with the peak of tectonic activity.

Attempts are being made in many parts of the Alps to relate individual movement and strain histories of the nappes to an overall kinematic pattern that is compatible with the plate tectonic model outlined earlier. This can be achieved less easily in the internal zones than in the external, owing to the more ductile deformation and more complex strain history of the former. However, in several areas, northwestwards transport directions (see e.g. Butler, 1983) appear to correlate with the main

NW–SE convergent phase in Upper Eocene–Lower Oligocene times. In the southern French Alps, Merle and Brun (1984) demonstrate that the Parpaillon nappe, a Pennine nappe that has overridden the external zone, exhibits an earlier movement to the northwest, followed by a southeastwards movement attributed to gravity sliding away from the uplifted belt to the northeast.

Butler *et al.* (1986) present a balanced crustal-scale section across the central (Franco-Swiss) sector of the Alps by restoring the *Frontal Pennine Thrust*, which marks the boundary between the external and internal zones (Figure 8.8). This section has been restored parallel to the main WNW-directed convergence direction in the external thrust belt (i.e., that of the Oligocene movements).

Shortening estimates from individual balanced sections demonstrate a minimum of 140 km displacement along the *Frontal Pennine*

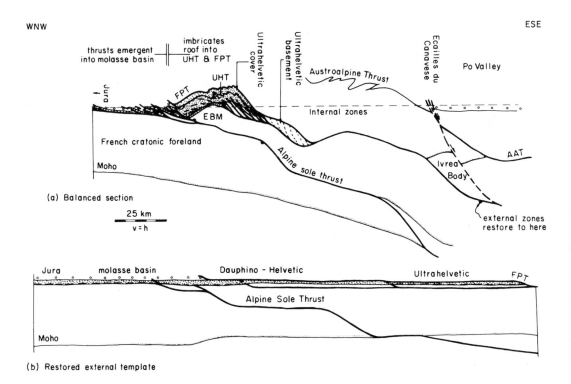

Figure 8.8 Balanced and restored sections illustrating the deep structure of the northwestern external Alpine thrust belt. Molasse, open circle ornament; other external cover sediments, stippled; *EBM*, external Belledonne massif; *FPT*, frontal Pennine thrust; UHT, Ultrahelvetic thrust; *AAT*, Austro-Alpine thrust (suture). From Butler *et al.* (1986)

Thrust (Figure 8.8). This implies an equivalent width of footwall in the form of a mid- to lower crustal wedge projecting beneath the internal zone of the Alps. However, the geometry of this wedge is dependent upon the pre-convergent crustal geometry: in particular, the extent of any basement thinning. The authors assume an original crustal thickness of 25 km, corresponding to the thickness on the un-deformed French craton.

The profile incorporates seismic refraction and gravity data (Perrier and Vialon, 1980) defining the depression of the Moho to a depth of c.50 km below the southern part of the internal zone (see Figure 5.26). However, the Moho has been displaced upwards and north-wards along a thrust regarded by Perrier and Vialon as the Alpine sole thrust. The dis-placement of the high-density Ivrea body to a relatively high crustal level (see gravity profile in Figure 5.23) is attributed by Butler *et al.* to the Frontal Pennine thrust. They regard this thrust as lying within the middle crust for most of the width of the internal zones, since no deep crustal rocks are exposed except at the eastern margin. Thus both the external and internal zones are attributed, according to this model, to thin-skinned thrust movements. The Austro-Alpine thrust, which transports the European–African suture over the European plate, restores to a position beneath the Po valley.

No precise estimate is available for the shortening across the Pennine nappes, which are difficult to restore because of their com-plex three-dimensional strain. According to Trumpy (1973), during the main convergent phase between late Eocene and early Oligo-cene time, at least 300 km of crustal shortening took place across the whole belt in a period of 5–6 Ma, giving an average deformation rate of 5–6 cm/year.

8.2 The Cordilleran orogenic belt of North America

At its simplest, the Cordilleran orogenic belt of North America may be divided into two major tectonic zones. The eastern zone comprises a Mesozoic to early Cenozoic foreland thrust belt produced by convergent deformation of the continental margin of the American plate. In the southern part of this zone, a later Cenozoic extensional regime has been super-imposed, giving rise to the Basin-and-Range province described in 4.3. The rocks and history of this zone can be related to each other and to the stable continental interior in a coherent and sensible manner, and will be described first. The western zone is a complex tectonic collage of suspect terranes (see 6.2), many of which are demonstrably alloch-thonous, and whose relationships with the continental margin to the east are either speculative or non-existent. This western zone was considered to represent the 'eugeosyn-clinal' belt, paired with the 'miogeosynclinal' eastern belt in the older literature, before the importance of strike-slip displacements and exotic terranes was discovered.

The structure of the orogenic belt is summar-ized in Figure 8.9. The eastern margin of the belt is represented by the thrust front, which possesses a sinuous course from the Alaska–Yukon border to the Gulf of Mexico, defining a belt that varies in width from 600 km to nearly 1400 km. The autochthonous part of the foreland thrust belt is defined by the proved limits of the N. American cratonic basement. The North Slope terrane is considered to be probably autochthonous, but all the terranes west of a line from the western boundary of the North Slope terrane, southwards along the western cratonic margin, are suspect. Several are involved in the thrust belt (e.g. the Eastern assemblage of British Columbia). Many of the terranes in the western belt have undergone large strike-slip displacements relative to North America.

The southwestern part of the orogenic belt is traversed by the San Andreas fault zone, along which dextral strike-slip motion is pre-sently taking place. This active zone is de-scribed in 6.3. Active subduction is also taking place at the western margin of the orogenic belt along the British Columbia–Washington–Oregon sector, and is responsible for the Cascades volcanic arc (see Figure 4.18).

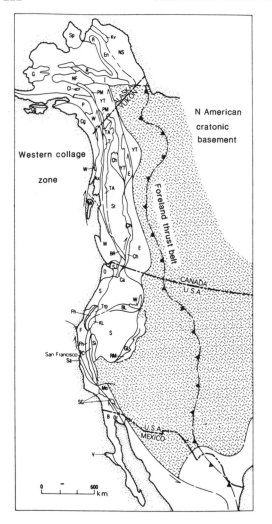

N American
cratonic
basement

Western collage

zone

Foreland thrust belt

CANADA
U.S.A.

San Francisco

U.S.A.
MEXICO

0   600
⌐──┬──┬──⌐ km

Figure 8.9 Tectonic summary map of the North American Cordilleran orogenic belt, showing the division between the western collage zone of suspect and displaced terranes, and the eastern parautochthonous fold-thrust belt underlain by North American cratonic basement. See source for a complete list of the terranes, and their descriptions. *NS*, North Slope and *E*, Eastern assemblage are terranes referred to in the text. The southern part of the map should be referred to Figure 4.17(*A*) for greater detail. From Coney *et al.* (1980)

The Cordilleran foreland thrust belt in the S. Canadian sector

The eastern part of the Cordilleran orogenic belt of N. America consists of an east-directed set of thrust sheets defining a belt up to 300 km wide, developed on Precambrian basement of the N. American craton. A detailed description of the southern Canadian Rocky Mountains sector (Figure 8.10) is provided by Price (1985). The sedimentary cover consists of an easterly-thinning wedge, divisible into 'pre-orogenic' and 'synorogenic' domains. The pre-orogenic cover consists of 'miogeoclinal' continental slope to continental shelf deposits ranging in age from Late Proterozoic to mid-Jurassic in the east, and in the west, of a 'eugeoclinal' assemblage of basic to acid volcanic rocks and immature clastic flysch deposits. The synorogenic cover consists of clastic deposits of late Jurassic to Palaeogene age, accumulated in a foredeep or foreland basin. The axis of this basin appears to have migrated north-eastwards during this period as the thrust-thickened load moved towards the craton.

The Late Proterozoic sequence comprises two quite different units. The Belt–Purcell Supergroup consists of up to 11 km of marine clastic sediments followed by carbonates deposited at a rifted continental margin. The overlying Windermere Supergroup consists of up to 9 km of coarse, immature sandstones and conglomerates, interbedded with shales and carbonates, lying unconformably on the Belt-Purcell rocks, and partly derived from them. The basin in which these beds accumulated appears to have been bounded by a major extensional fault, which subsequently reversed its displacement direction during the thrusting.

The Lower Palaeozoic strata consist of a thin sequence, up to *c*.1 km thick, of shelf carbonates and shales that thicken westwards to about 5 km along a carbonate bank margin. West of this line, there is a facies change in most units to dark mudstones with interbedded basic volcanic rocks and coarse clastic deposits. This facies change is interpreted as evidence of a back-arc or marginal basin west of the continental shelf, with a volcanic arc on its western side.

The Upper Palaeozoic strata are unconformable on the Lower. In the basin domain, the Lower Carboniferous deposits are similar to those of the Lower Palaeozoic, and rest on them with angular discordance. In the platform sequence, Upper Devonian carbonates overstep their Lower Palaeozoic equivalents.

This distinction between shelf and miogeoclinal slope environments is maintained in the Mesozoic sequences. 100 m of Jurassic marine shales on the platform are the lateral equivalent of a 1 km-thick sequence of Triassic to Jurassic shallow-marine deposits, consisting of assorted sandstones, carbonates and evaporites. The basin assemblage of equivalent age is 10 km thick and contains a high proportion of volcanic and volcanogenic deposits of mainly andesitic derivation. Deformed mid- to late Jurassic granitic plutons are associated with these rocks, which are also cut by younger, Cretaceous granites.

The changes in depositional environment recognized in these facies variations mark the boundaries of major tectonic provinces within the Rocky Mountains belt. Important changes in thickness and mechanical properties of the cover have controlled the nature of the fold-thrust deformation. Consequently, abrupt changes in tectonic pattern take place both across and along strike, due to these lateral changes in facies and thickness.

The belt is interpreted as a typical, thin-skinned, foreland thrust belt, where many of the ideas incorporated into modern thrust-tectonic theory were developed (see e.g. Dahlstrom, 1970). The structure is dominated by thrust faults that are primarily west-dipping and eastwards-verging (Figures 8.10, 8.11). Flexural-slip folds are developed in association with the thrusts, many of which die out in the cores of anticlines. Thrusts in the eastern part of the belt commonly detach on weak horizons such as the Upper Cretaceous Alberta Group, the Jurassic marine shales of the Fernie Group, and at the base of the thick Upper Permian Paliser limestone.

The thrusts in the eastern part of the belt, from the thrust front to the Rocky Mountain Trench, appear to have developed in piggyback sequence, by eastward displacement of the cover over the autochthonous platform (Figure 8.11). Precambrian basement with similar geophysical properties can be traced at least as far as the eastern part of the Purcell anticlinorium (Figure 8.11) without any disruption of its distinctive NE–SW magnetic fabric. In the western part of the belt, the Proterozoic strata of the Belt–Purcell and Windermere Supergroups show variable penetrative strain and metamorphic grade, both of which increase progressively westwards. Accurate section balancing is therefore confined to the eastern section.

A major change in structural level, amounting to about 20 km of stratigraphic succession, takes place across the Kootenay Arc, where Belt–Purcell strata in the core of the Purcell anticlinorium are adjacent to Triassic–Jurassic basin deposits to the west. The Kootenay Arc is therefore interpreted as a major west-facing monocline marking the western edge of the North American craton. This structure corresponds to a change in crustal thickness from $c.50–55$ km at the Purcell anticlinorium to 30–40 km beneath the interior part of the Cordillera, and also to a corresponding increase in the negative Bouguer gravity anomaly. The size of this anomaly can be explained by the calculated increase in crustal thickness of 10 km. However, in order to balance the thin-skinned crustal shortening in the east, the 40 km-thick crust of the autochthonous platform is required to extend westwards beneath the Purcell anticlinorium to the edge of the Kootenay Arc (Figure 8.11, 8.12). This crustal structure confirms the view, based on the study of the sedimentary facies, that the thick miogeoclinal strata of the Proterozoic and Lower Palaeozoic sequences accumulated on thinned continental crust at the continental margin. The minimum total shortening achieved by the thin-skinned part of the belt is estimated to be 170 km.

Deformation in the thrust belt appears to have spanned a period of almost 100 Ma from late Jurassic to Palaeocene time. Uplift and erosion of shelf assemblages is first recorded in Upper Jurassic deposits, and early reverse/transfer faults in the Purcell anticlinorium are cut by early to mid-Cretaceous batholiths.

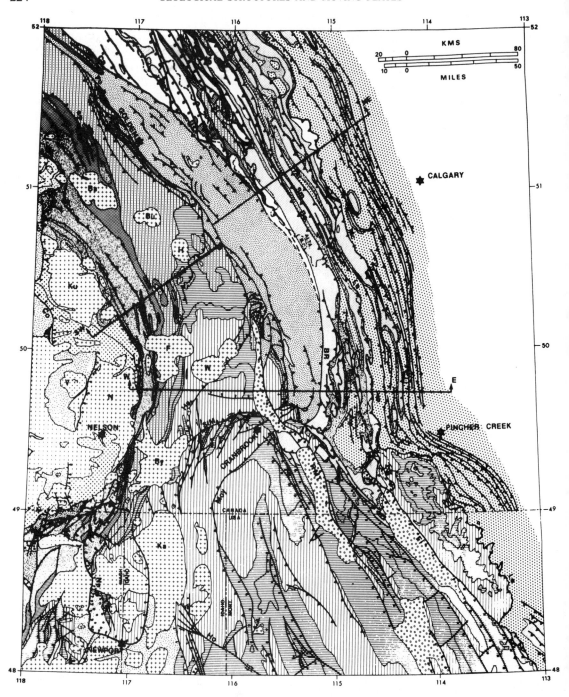

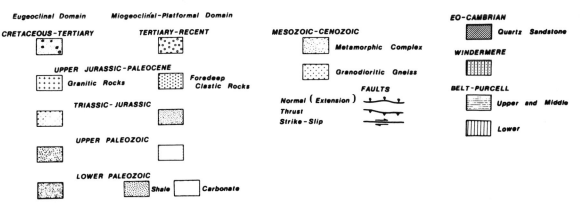

Figure 8.10 Geological map of the foreland thrust-fold belt of the southern Canadian Rocky Mountains, in the central sector of the N. American Cordilleran orogenic belt. For names of key faults and batholiths distinguished by letters, see source. From Price (1981)

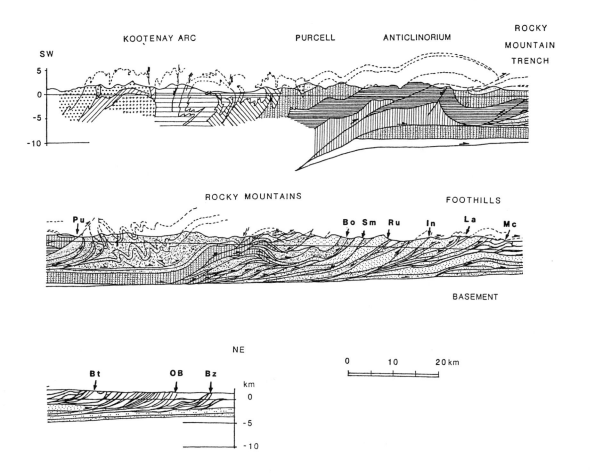

Figure 8.11 SW–NE structural profile across upper line on Figure 8.10. Faults identified on the section are: *Pu*, Purcell; *Bo*, Bourgeau; *Sm*, Sulphur Mountain; *Ru*, Rundle; *In*, Inglismaldie; *La*, Lac des Arcs; *Mc*, McConnell; *Bt*, Burnt Timber; *OB*, Old Baldy; *Bz*, Brazeau. Note different ornament in Kootenay are (ruled) and in foredeep clastics (blank) for clarity. From Price (1981)

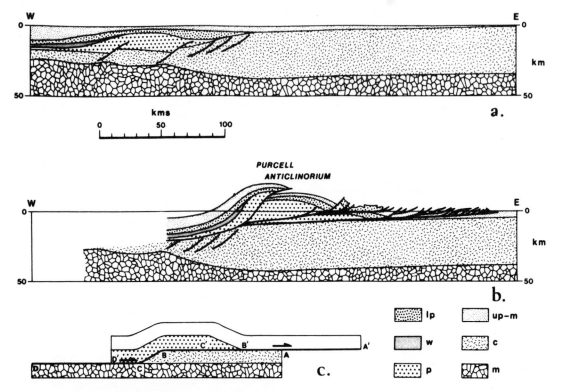

Figure 8.12   Diagrammatic sections to illustrate an interpretation of the evolution of the Purcell anticlinorium (see Figure 8.11) (*a*) Restored original section; (*b*) present section drawn to eliminate effects of erosion pre- and post- thrusting; (*c*) schematic representation showing displacement (*A* to *A'* etc.). *m*, mantle; *c*, continental crust; *p*, Belt–Purcell assemblage; *w*, Windermere assemblage; *Lp*, lower Palaeozoic assemblages; *up–m*, Upper Palaeozoic and Triassic–Jurassic assemblages. From Price (1981)

The late Cretaceous granite plutons of the Kootenay Arc are post-tectonic in relation to the deformation fabric. Further east, the McConnell and Lewis thrusts post-date Upper Cretaceous slope deposits but pre-date late Eocene–early Oligocene foredeep-basin molasse. Price estimates that at least 100 km of horizontal displacement occurred across the thrust belt during this period of less than 30 Ma, corresponding to a rate of 3 km/Ma, or 3 mm/year.

The Purcell anticlinorium is interpreted as the geometric consequence of lifting the thick sedimentary prism, originally deposited on the attenuated crust of the continental margin, on to the main part of the craton (Figure 8.12). Crustal convergence of possibly 200 km over the whole width of the belt is thus accommoda-ted to a large extent by overlapping of the already thinned continental crust along the continental margin, and is viewed by Price as an example of intraplate convergence, invol-ving the destruction of a marginal basin situa-ted behind (east of) the main eastward-dipping Cordilleran subduction zone.

The western collage zone of suspect terranes

The concept of displaced or suspect terranes (see 6.2) was developed in this region (see Wilson, 1968; Monger *et al.*, 1972; Jones *et al.*, 1972). In Figure 8.9 the distribution of more than fifty suspect terranes identified by Coney *et al.* (1980) is shown. The principles governing their recognition are discussed in 6.2. Adjoin-ing terranes may be distinguished by discontinuities of structure or stratigraphy across their boundaries, that cannot be explained on the basis of normal facies or tectonic changes. Many terranes contain palaeomagnetic records that differ strongly from those of the stable craton, or of adjoining terranes. Terranes are regarded as allochthonous or exotic if their faunal or palaeomagnetic signatures indicate that they originated a considerable distance from their present position relative to the craton. Many terranes show evidence of an origin far to the south of their present latitude, and many also have undergone translations of hundreds of km after collision. Palaeomagnetic evidence also indicates significant rotations about the vertical in many cases (e.g. the large terrane in Oregon, labelled *S* in Figure 8.9).

The history of the western zone can be pieced together by comparing the stratigraphy of the autochthonous and parautochthonous foreland sequences with those in the suspect terranes. As we have seen, the western boundary of North America was a passive continental margin throughout late Precambrian and early Palaeozoic time, during which a broad miogeoclinal terrace developed. Apart from a brief period of convergence and collision in the mid-Palaeozoic, this situation continued into the late Palaeozoic. In late Triassic to mid-Jurassic time, however, a subduction zone became established which eventually consumed the Palaeozoic proto-Pacific ocean. All the Palaeozoic terranes now found outside the Palaeozoic passive continental margin must therefore be suspect, and must have accreted to that margin during Mesosozic–Cenozoic time. Younger terranes outside that margin must also be suspect, although their allochthonous nature may be more difficult to prove unless they include Palaeozoic basement.

Most of the suspect terranes listed by Coney *et al.* contain sedimentary and volcanic sequences of oceanic affinity, and rocks older than mid-Palaeozoic are rare. A few contain pieces of oceanic crust (e.g. the Cache Creek terrane of Western Canada, and the Klamath Mountains terrane of California — see Figure 8.9). The Cache Creek terrane contains Per-

mian Tethyan faunas quite distinct from those found in adjoining blocks.

Other terranes represent fragments of island arcs of late Palaeozoic to Jurassic age. The large Stikine terrane of Western Canada (Figure 8.9) contains a Lower Carboniferous to Permian volcanic sequence overlain by Upper Triassic to mid-Jurassic volcanogenic strata. This terrane has no continental basement. Other terranes represent volcanic arcs formed on older basement sliced from a distant continental margin.

Several terranes can be shown to have amalgamated before their final accretion to the North American craton. For example Jones *et al.* (1977) demonstrate that Wrangellia collided with the Alexander terrane before final accretion to Western Canada and Alaska. These terranes contain different Palaeozoic basement rocks originating far to the south, but display similar Upper Jurassic to Cretaceous sequences and evidence of volcanic arc activity. The combined terrane accreted to the continental margin in mid-Cretaceous times. Since its accretion, further fragmentation has occurred, and the terrane now extends in several detached pieces over 2000 km from Oregon to Alaska.

The process of strike-slip terrane accretion appears to have extended over a period of at least 120 Ma from mid-Jurassic to early Cenozoic time. During most of this period, the continental margin was a subduction zone, so that accretion took place by a process of oblique convergence combining underthrusting with strike-slip movements. The former presence of subduction zones is attested by the belts of highly deformed chert, ophiolite and greywacke sequences, metamorphosed in blueschist facies, such as the Franciscan complex of California.

The strike-slip component appears to have been dextral throughout, so that the accreting material seems to have originated consistently to the south of its final resting place. Many of the fragments of volcanic arcs may be totally foreign to North or even South America, and may have travelled from the far side of the

Pacific Ocean. Ernst (1984) provides a quantitative analysis of the process. By assuming symmetrical spreading at the East Pacific ridge, a figure of about 10 000 km of western overriding of Pacific ocean plate is derived. To this E–W convergence should be added several thousand km of northward drift of the Pacific plate.

8.3 The Hercynian orogenic belts of Western Europe and North America

An orogenic belt of Hercynian age, often termed the *Variscan* belt, occupies most of Western Europe south of a line through the southern British Isles and northern Germany, and west of the Tornquist line marking the edge of the Russian platform (Figures 8.13, 8.16). On the eastern side of the Russian platform, the Urals belt formed during the same period. In North America, the equivalalent orogeny is termed the *Alleghenian*, and in North Africa, the *Mauritanian*.

The orogeny spans mid-Devonian to early Permian time, and immediately follows the Caledonian orogeny. In Europe, the Hercynian belt is oblique to the earlier Caledonian belt, but in North America, the two belts are parallel, and partly superimposed, and are difficult to distinguish from each other in many areas. Useful general descriptions of the belt are provided by Windley (1977), Ziegler (1975) and Weber (1984). The preferred name for the European orogenic belt is the *Variscides* (Hutton and Sanderson, 1985), but *Hercynian* is probably the more internationally acceptable name for the orogeny world-wide.

The width of the belt in Europe is about 2000 km, and the structural and stratigraphic pattern is difficult to interpret because the various outcrops are separated by post-Hercynian cover and, in the south, by the overprinting effects of the Alpine orogeny. The European Hercynides, or Variscides, have traditionally been regarded as a different type of orogenic belt to both the Caledonides and the Alps. Thus Zwart (1967) classifies orogenic belts into Hercyno-type and Alpino-type on the basis of their structural, metamorphic and igneous characteristics. The Hercyno-type, of which the West European Variscides are the type example, were differentiated from the Alpino-type by (i) large volumes of granitoid pluton, (ii) regional low-pressure, high-temperature metamorphism, and (iii) poorly-developed fold-thrust tectonic shortening. However these characteristics do not apply to the whole Hercynian belt. In North America the Hercynian orogeny is represented by a linear fold-thrust belt containing Barrovian metamorphic rocks and few granites. Moreover linear fold-thrust belts exist also in the marginal zones of the European Hercynides, in SW England and S. Wales, and in the Cantabrian–Asturian chain, for example (Figures 8.17, 8.19).

The regional context of the Hercynian belts is summarized in Figure 8.13. Following the Caledonian orogeny, the continents of Laurentia and Baltica had become welded together as far south as the northern Appalachians. To the south lay the proto-Tethys Ocean, with Gondwanaland on its southern side. At the end of the Hercynian orogeny, Africa had collided with Laurentia to form the Alleghenian sector of the Hercynides. Many authors have pointed out the importance of dextral shear within the European Hercynides (e.g. Arthaud and Matte, 1977). A generally northwestwards movement of Africa in relation to Laurentia–Northern Europe explains both convergent movement in the Alleghenian sector and dextral strike-slip effects in Western Europe, and forms the basis of most plate tectonic reconstructions.

A simple subdivision of the Hercynian belt (Figure 8.13) is made by Dewey and Burke (1973). The outermost zone is part of the Hercynian foreland on which formed basins of continental deposits during the Devonian, shelf deposits in the Lower Carboniferous, and coal basins in the Upper Carboniferous. This zone is represented in S. Wales and in the western side of the Alleghenian belt. The middle zone contains both marine and non-marine early Devonian sediments, mid-Devonian basic volcanic rocks, and mainly

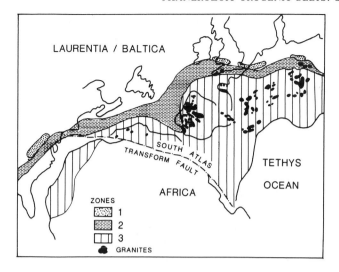

Figure 8.13 Outline map of the tectonic setting and principal subdivisions of the Hercynian orogenic belt system of Western Europe and North America. Zones: 1, discontinuous foreland basins; 2, external zone characterized by Upper Carboniferous flysch basins and fold-thrust belts; 3, internal zone characterized by basement massifs, high-temperature, low-pressure metamorphism and abundant granite plutons. After Windley (1977) and Dewey and Burke (1973).

shales in the early Carboniferous. Flysch basins, exhibiting the 'Culm' facies, formed in the mid-Carboniferous, and were subjected to northward-directed thrust movements. The inner zone contains a number of Precambrian basement blocks, such as the Bohemian (Moldanubian), Armorican, and central Iberian massifs. Devonian sedimentation in this zone was largely controlled by the distribution of the basement blocks. Sedimentary sequences are generally thin, and carbonates are typical. In the Lower Carboniferous, sedimentation was interrupted by tholeiitic vulcanism. The zone is characterized by high-temperature, low-pressure regional metamorphism, and by abundant granitic plutons and local migmatites. In the uppermost Carboniferous, a number of intermontane sedimentary basins developed, together with potassic ignimbritic vulcanism.

Three main phases of deformation are recognized within the period occupied by the Hercynian orogeny in the West European belt, each of which can be detected over most of the belt. These phases are the *Bretonic* (*c*.345 Ma BP), the *Sudetic* (*c*.325 Ma) and the *Asturic* (290–295 Ma). The Bretonic phase is responsible for the widespread Devonian–Carboniferous unconformity. According to Ziegler (1975), significant shortening occurred across the belt at that time. The Sudetic deforma-

tion corresponds to the main uplift phase of the interior of the Hercynian belt, and was associated with the main episode of granitic emplacement and acid to intermediate vulcanicity. The Asturic phase, in the uppermost Carboniferous, produced the marginal belts of fold-thrust deformation as well as further deformation in the interior zone.

The Alleghenian belt

The Phanerozoic orogenic system of eastern North America is divided into three separate sectors: the Northern Appalachians, extending from Newfoundland to the Hudson River; the Central–Southern Appalachians from there to Central Alabama; and the Ouachita–Marathon belt from northern Mississippi to Texas. The Northern Appalachians are primarily Caledonian in age (Acadian and Taconic), but in addition suffered Hercynian deformation in the south-eastern part of the belt.

The Central–Southern Appalachian belt is the type area of the Alleghenian orogeny. The belt here is about 2000 km long and 500 km across (Figure 8.14). It consists of four main zones bounded on the Atlantic side by younger deposits of the coastal plain. The outermost, foreland, zone comprises the Appalachian and Black Warrior basins, which contain undeformed or weakly-deformed Upper

Palaeozoic (mainly Carboniferous) strata. Lower Carboniferous (Mississippian) marine carbonates are overlain by Upper Carboniferous (Pennsylvanian) fluvial or deltaic deposits, with an overall thickness of generally under 1 km. These strata are affected by folding near the southeast margin of the zone.

The three zones making up the Alleghenian orogenic belt are known as the *Valley-and-Ridge*, the *Blue Ridge* and the *Piedmont* provinces (Figure 8.15). The Valley-and-Ridge province contains a thick Palaeozoic succession without appreciable break between Silurian and Devonian, or between Devonian and Carboniferous. The facies of the Carboniferous are similar to those of the foreland. Important coal-bearing deposits occur in the Upper Carboniferous. This province has long been considered to be an example of a major thin-skinned thrust belt (see e.g. Gwinn, 1964).

The eastern boundary of the province is marked by a major fault, southeast of which lies the Blue Ridge province, consisting of an upthrust block of Precambrian (Grenville) crystalline basement together with late Precambrian to early Palaeozoic sedimentary cover. The Piedmont belt consists of metamorphic rocks of probably pre-Carboniferous age, cut by abundant granite and gabbro intrusions of Carboniferous age (330–260 Ma), some of which are strongly deformed and gneissose. This belt is interpreted as a Carboniferous island arc.

The Alleghenian structure of the Central-Southern Appalachians is dominated by westwards overthrusting towards the foreland. A major décollement horizon within Silurian salt deposits forms a relatively shallow detachment surface for thin-skinned thrusting in the Valley-and-Ridge province. The COCORP deep-

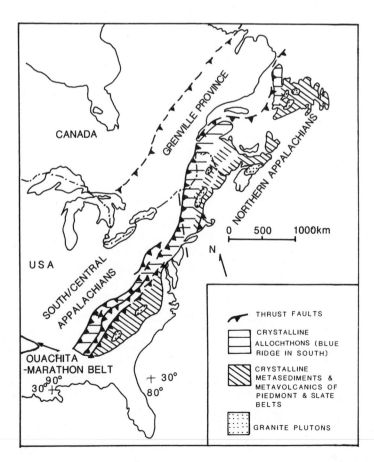

Figure 8.14 Tectonic summary map of the Appalachian orogenic belt of North America. Note the subdivision into external thrust-fold belts and internal Piedmont and Slate belts. The eastern end of the Ouachita–Marathon belt is shown in the extreme SW. After Cook *et al.* (1981).

seismic profile across Georgia (Cook *et al.*, 1981) appears to confirm this model (Figure 8.15*B*) in respect of the Valley-and-Ridge, Blue Ridge and inner Piedmont belts. Two alternative models are proposed for the eastern Piedmont belt: one envisages a mid-crustal detachment extending to the edge of the Coastal Plain, then descending to the Moho; and the other a zone of deep thrusts descending to the Moho beneath the eastern Piedmont, along the King Mountain belt. In both models, the continental crust of the eastern Piedmont and Coastal Plain is shown to be substantially thinner — about 30 km, compared with *c.*45 km in the main Appalachian belt. Thus the major part of the belt is allochthonous, involving displacements of up to several hundred km. The age of the deformation appears to span a long period of time. Earlier thrusts have been dated at *c.*380 Ma and 356 Ma, but the main Alleghenian deformation appears to relate to post-metamorphic displacements of late Carboniferous to Permian age (270–240 Ma BP). The main deformation is generally attributed to collision with North Africa.

The Ouachita–Marathon belt to the south (Figure 8.14) is thought to be related to a quite separate collision with a different microcontinent, which took place in mid-Upper Carboniferous times. Both orogenic belts involve sequences of shelf–slope sediments of the North American plate, together with portions belonging to the advancing Gondwanaland plates. An earlier collision took place in the Northern Appalachian belt (see later) where a continental fragment known as *Avalonia* collided with the North American craton in mid-Devonian time, giving rise to an Acadian orogenic phase there. Thus both to the north and to the south of the main Alleghenian sector of the North American Hercynian belt, collision with microplates preceded the main African–North American collision in end-Carboniferous time.

The West European sector

Weber (1984) summarizes the evidence for the nature of the pre-Hercynian basement in the West European Variscides (Figure 8.16), and concludes that, over most of the region, the basement is no older than about 700 Ma BP (i.e. derived in the Cadomian orogeny of Late Proterozoic age). Exceptions are the Armorican and Bohemian massifs, which are founded on older Precambrian blocks. The evidence for the nature of the basement comes mainly from a study of $^{87}Sr/^{86}Sr$ initial ratios indicating that the Hercynian granites are derived from melts of relatively young continental crust (Vidal *et al.*, 1981). The Cadomian orogeny appears to have succeeded a period of generally oceanic sedimentation over most of the West European region. Weber also discusses the evidence relating to the existence of the Caledonian orogeny within the Variscan belt. Although there has been no severe regional deformation, involving significant crustal shortening, a widespread suite of granite plutons was emplaced in Ordovician to Silurian times. This Lower Palaeozoic magmatism is broadly coeval with a high-grade metamorphic event represented for example in the granulite-facies rocks of the Saxon Granulitgebirge. The stratigraphic record suggests that this high-grade event took place at depth during continuous sedimentation at the surface, since a complete stratigraphic sequence from late Precambrian to Carboniferous occurs within the adjacent Saxothuringian zone. Weber suggests, following Calsteren *et al.* (1978), that both the granite emplacement and the subsequent high-temperature metamorphism were produced by extensional crustal thinning and rifting, enabling the warmer asthenosphere material to rise to high levels within the lithosphere (see 4.2). If these ideas are correct, the implication is that the nature of the Caledonian 'orogeny' changes dramatically from northern to southern Europe, from an essentially convergent regime to a divergent one.

Another important orogenic event that is usually regarded as pre-Hercynian is a pre-Upper Devonian phase of deformation and granite emplacement recognized in the basement complexes of the Saxothuringian zone, the Bohemian massif, and the Massif Central, for example, where metamorphic rocks with

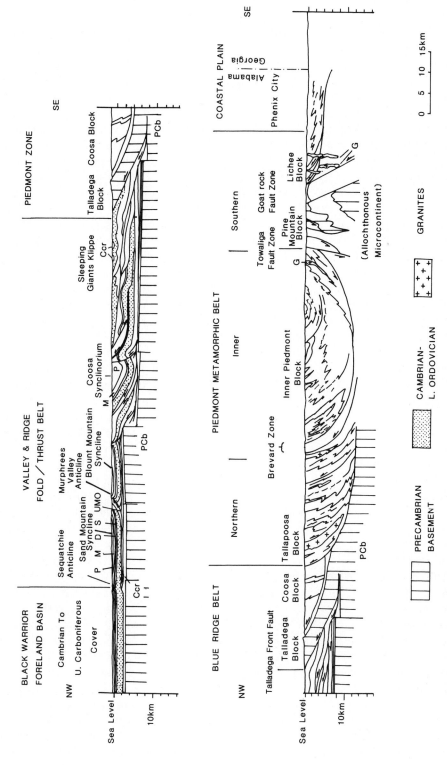

Figure 8.15 (*A*) Structural profile across the Southern Appalachian orogenic belt, showing the main tectonic units and structures. Pcb, Precambrian; Ccr, Cambrian; S. Silurian; D, Devonian; M, Mississippian; P, Pennsylvanian.

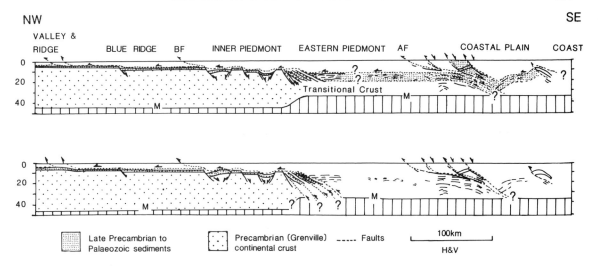

Figure 8.15 (*B*) Schematic restored cross-sections based on COCORP deep-seismic reflection data showing two possible interpretations: in the upper section a sub-horizontal detachment extends to beneath the coastal plain; in the lower, the inner Piedmont–Blue Ridge allochthon is shown as rooting at the western edge of the inner Piedmont zone. After Cook *et al.* (1981)

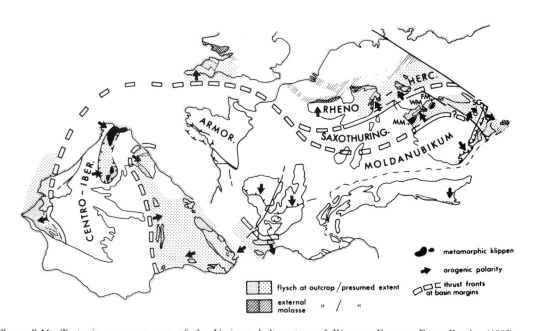

Figure 8.16 Tectonic summary map of the Variscan belt system of Western Europe. From Franke (1985)

intrusive granites, yielding ages in the range 370–400 Ma, are overlain by Upper Devonian clastic sediments. Unlike the earlier rifting phase, this mid-Devonian (*Acadian*) event was associated with significant crustal shortening, involving folding and thrusting of the metamorphic basement, and producing nappes of granulite-facies rocks resting on lower-grade material. In the Saar 1 borehole, within the Mid-German Crystalline Rise (Figure 8.17*A*), unmetamorphosed mid-Devonian sediments rest on Lower Devonian crystalline basement, indicating rapid uplift during this episode. However other basement complexes show much younger K–Ar cooling ages, indicating that uplift to high crustal levels did not occur until the Lower Carboniferous.

Although the major folding and metamorphism of the basement complexes is of early Devonian age, the weakly metamorphosed or unmetamorphosed Upper Palaeozoic cover was not folded until the Carboniferous. Weber

suggests that continued uplift of the basement massifs took place under conditions of general crustal shortening throughout Upper Devonian and Lower Carboniferous time, documented by repeated influxes of turbiditic flysch in the neighbouring basins, such as the Rhenisches Schiefergebirge. These movements took place along shear zones, producing belts of mylonitic gneisses in the basement crystalline complexes.

In Figure 8.17*A* a section across the Rhenohercynikum (Rhenisches Schiefergebirge) and northern Saxothuringian zones of northern Germany is shown. The profile illustrates consistently NW–verging overfolds and reverse faults in the greywacke cover throughout the Rhenohercynikum. These structures are interpreted as a thin-skinned thrust complex detaching at a shallow depth. Weber estimates that the basement of the Mid-German high has travelled up to 100 km to the northwest. In the narrow Northern Phyllite zone, between the Rhenohercynikum and the Mid-German

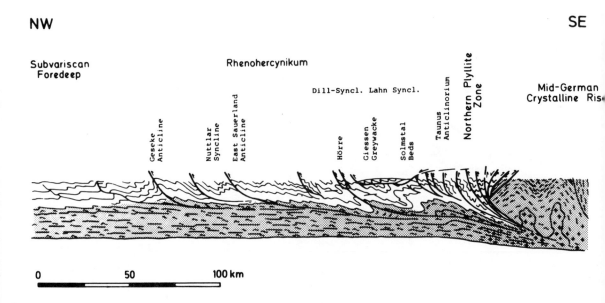

Figure 8.17 Schematic structural profiles across the Variscan belt of Western Europe. (*A*) NW–SE traverse across the Rhenohercynikum (Rhenisches Schiefergebirge) to the Saxothuringian zone. From Weber (1985)

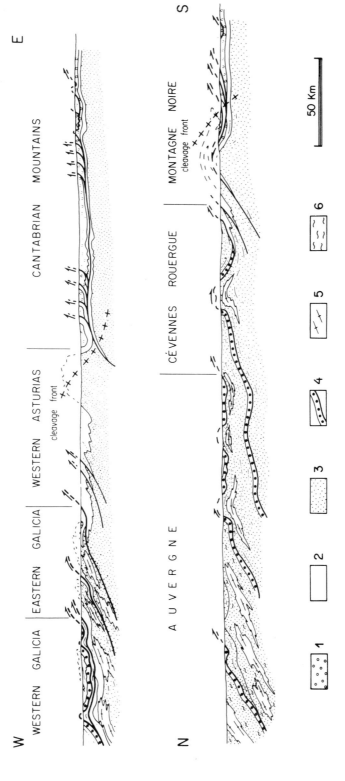

Figure 8.17 Schematic structural profiles across the Variscan belt of Western Europe. (*B*) W–E traverse from western Galicia to the Cantabrian mountains (N. Spain); and N–S traverse across the Massif Central (France). From Matte and Burg (1981)

Crystalline Rise, metamorphic temperatures reached 400–450°C, but elsewhere in the cover of the Rhenohercynikum, temperatures were typically in the range 200–300°C. The age of the NW-directed fold-thrust deformation is late Devonian to early Carboniferous (Bretonic) and there is evidence of a northward progression of the deformation from about 330 Ma in the south to c.300 Ma in the north.

Weber notes that there is no evidence of the development of oceanic crust, or of its subduction, in the German Variscides. He explains the orogeny as an initial phase of intracratonic extension and rifting in the Lower Palaeozoic, followed by intracontinental crustal shortening by A-subduction, or intra-crustal slicing, of the kind suggested in the Himalayas (see 5.4). This convergent deformation continued into the early Carboniferous (Bretonic).

He points out that the structures on the southern side of the West European Variscides verge southwards (see e.g. profile across the Massif Central in Figure 8.17B), giving the belt as a whole a bilateral structural symmetry. A N-dipping subduction zone on the south side of the Variscan belt, along the northern margin of the proto-Tethys Ocean, is considered to be a possible explanation of the structural pattern (see Figure 8.20). However evidence as to the nature of the Variscan structure of the southern part of the belt is difficult to assemble owing to the effects of the Alpine orogeny.

The SW British Isles

The western extension of the Rhenohercynian zone of north Germany (zone 2 of Figure 8.13) occurs in SW England and SW Ireland (Figure 8.16A). To the north is the foreland zone (zone 1) of Dewey and Burke (1973) represented in S. Wales. The rocks of zone 2 show only low-grade metamorphism (up to greenschist facies) and are cut by a major post-tectonic granite pluton of probable Permian age, the Cornubian batholith. The sedimentary sequence involved in the deformation consists mainly of Devonian to late Carboniferous flysch, grading laterally northwards (in the Devonian) into shallow-marine shelf deposits and continental red-beds. Lower Palaeozoic rocks occur in

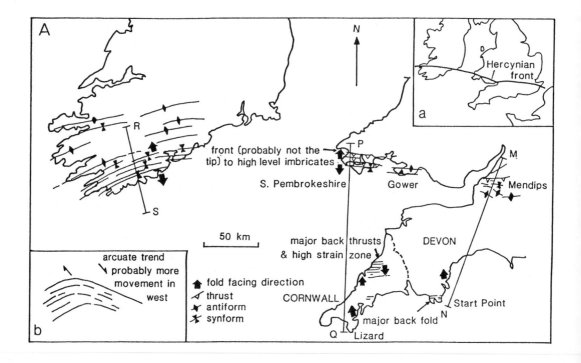

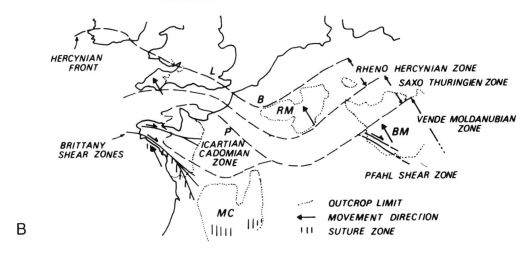

B

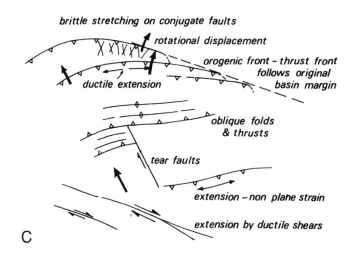

C

Figure 8.18 The external fold-thrust belt of the SW British Isles. (*A*) Tectonic summary map showing the trends of the main folds and thrusts in SW Ireland, S. Wales, the Mendips, and SW England, with the positions of the profiles *M–N*, *P–Q*, and *R–S* of Figure 8.19. After Coward and Smallwood (1985). (*B*) Summary map of the Variscan belt of W. Europe. *B*, Brussels; *L*, London; *P*, Paris; *BM*, Bavarian massif; *RM*, Rhenish massif. (*C*) Schematic diagram showing structures associated with postulated oblique closure: major thrusts, pinned at one lateral tip, suffer some rotational displacement and develop extensional strains along their traces. (*B*), (*C*) from Coward and Smallwood (1985)

anticlinal fold cores in S. Wales, along the margins of the belt. However, there are no large nappes carrying Lower Palaeozoic or basement rocks such as those in the Alleghenian.

Dewey (1982) and Leeder (1982) attribute the formation of the Upper Palaeozoic basin of SW Britain to back-arc crustal extension, related to a subduction zone through southern France, and estimate a stretching factor β of about 2.

The structure of the fold-thrust belt of SW Britain is summarized by Coward and Smallwood (1985). They point out that the belt shows many characteristic features of a thin-skinned foreland thrust belt, but that in many areas the details of the thrust geometry are obscured by the absence of a well-dated layer-cake stratigraphy. In Figure 8.18 the main elements of the regional structure are shown. There are considerable changes both along-strike and from north to south. SW Ireland is dominated by rather upright folds which face south in the south (see balanced section by Cooper *et al.*, 1984, and Figure 8.19). In S. Wales and the Mendips, good layer-cake stratigraphy defines a series of N-verging overfolds and thrusts, and enables balanced sections to be constructed across the belt (see Williams and Chapman, 1985; Hancock *et al.*, 1983).

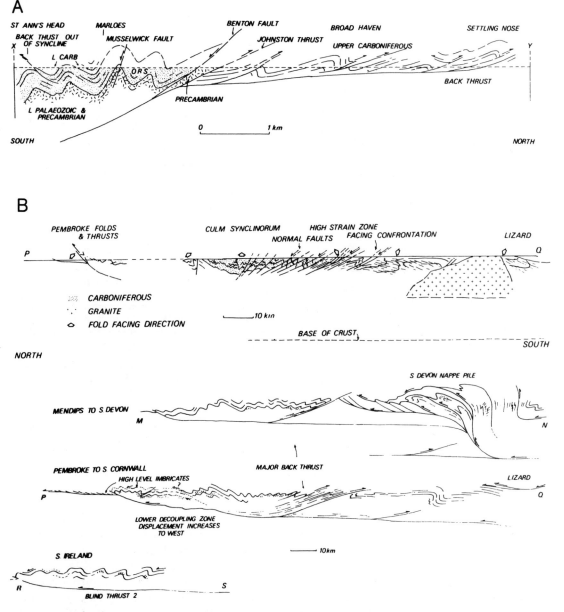

Figure 8.19 Simplified structural profiles across the Variscan belt of the SW British Isles. (*A*) S–N section along the west Pembrokeshire coast. Vertical and horizontal scales equal. (*B*) Synoptic N–S sections: (*a*) from Pembroke to SW Cornwall (*P–Q*); (*b*) the Mendips to S. Devon (*M–N*); (*c*) an interpretation of (*a*); and (*d*) across SW Ireland (*R–S*). From Coward and Smallwood (1985)

Coward and Smallwood show two sections across the south Pembrokeshire sector that illustrate the general nature of the deformation (Figure 8.19). The northern part of the section crosses an imbricate zone developed in Upper Carboniferous strata. These structures detach on a shallow sole thrust dipping gently south. This zone is bounded on its southern side by the Johnston and Ritec thrusts, which define the northern limit of a zone of major folds in

Devonian and Lower Carboniferous cover. Some of the major anticlines expose Lower Palaeozoic strata in their cores, and Precambrian basement is brought up along the Johnston thrust. The structures verge northwards in the northern belt but fan through the vertical in the southern belt to verge south in the southern part of the sector. The authors suggest a rather deeper detachment level for the structures in the southern belt. The folds are considered to have been formed by buckle-shortening and subsequently thrust northwards along the Ritec thrust, which is out of sequence. Restored sections indicate about 45% shortening in the southern belt, and 25% in the northern. The structures in the imbricate northern belt are considered to have formed in piggyback manner, but the major Johnston and Ritect thrusts appear to have climbed up, out of sequence, from a deeper detachment level. The authors suggest that this later movement has folded and uplifted the higher-level detachment, at the base of the Upper Carboniferous, above the erosion surface in the southern belt (Figure 8.19A). The rocks in the south show internal strain and cleavage indicating compression across the strike of the belt, and some extension along it.

The structure of Devon and Cornwall presents a more complex geometrical problem. The Culm synclinorium in the north exhibits a fan-like arrangement of folds, and is separated by a zone of normal faults from a central zone of nappes that face northwards (Figure 8.19B). The southern belt consists of NW-directed thrusts, one of which carries the high-grade metamorphic–igneous complex of the Lizard (see Rattey and Sanderson, 1984). Cleavage throughout the area is well-developed, showing high strains. A major high-strain zone, 25 km wide, marks the Tintagel decoupling zone, for which a displacement of over 20 km is estimated (Shackleton et al., 1982). This zone is considered to mark the detachment below the Culm structures, and to continue beneath the southern belt.

Although the primary direction of thrust transport is to the north, major back-thrusts are recognized along the south side of the Culm synclinorium, and may explain the steep zone affecting the south Devon nappes (Figure 8.19B). Throughout south Devon and Cornwall, the extension lineations indicate a NNW transport direction, approximately normal to the thrust outcrops. The back-thrusts, however, are E–W, oblique to the earlier structures. Coward and Smallwood suggest that the obliquity reflects differential displacement with a dextral sense. They relate this observation to the pattern of structures over the region as a whole (Figure 8.18B) and observe that the slightly arcuate pattern of the fold trend in SW Ireland is normal to the NNW transport direction, whereas the eastern extension of the frontal belt through S. Wales and the Mendips exhibits an E–W to WNW–ESE trend, oblique to the inferred transport direction. This prompts the authors to suggest a bow-shaped displacement with the maximum displacement in SW Ireland, becoming less to the east. This variable displacement model could explain the extension parallel to the fold axes in S. Wales by rotation of the thrust sheet around its lateral tip (Figure 8.18C).

The age of the deformation appears to vary from south to north. Uplift of south Devon and Cornwall occurred in late Devonian to early Carboniferous times, from K–Ar dating of slates. However, stratigraphic evidence points to mid-Carboniferous movements on the major thrust sheets contributing flysch sediments to the Culm basin. Late Carboniferous ages are indicated for the uplift of the Culm synclinorium and the fold-thrust belts of S. Wales and the Mendips. An Upper Carboniferous age is also indicated for the back-thrusting in south Devon. Coward and Smallwood estimate the total shortening across the belt to be about 150 km within a total time of about 90 Ma, giving a rather slow average rate of 0.4 cm/year.

The lack of angular unconformities throughout the main Devonian and Carboniferous outcrops of the region suggests submarine deformation, without the development of major uplifted landmasses. The presence of

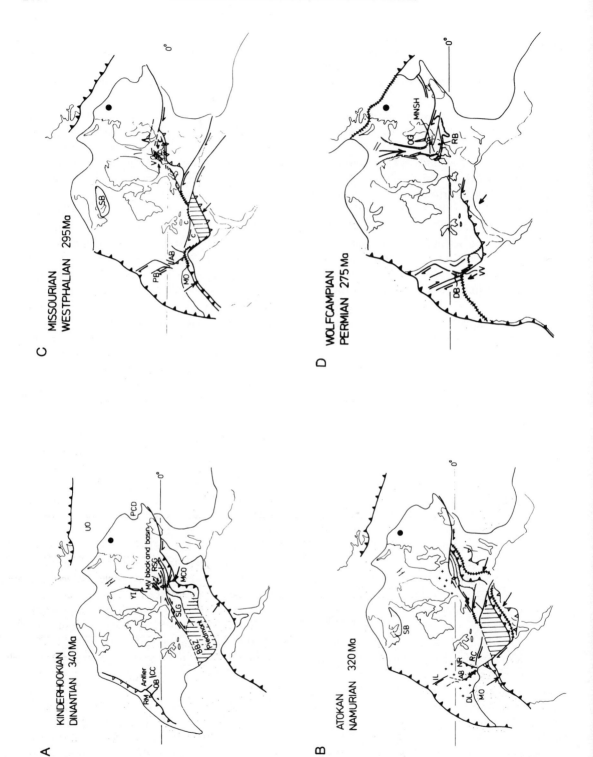

Devonian olistostromes and the Culm flysch deposits, however, suggest submarine tectonic instability related to the migrating thrust sheets.

The structure of the SW British Isles represents a westward continuation, therefore, of the outer thin-skinned Variscan zone, and shows obvious similarities to the Rhenohercynian zone of north Germany. The transport direction there and elsewhere in Western Europe is also to the northwest, suggesting a continuity throughout the Variscides from the Massif Central and Bohemian Massif northwards. The bend in the Variscan front from Ireland to Germany is apparently determined by the limits of the original basin. This shape, markedly oblique to the transport direction of the nappes, coupled with the presence of the less easily deformable Armorican Massif, may be responsible, according to Coward and Smallwood, for the lateral displacement of the main outer compressional zone from SW Britain to the Rhenohercynian, and suggests a strongly transpressive belt between (Figure 8.18B). An important implication of the estimated 150 km shortening across the British sector is that the original continental crust of the southern British Isles must extend beneath

the Armorican Massif, which therefore should be regarded as allochthonous.

Plate-tectonic interpretation of the Atlantic Hercynian region

The evolution of this region is discussed by Dewey (1982), who points out that the best approach to the problem of Upper Palaeozoic plate movements is to work backwards from the relative certainty of the early Permian continental fit, using palaeomagnetically derived movement vectors (Figure 8.20). This exercise confirms previous views expressed by Arthaud and Matte (1977) and others that the West European Variscides are controlled by an important component of dextral strike-slip motion during Upper Carboniferous times. Arthaud and Matte had suggested that the European Variscides formed a belt of complex strike-slip movement joining the northern end of the Alleghenian collision suture with the southern end of the Uralides suture. Many of the complexities of the West European belt can be attributed to the effects of a strike-slip regime: (i) varying and locally strongly-curved structural trends; (ii) rapid changes in metamorphic grade across major steep shear zones; (iii) low-angle thrusts emplacing thin flakes of crustal and locally upper-mantle rocks; and (iv) obducted ophiolites generated in small intracontinental basins. This tectonic pattern contrasts markedly with the linear Alleghenian belt, attributed to relatively straightforward collision normal to its strike. A minimum shortening of 200 km across the belt gives a minimum strike-slip displacement for the West European belt.

According to Dewey's model, in the early Lower Carboniferous (Dinantian), closure of a minor ocean basin produced by late Devonian crustal stretching (see above) led to collision in the Massif Central, and the southwestwards emplacement of nappes (Figure 8.20A, and Figure 8.17). Probably at the same time, collisional shortening occurred in the Piedmont zone of the Alleghenian.

In mid-Carboniferous times, the Sudetic

Figure 8.20 Plate tectonic evolution of the Appalachian–Hercynian system. (A) Early Dinantian tectonic setting: *BBZ*, Brevard-Blue Ridge zone; *CC*, Cedar Creek uplift; *MCO*, Massif Central ocean; *MV*, Midland Valley; *OB*, Oquirrh basin; *PCD*, PreCaspian depression; *RM*, Roberts Mountain thrust complex; *RSG*, Rhenisches Schiefergebirge graben; *SLG*, St Lawrence graben; *UO*, Urals ocean; *YI*, Ymer Island deformation; coarse stipple, oceanic crust; vertical lines, deformation; arrows, movement directions; black circle, average Laurasia–Gondwana rotation pole. (B) Namurian; *AB*, Anadarko basin; *DL*, Dimple limestone; *IL*, Idaho lineament; *MO*, Marathon ocean; *NR*, Nemaha ridge; *RC*, Rough Creek fault zone; *SB*, Sverdrup basin; crosses, area of uplift; other ornament as in (A). (C) Westphalian (Asturic phase): *C*, coal basins of the Hercynian foreland; *PB*, Paradox basin; *V*, mafic volcanism and sill intrusion; other symbols as for (A) and (B). (D) Early Permian; *DB*, Delaware basin; *MNSH*, mid-North Sea high; *OG*, Oslo graben; *RB*, Rotliegendes basin; *VV*, Val Verde basin; light hachured line, limits of Zechstein marine transgression; broad arrow, boreal provenance of Zechstein transgression; other symbols as for (A)–(C). (A)–(D) from Dewey (1982)

phase in Europe is attributed to continued intracrustal shortening of the flysch basins generated north of the collisional uplifts of the previous phase (Figure 8.20*B*). Northward migration of thrust stacking led to flexural crustal depressions on the margins of the foreland, giving rise to the coal basins of S. Wales and the Ruhr. Similar effects of the progressive westwards migration of the thrusting are seen in the Alleghenian belt. The main collisional deformation and uplift appears to have been completed in Westphalian times in Western Europe, during the Asturic phase (Figure 8.20*C*) In late Carboniferous time, and continuing into the Permian, changes in plate vectors related to the collision are expressed in the formation of rifts in the North Atlantic, between Greenland and Norway, and in the Oslo graben. Meanwhile, continued collision took place across the Central Appalachians until well into the Permian period (Figure 8.20*D*).

8.4 The Caledonian orogenic belt of the North Atlantic region

Regional setting

The Caledonides of the British Isles, together with their extension northwards into Scandinavia and Greenland, and southwards into Nova Scotia and New Brunswick, are probably, after the Alps, the most intensively studied and best known of the Phanerozoic orogenic belts. Much of the early work on structural and metamorphic geology took place in the Scottish Highlands, which has experienced periodic invasions by geologists, in order to test out new structural or tectonic ideas, since the early mapping was completed.

The extent of the belt before the opening of the Atlantic is shown in Figure 8.21. It occupies a coastal belt in East Greenland and Western Scandinavia, extending northwards to include Spitzbergen and Franz Josef Land in the Arctic Ocean. Southward, the belt embraces most of the British Isles, and extends to Newfoundland and the Northern Appalachians

in North America. To the east and southeast, the belt is truncated by the younger Variscan belt. Ziegler (1985) summarizes the available evidence from Western and Central Europe, which indicates that a complex system of late Caledonian fold belts occupied much of this region. The eastern margin of the belt in southern Norway crosses beneath younger cover through Denmark and Poland. The southwestern branch of the Caledonian belt is usually known as the Acadian belt, and is characterized by a rather later orogenic climax (mid- to late Devonian). The formation of the main Caledonian belt was completed by the early Devonian.

The present width of the main Caledonian belt, after removing the intervening oceanic crust, is about 1000 km, but this overestimates the Devonian width by probably about 300 km because of early Mesozoic crustal extension along the Atlantic margins. The foundations of the belt are to be seen in the Precambrian shield regions of North America–Greenland to the west, and Fennoscandia to the east. These regions are mainly composed of large Archaean and Early Proterozoic cratons, but are also crossed by linear Mid- and Late Proterozoic orogenic belts that form an important component of the Caledonian basement. The earlier of these belts is the Grenville–Sveconorwegian belt, which ceased to be active about 1000 Ma ago. The Grenville sector lies, parallel to the Acadian sector of the Caledonides, along the southeastern side of the Canadian shield, and crosses the British Isles to join the Sveconorwegian branch in southern Norway and southwest Sweden, where it forms a N–S belt along the western side of the Fennoscandian shield. Most of Scotland appears to be formed from a basement of Grenville age, reworked during the Caledonian orogeny. The later of the Precambrian belts that influenced the Caledonian basement is the Timanides of northern Russia, which form the northeastern boundary to the Fennoscandian shield, and has a possible counterpart in northeast Greenland. This belt remained active until the mid-Cambrian. A belt of similar age in south-

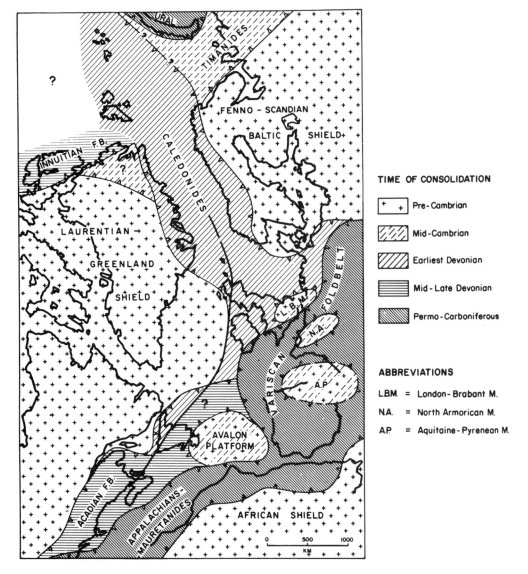

Figure 8.21 Location and extent of the Caledonian orogenic belt of the North Atlantic region prior to Mesozoic opening of the Atlantic ocean, showing its relationship to the older Precambrian shields and to other Palaeozoic orogenic belts. From Ziegler (1985)

ern Britain and Western Europe is known there as the *Cadomian*. Rocks affected by the Cadomian orogeny form a number of isolated cratonic areas within the younger orogenic belts in the south of the region. Of these, the London–Brabant massif and the Avalon platform are important in defining the southeastern margin of Caledonian tectonic activity

in England and Newfoundland respectively.

The relative positions of each side of the Caledonian belt have changed considerably due to large sinistral strike-slip movements (*c.*1800 km according to Dewey and Shackleton, 1986) of mid-Devonian age. When this movement is restored, we find that northern Newfoundland is contiguous to S. Britain, the

Greenland craton faces the Scandinavian Caledonides, and the East Greenland Caledonides face the Arctic Ocean.

We shall now discuss the British Isles and Scandinavian sectors of the belt in some detail, referring briefly to the Appalachian sector as well.

Tectonic subdivision of the British Isles

The principal tectonic units of the Caledonides of Britain are shown in Figure 8.22. They are: (1) the *NW foreland*; (2) the *Northern Highlands*; (3) the *Grampian Highlands*; (4) the *Midland Valley*; (5) the *Southern Uplands*; (6)

the *Lake District*; (7) the *Irish Sea block*; (8) the *Welsh basin*; and (9) the *Midlands platform* (the British part of the London–Brabant massif of Figure 8.21). Most of these zones, north of the Welsh basin, can be traced into Ireland. A fundamental distinction has been recognized for many years between the northern 'metamorphic' Caledonides (Read, 1961) and the southern Caledonides, which exhibit at most the lowest metamorphic grades and are characterized by slates. Following the original suggestion of Wilson (1966) that a suture through the middle of the British Caledonides represented the line of closure of the 'proto-Atlantic' Ocean, Dewey (1969, 1971) presented the first plate tectonic model for the Caledonides of the region. Dewey (1971) highlighted the significance of Read's subdivision, and termed the northern part (zones 2–3) the *orthotectonic belt*, interpreted as part of a North American plate, and the southern part (zones 4–8) the *paratectonic belt*, most of which was interpreted as part of a southern, European, plate. The suture is now universally regarded as lying between zones 5 and 6, through the Solway Firth.

Since the pioneering work of Dewey, many plate tectonic models have been proposed, and the British Isles sector of the Caledonian belt is now regarded as a college of terranes, all displaced to a greater or lesser extent from their original positions relative to the North American craton. Major strike-slip boundaries separate zones 2 and 3 (the *Great Glen Fault*), zones 3 and 4 (the *Highland Boundary Fault*) and zones 4 and 5 (the *Southern Upland Fault*). Strike-slip displacements are also recognized between zones 6 and 7, and zones 7 and 8, and probably exist along the Solway suture as well. Thus all six internal zones are suspect terranes (see 6.2).

Much of the evidence required in any analysis of the tectonic pattern and history of this region comes from the study of the basement in the different zones (Watson and Dunning, 1979). Dewey pointed out the major difference between the gneissose basement of the orthotectonic Caledonides (i.e. the northern plate)

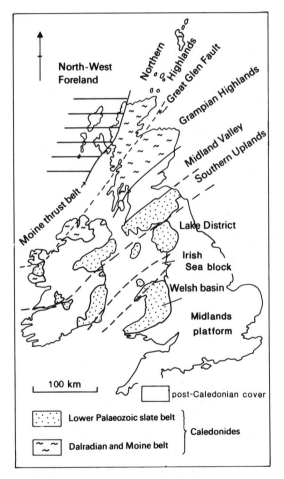

Figure 8.22 Location of the main tectonic zones of the Caledonian orogenic belt in the British Isles.

and the low-grade rocks cut by acid plutonic complexes that formed the basement of the southern plate. These differences correspond to important contrasts in geophysical characteristics of the deeper crust. The deep seismic LISPB profile (Bamford *et al.*, 1977) indicates a layer with high seismic velocities at depths of 6–12 km below zones 2–4. This layer is absent south of the Solway Firth (Figure 8.26*A*). Studies of crustal heat production by Richardson and Oxburgh (1978) suggest that much of the upper continental crust of England and Wales consists of low-grade metamorphic rocks with acid plutons.

There are important differences also in the age of the basement in each case. The last major orogeny to affect the southern basement is the Cadomian, yielding dates ranging from *c.*800 Ma downwards. In the north, the basement is either Lewisian (*c.*1700 Ma BP) or Grenvillian (*c.*1000 Ma). These differences in the nature of the basement are complemented by differences in the age of the cover. The Caledonian orogeny is, by general agreement, post-Grenvillian in age, and in the north involves sedimentary cover at least as old as 800 Ma (the Torridon Group and, possibly, the Grampian Group). In the south, the oldest cover rocks to be affected are usually Cambrian, or very late Precambrian, in age. We shall describe the general stratigraphic and structural features of each of the zones in turn, before discussing their possible plate tectonic context.

The NW foreland (zone 1)

This region, often termed the 'Hebridean craton', is neither a true foreland nor a true craton. It is transected by a major low-angle fault zone (the Outer Hebrides fault) which may have acted as a Caledonian thrust, and another major low-angle structure, the Flannan fault zone, is identified by deep seismic refraction at depth, and projects to the surface on the continental shelf west of the Outer Hebrides. It is likely therefore that all the northern zones are allochthonous with respect to the 'true' foreland of North America–Greenland. However, on either side of the Outer Hebrides fault zone, the rocks of zone 1 are virtually unaffected by Caledonian deformation except near the margin of the Moine thrust zone, where the cover is tilted gently towards the southeast.

The basement consists of the Lewisian complex (see 9.5), which comprises late Archaean crust formed about 2900 Ma BP, and reworked during the Laxfordian orogeny about 1700 Ma BP. The rocks consist predominantly of granulite- or upper amphibolite-facies gneisses. The Lewisian basement is overlain by three distinct units of unmetamorphosed pre-Caledonian sedimentary cover, the Stoer Group (*c.* 1000 Ma old), the Sleat and Torridon Groups (*c.*800 Ma old) and the Cambro-Ordovician sequence. The Stoer, Sleat and Torridon Groups are largely continental, fluviatile redbeds, whereas the Cambro-Ordovician is a thin marine shelf sequence consisting mainly of orthoquartzites and carbonates.

This foreland zone is involved in the major Moine thrust belt, in which a number of distinct nappes or thrust sheets can be recognized, usually with complex internal deformation. However, the typical zone 1 stratigraphy can be recognized throughout this thrust belt except in the uppermost thrust sheet, the Moine nappe. The Moine thrust, which underlies the Moine nappe, marks the western boundary of the Northern Highland zone (zone 2), whose stratigraphic and structural characteristics are quite distinct from those of zone 1.

Structure of the Moine thrust belt

The structure of this classical foreland thrust belt is summarized by Elliott and Johnson (1980) and McClay and Coward (1981). The first mapping and comprehensive description of the belt were carried out by Peach *et al.* (1907) of the UK Geological Survey, and a considerable amount of detailed research has been directed at the belt subsequently. The belt extends along strike for over 190 km (Figure 8.23) and is up to 11 km in width. A

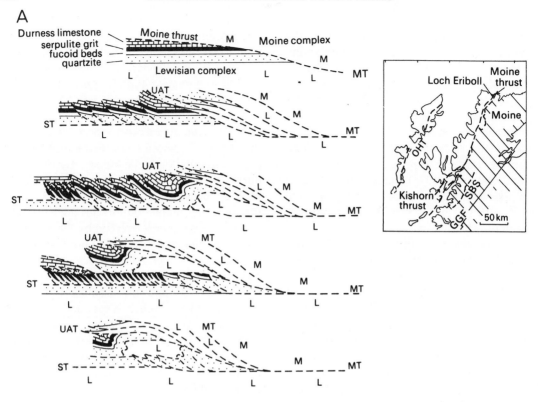

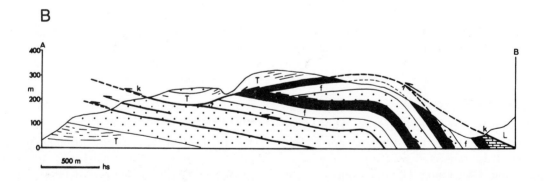

Figure 8.23 Structure of the Moine thrust zone of NW Scotland. (*A*) Location map (inset) and diagrammatic sections across the northern end of the belt at Loch Eriboll, showing stages (1–4) in its evolution. Profile (5) shows a later stage of the central part of (4). Note the propagation of the thrusting northwestwards, in general, (i.e. piggyback) but with some out-of-sequence movements along the Arnaboll thrusts. *MT*, Moine thrust; *OHT*, Outer Hebrides thrust (fault); *ST*, sole thrust; *UAT*, upper Arnaboll thrust; *SBS*, Sgurr Beag slide; *GGF*, Great Glen fault; *M*, Moine complex; *L*, Lewisian complex. After McClay and Coward (1981). (*B*) Structural profile across the Kinlochewe area, towards the southern end of the Moine thrust zone, illustrating out-of-sequence movement of the Kinlochewe thrust (*k*) which cuts downwards across lower thrusts, and is interpreted as extensional, due probably to gravitational gliding. From McClay and Coward (1981)

series of nappes or thrust sheets are recognized, resting on the basal or sole thrust. The nappes appear to be lensoid in character, wedging out laterally to be replaced by other nappes along strike. The uppermost nappe, termed the Moine nappe, resting on the Moine thrust, marks the western limit of the Moine Complex and is discussed below. The nappes are divided into a lower series of imbricate sheets that are essentially thrust duplexes of stacked Cambro-Ordovician cover, around 300 m thick. Above the imbricate nappes are several large nappes carrying Lewisian basement together with Cambro-Ordovician cover. South of Ullapool, Torridonian cover also appears in the higher nappes. These upper nappes are up to 500 m thick and usually deformed internally by minor thrusts and overfolds. The main nappes, from north to south, are the Arnaboll, Glencoul, Ben More, Kinlochewe–Kishorn and Tarskavaig nappes. The nappe zone is widest at Assynt, where the Glencoul and overlying Ben More nappes result in an antiformal bulge of the overlying Moine thrust, and in southeast Skye, where the thick Tarskavaig nappe overlies the Kishorn nappe. Elsewhere, for example between Ullapool and Kinlochewe, large nappes are absent at the surface outcrop.

In Figure 8.23A a series of diagrammatic sections across the thrust belt at Loch Eriboll in the north is shown. The thrust sequence here consists of lower and upper imbricate duplexes carrying only Cambro-Ordovician cover, overlain by the Arnaboll nappe. This nappe is a double duplex structure with internal thrust stacking of rather thicker slices of Lewisian basement and Cambrian cover. The Moine thrust forms the roof thrust to this complex duplex.

The transport direction of the thrust sheets is towards WNW and the thrusts appear to have propagated from east to west, since the upper thrusts are deformed by the lower. However, in some parts of the belt the higher nappes appear to have undergone later reactivation (in some cases involving local extension) that resulted in their cutting across structures in the underlying imbricate nappes. The Kinlochewe nappe, for example, locally cuts down the stratigraphic sequence in the underlying nappe (Figure 8.23B). McClay and Coward suggest that late re-activation took place on some of the upper thrusts due to gravitational sliding. If the Cambro-Orodovician strata are returned to their original horizontal orientation, the Kinlochewe thrust has an appreciable dip towards the foreland. It is possible that several of these higher roof thrusts, deformed into hangingwall antiforms by movements on underlying thrusts, may have become gravitationally unstable and slid towards the foreland, cutting down through the underlying nappes. It is also possible that renewed movements on thrusts at a deeper level caused out-of-sequence movements near the surface.

Displacements on the lower thrusts, estimated by section balancing, vary from 3.5 to 30 km; however, movement on the Moine thrust is much greater: a minimum displacement of 40 km has been estimated by various authors but the actual displacement may be nearer 100 km (Elliott and Johnson, 1980).

The date of the movements has been established at 430 Ma BP (mid-Silurian) from the age of intrusive igneous rocks emplaced during the thrust sequence (Van Breemen et al., 1979). How this relates to the movements in other major thrusts to the south, and to the collisional orogeny as a whole, will be discussed below.

The Northern Highlands terrane (zone 2)

The Northern Highlands extends from the Moine thrust in the west to the Great Glen fault (Figure 8.24) and varies from 30 to 50 km in width. The Caledonian part of this terrane consists of the Moine Complex, together with its interleaved Lewisian basement. The psammites and pelites of the Moine succession represent a thick sequence of fluviatile or deltaic sediments that rest unconformably on the Lewisian basement. The Moine Complex and its Lewisian basement were highly deformed and metamorphosed in the Grenville

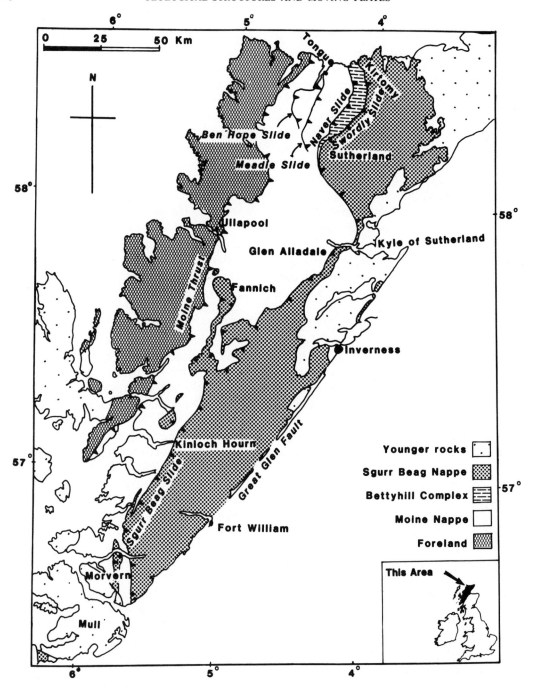

Figure 8.24 Outline tectonic map of the foreland and Northern Highlands zones of the Caledonides in north Scotland. Note the two major nappes, Moine and Sgurr Beag, making up the Northern Highlands. From Winchester (1985)

orogeny at $c.1000$ Ma BP, and were later reworked in the Caledonian orogeny. The original margin of the Grenville belt probably lay a short distance east of the present Moine thrust outcrop, and parallel to it, but has been transported westwards above the Moine thrust, so that the margins of Grenvillian and Caledonian deformation are now almost coincident.

Caledonian effects within the Moine complex are difficult to distinguish from the earlier Grenvillian. Pervasive ductile deformation has appreciably affected the whole terrain, but effects are most intense along several major ductile shear zones or *slides*. The Moine thrust itself is the most westerly of these. It is a deep-level structure, as shown by the development of a wide zone of mylonite (over 600 m thick in places), that possesses a well-developed elongation fabric in the direction of thrust transport. The mylonite sheet is cut by later, brittle, high-level thrusts that have transported the deeper material to the surface. It is considered that the high-level brittle thrusts pass downwards into mylonites and ultimately to broad ductile shear zones, with increasing depth.

The superimposition of Moine complex above Lewisian appears to disobey the usual rule that thrusts place older rocks above younger. The explanation seems to be that the Lewisian–Moine unconformity had been highly deformed in the Grenville orogeny prior to the thrust movements.

The Glenelg area (Barber and May, 1976) provides a link between the complicated structure of the Moine Complex and the rather simple structure of the thrust belt. Immediately east of the Moine thrust in this area, the Lewisian basement and Moine cover are interfolded. However, the degree of deformation in the Lewisian in the western and eastern parts of the area shows a marked contrast. In the Lewisian of the western outcrop, the deformation is similar to that seen in the basement nappes of the thrust zone to the west. In the eastern part of the area, both Lewisian and Moine rocks are highly deformed. Three phases of deformation can be distinguished,

associated with high-grade metamorphism of Grenville age, prior to the development of the mylonite fabric, which is the first Caledonian structure seen here, and is affected in turn by two further phases of Caledonian deformation. These latter three phases are the only deformations seen in the cover west of the Moine thrust.

Further to the east, the situation becomes even more complicated. The grade of Caledonian metamorphism increases to amphibolite facies, and complex interference patterns are formed on both minor and major scales by the superimposition of ductile deformations of Grenville and Caledonian age. In the region between Strathconon and Glen Affric, Tobisch *et al.* (1970) summarize the results of detailed structural mapping over a region, 450 km^2 in extent, of interfolded Lewisian and Moinian rocks. The authors recognize seven distinct sets of folds, of which the last four are Caledonian and the first three probably Grenvillian. The second Caledonian phase (termed the Monar folds by the authors) produced the typical 'Caledonoid' NE–SW folds, with generally steep axial planes, found over wide areas of the Scottish Highlands. The Caledonian folds affect a major shear zone, termed the Sgurr Beag slide (Figure 8.24; Tanner, 1970; Rathbone *et al.*, 1983; Kelley and Powell, 1985) which is a deeper-level counterpart to the Moine thrust zone. This slide extends for the whole length of the Northern Highlands terrane (it is known as the Naver slide in Sutherland) and divides the western part of the Moine Complex (the Morar Division) from the central part (the Glenfinnan Division). The slide consists of a zone of highly deformed rocks in which the strain increases over a distance of about 500 m to a maximum along a narrow zone that usually corresponds to a Moine–Lewisian junction. Linear shape fabrics in the slide zone indicate a WNW transport direction similar to that of the Moine thrust. Although now usually steeply-dipping, the slide is considered to be an originally low-angle displacement plane, equivalent to the Moine thrust but at a mid-crustal level.

Winchester (1985), by matching amphibolites on either side of the slide, estimates a westwards displacement of more than 140 km. Kelley and Powell (1985) conclude that it must have formed at depths greater than 15 km at around 450 Ma BP, during the period of Caledonian metamorphism. They suggest that a period of c.25 Ma may have elapsed between the ductile movements on this slide and the movements along the Moine thrust zone.

The large displacements in both the Moine and Sgurr Beag structures, coupled with the lack of any deep crustal rocks brought up along them, and with the general uniformity of grade across the nappes, suggested to Coward (1980) and Rathbone et al. (1983) that the displacements initially had a very low angle of inclination, perhaps following major horizontal zones of weakness within the middle crust for large distances (see Figure 8.26B). On the other hand, Soper and Barber (1982) argue in favour of a 'thick-skinned' model for the Northern Highlands (Figure 8.25), where both the Moine thrust and the upper slides steepen at depth then become listric and detach at the Moho. This model is similar to the one suggested for the Himalayas (see Figure 5.26). A comparison with the deep seismic reflection profile (Figure 8.26B) suggests that the thrusts in fact detach at a level about 8 km above the Moho. Soper and Barber attribute most of the displacement on the Moine thrust to an early phase of deformation (D1–D2) prior to the peak of Caledonian metamorphism at c.450 Ma BP, and only about 30 km to the late post-metamorphic phase at c.420 Ma seen in the thrust belt.

An interesting geometrical result of this model is the requirement for an exotic 'roof nappe', now completely removed by erosion, to provide the roof of the crustal duplex structure and to bury the Moine Complex to a suitable depth during the period of Caledonian metamorphism. The authors point out that only crustal-scale overthrusting produces the degree of crustal thickening required in major orogenic belts. The sigmoidal nature of the thrust/shear zone displacement surfaces is at-tributed to deep-crustal ductile shortening, and upper-crustal gravity spreading, acting on a plane with an initially even dip of around 30°.

Coward et al. (1983), in a discussion of Soper and Barber's model, point out that the measured displacements on the thin nappes of the thrust zone require a low-angle displacement extending at least 60 km back from the present Moine thrust outcrop. Thus the major ramp must be at least 60 km behind the thrust front.

Soper and Barber, following Mitchell (1978), attribute the Caledonian orogeny of the Northern Highlands to the effects of a collision, during the period 470–450 Ma BP, of the North American continental margin with a subduction zone and island arc system lying to the south. The postulated upper exotic nappe could be an obducted ophiolite nappe derived from the subduction zone as a result of this collision. The later (D4) events at the thrust front (at 420 Ma BP) are attributed to the final continental collision along the Solway suture.

The concept of a missing high-level ophiolite nappe is also a feature of the Dewey and Shackleton (1984) model for the plate tectonic evolution of the whole Caledonian belt (see below). They point out that the Unst ophiolite complex in Shetland may be part of such a nappe and that ophiolite nappes are well known in the Appalachian belt.

The Grampian Highlands terrane (zone 3)

This terrane (50–75 km in width) is separated from the Northern Highlands terrane by the Great Glen Fault, which has long been recognized to have a major sinistral strike-slip displacement, originally estimated at 100 km (Kennedy, 1946) and more recently at 160 km (Winchester, 1973) on the basis of the apparent displacement of metamorphic zones. A much smaller Mesozoic dextral displacement of 30 km has also been recognized (Holgate, 1969). Sinistral displacement of the North American continent in relation to Europe, based on palaeomagnetic data, is much larger, approximately 2000 km (Van der Voo and Scotese, 1981). Much of this displacement,

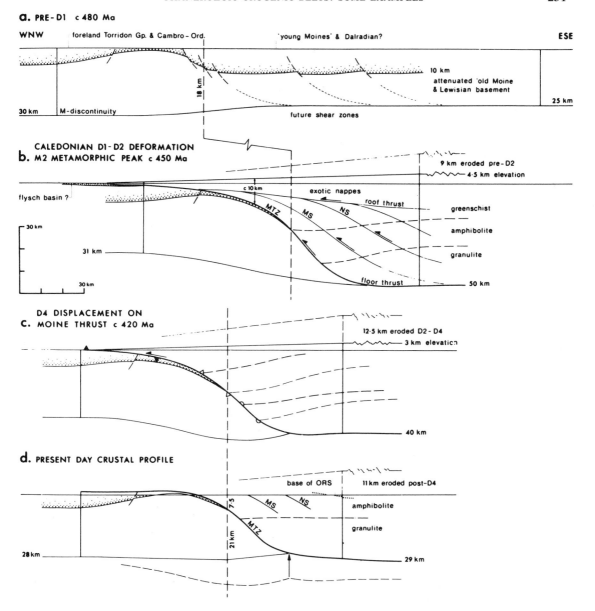

Figure 8.25 Schematic profiles across the Northern Highlands illustrating a thick-thinned model for their tectonic evolution. Note the importance of the postulated exotic roof nappe, now eroded, in the model. From Soper and Barber (1982)

however, may be taken up along other major strike-slip faults, and possibly along a hidden fault near the continental margin.

Consequently, the Grampian Highlands terrane is displaced in relation to the Northern Highlands and, although similarities in strati-graphy and structure exist between them, it is difficult to make direct connections. Reconstructions involving profiles across the Scottish Highlands as a whole should be therefore be viewed with caution. With the Winchester fit, at the D1–D2 period, there is an overlap of

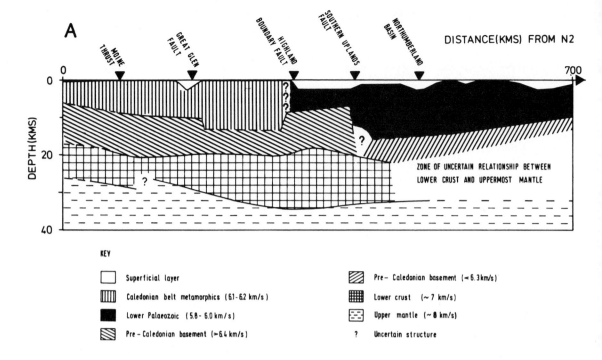

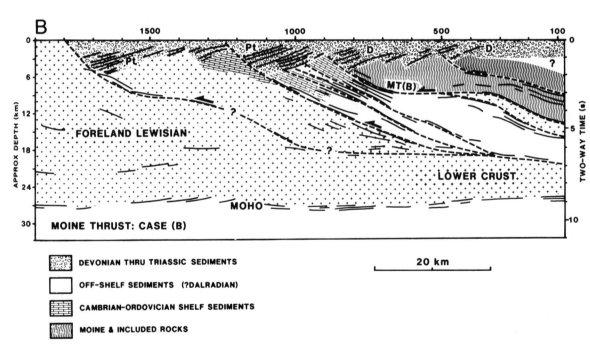

Figure 8.26 (*A*) Model of the crustal structure of northern Britain based on the LISPB deep-seismic reflection survey profile. Seismic P-wave velocities of the different layers are shown in the key. From Bamford *et al.* (1978). (*B*) Possible interpretation of the MOIST deep-seismic reflection profile across northern Scotland. Reflectors indicated by solid lines; faults (dashed) inferred; *D*, Devonian; *Pt*, Permo-Triassic. The depth conversion exaggerates the depths of sedimentary basins by a factor of 1.5–2. *MT(B)* is the preferred position of the Moine thrust, from two possible solutions given in the source reference. Note that the major faults are extensional, but are inferred to be reactivated Caledonian thrusts. From Brewer and Smythe (1984)

only around 50 km of presently-exposed ground across the fault.

The major difference between the Grampian and Northern Highlands lies in the nature of the pre-Caledonian cover. The Grampian Highlands is made up of an extremely thick sequence of late Precambrian to Cambrian sediments forming the Grampian Group and the overlying Dalradian Supergroup. The Grampian Group, often termed the 'young Moines' consists of a very similar sedimentary association to the Pre-Grenville Moine rocks of the Northern Highlands. An inlier of rocks correlated with these older Moine rocks has been identified in the northeastern Grampian Highlands by Piasecki and Van Breemen (1979) who gave them the name 'Central Highland Division'. After restoring the postulated movement on the Great Glen fault, these rocks lie along-strike from the Glenfinnan Division of the Moine complex. They are separated by a major slide, the Grampian slide, from the Grampian–Dalradian cover sequence. The main movement on this slide is dated at c.750 Ma BP in an event widely known as the Morarian (Lambert, 1969). Since this event affected both basement and cover, the deposition of the Grampian Group must predate 750 Ma BP. No major unconformities have been identified in the thick Grampian–Dalradian sequence, which implies that movements on the slide were taking place at depth during continued sedimentation at the surface. This in turn suggests that the Grampian slide, in common with several of the other major displacement zones in the Scottish Highlands, originated as an early extensional fault and was later reactivated in the Caledonian orogeny as a compressional thrust (cf. Soper and Barber, 1982).

Lewisian basement appears at the southwestern end of the Grampian Highlands terrane, in Islay, where it is overlain by Dalradian rocks above the Loch Skerrols thrust, tentatively correlated with the Moine thrust. Plant et al. (1984) suggest that the Moine complex (Grenvillian) basement ends along the NW–SE Cruachan line in the south-west of the terrane, allowing Dalradian sediments to rest directly on Lewisian basement. Several lines of evidence indicate that the nature of the basement of the Grampian Highlands is the same as that of the Northern Highlands. The LISPB seismic profile (Figure 8.26A) shows that the two lower layers recognized by their different seismic velocities can be traced across the Highland Boundary Fault. An analysis of inherited zircons from Caledonian granites in both terranes shows a similar pattern, suggesting a derivation by partial melting of mid-Proterozoic crust in the age range 1800–1000 Ma. The Grampian basement is inferred therefore to be composed of Lewisian crust modified and added to during the Grenville orogeny. A regional geochemical analysis of the Caledonian rocks of zones 1–3 (Plant et al., 1984) brings out a major geochemical discontinuity within the Grampian terrane, following the Moine–Dalradian boundary between the Grampian Group and Dalradian Supergroup. The Torrodinian, Moine, and Grampian Group sediments indicate derivation from a common source dominated by intermediate to acid calc-alkaline rocks of Archaean and Proterozoic age, resembling the present Ketilidian belt of S. Greenland. The geochemistry of the Dalradian sediments in contrast, points to significant contributions from basic to ultrabasic vulcanicity and from Ba, Pb and Zn mineralization.

These geochemical features confirm previous interpretations (e.g. Harris and Pitcher, 1975) suggesting that the Dalradian succession accumulated in a marine, intracontinental basin formed by stretching and rifting of the continental crust. The original margin of the basin, where much thicker Dalradian sequences formed on thinned Moine crust, probably ran obliquely across the present Grampian terrane in such a way that the main Grampian division outcrop represents the edge of the old continent, and the Dalradian outcrop of the SW Highlands and Highland borders represents the basin.

The structure of the Grampian Highlands has occasioned considerable debate over the years. The original interpretation in terms of a series of large nappes was put forward by Bailey (1922). This model has been considerably modified but, in its fundamental aspects, still remains the basis for modern interpretations. A regional synthesis of the structure in the SW Highlands is presented by Roberts and Treagus (1977). The major nappes in this region (Figure 8.27A) are: (i) the *Appin nappe*, resting on the Fort William slide, which marks the contact between the Grampian Division and the overlying Dalradian; and (ii) the *Ballachulish–Tay nappe* which rests on

the Ballachulish slide. The latter slide passes southwards into the Iltay boundary slide that underlies the Tay nappe to the south. In the northeastern part of the Grampian Highlands terrane, these nappes are overlain by the *Banff nappe* which includes a sheet of gneissose basement dated at *c*.700 Ma BP (Stuart *et al.*, 1977). This nappe is regarded as allochthonous, and possibly derived from a southern continental block of Cadomian age. An interesting aspect of the major structure of the region is that the nappes in the northwest face to the northwest, whereas those in the south-

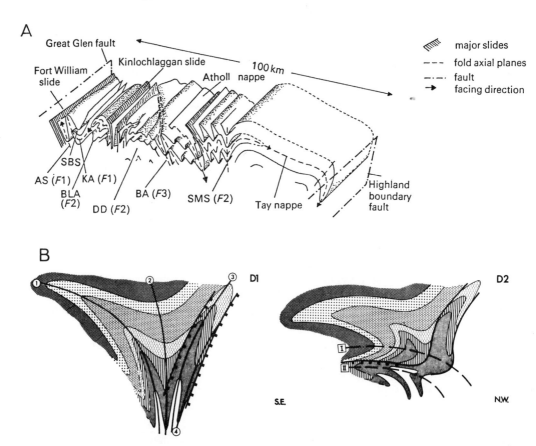

Figure 8.27 Structure of the Grampian Highlands. (*A*) Interpretative model showing the principal major structures of the Grampian Highlands. D1 structures: *AS*, Appin syncline; *KA*, Kinlochleven anticline. D2 structures: *SBS*, Sgurr Beag synform; *BLA*, Ballachulish antiform; *DD*, Drumochter dome; *SMS*, Stob Mhor synform. D3 structure; *BA*, Bohespic antiform. After Thomas (1979). (*B*) Interpretative profiles illustrating a possible two-stage model (D1, D2) for the tectonic evolution of the Grampian Highlands. 1, Aberfoyle–Ardrishaig anticline; 2, Ballachulish–Loch Awe syncline; 3, Kinlochleven–Islay anticline; 4, Appin syncline; I, Ben Lui–Stob Bhan fold; II, Kinlochleven–Ben Chuirn fold; square-toothed lines, major slides. Ornament indicates different subgroups: from bottom to top: Grampian Group (blank), and the Lochaber, Islay, Easdale, Crinan subgroups, and the Southern Highland Group (outermost). From Roberts and Treagus (1977)

east face southeastwards (Figure 8.27*A*). This gave rise to the 'fountain of nappes' hypothesis that visualized the Dalradian cover being squeezed laterally by converging basement blocks on either side and escaping sideways by gravitational flow.

Roberts and Treagus showed that three deformation phases controlled the major structure, and that the D1 phase was associated with major isoclinal folds and with a penetrative fine-grained fabric. Several major recumbent folds, formerly thought to be D1, were shown to be of D2 age. These folds developed during the metamorphic climax. The later upright D3 folds refold the earlier recumbent structures into locally steep attitudes. Their model to explain this structure (Figure 8.27*B*) attributes the originally upright D1 folds to the lateral squeezing effect, and the recumbent D2 folds to the consequential gravitational spreading, followed in D3 by a further horizontal contraction. The major slides appear to have been initiated in D1 times but have clearly also been re-activated in D2. Harris *et al.* (1976) suggest that the southeastwards transport of the large Tay nappe was largely accomplished during the D2 deformation, when about 6 km of horizontal transport took place. Both the Fort William and Ballachulish slides emplace younger strata on older, and are lags rather than thrusts. Like the Grampian slide to the north, these slides may be primarily extensional in origin.

The age of the main D1–D2 deformation overlaps the main Caledonian metamorphism in this region. This is the type area for the 'Grampian orogeny' (Lambert and McKerrow, 1976) which is dated at *c.*490 Ma BP (Arenig) by a variety of methods. Note that there is a progressive change in the apparent age of initial movement on the main slide zones from 490 Ma in this region to 450 Ma at the Sgurr Beag slide, and 420 Ma in the Moine thrust zone.

An alternative to the 'fountain-of-nappes' hypothesis is provided by Dewey and Shackleton (1984), who believe that all the early (D1–D2) structures were NW-facing and that

there is no 'root zone' within the present boundary of the terrane (Figure 8.34). They attribute the NW-directed nappe pile to an obduction orogeny in which the highest nappe is the lost ophiolite nappe of Soper and Barber (1982). The obduction is thought to have taken place by detachment at a ridge-fracture zone interaction (Figure 8.34*C*) producing an overriding young oceanic plate that progressively obducts on to continental shelf and ultimately over the continent itself. This major nappe is represented in the Betts' Cove and Bay of Islands ophiolite complexes in Newfoundland, at Trondhjeim in central Norway, and steep slices are preserved along the faulted southern boundary of the terrane, for example along the Highland Boundary fault. Later SE-facing structures are attributed to 'retrocharriage' or back-folding, resulting from northward subduction commencing at *c.*460 Ma BP. The latter process resulted in uplift of the Scottish Highlands during the period 460–440 Ma BP. The Southern Uplands accretionary prism, or its lateral equivalent, is thought to represent the southern margin of this subduction zone.

The Midland Valley (zone 4)

Major strike-slip displacements have occurred along both the Highland Boundary and Southern Uplands faults that form the north-western and south-eastern margins of the 40 km-wide Midland Valley terrane (Figure 8.22). Sinistral displacements during the Devonian have been established by Bluck (1980), and dextral displacements during the Carboniferous are described by Read (1987).

This zone is therefore a displaced terrane with no direct relationships to the zones to the north and south. The terrane consists of a hidden basement of probable Grenville type (Watson and Dunning, 1979), similar to that underlying the main part of the Grampian Highlands (Figure 8.26*A*). The upper layer of metamorphic Grampian–Dalradian sediments, however, is missing, and in its place is a thick, gently folded sequence of Ordovician to Permian sediments and volcanics. Over most of the zone, the Caledonian history is obscured by the younger Devonian and Carboniferous cover. The absence of an unconformity at the base of the Devonian in the Silurian inlier of Lesmahagow indicates that the end-Silurian collision did not result in deformation of the Lower Palaeozoic cover here. In the southwestern part of the terrane, however, a major unconformity separates the obducted ophiolite sheet of Girvan–Ballantrae from the overlying mid-Ordovician strata.

The Upper Devonian tectonic history of the Midland Valley is discussed by Bluck (1980) who demonstrates that, during this period, it formed a strike-slip extensional graben or pull-apart basin (see 6.1) bounded by sinistral boundary faults on each side. Over 3 km of continental red beds formed in this basin, fed by rising Caledonian mountain belts on both sides. The Upper Devonian sediments rest unconformably on a gently-folded Lower Devonian sequence that accumulated in two basins, separated by a volcanic chain. Bluck considers that the Midland Valley acted as a sinistral strike-slip-controlled graben basin during both the Lower and Upper Devonian, but his detailed account is confined to the Upper.

An analysis of the deformation pattern in the Lower Devonian rocks indicates a preferred orientation of folds in a ENE–WSW direction, 35° clockwise from the orientation of the boundary fault. This is consistent with the relationship predicted for a sinistral strike-slip region (see Figure 6.1).

The Upper Devonian (Upper Old Red Sandstone) sequence begins with alluvial fan deposits and ends with coastal sediments. Palaeocurrent directions are generally eastwards or northeastwards. To explain the disposition of the coarse conglomerate wedges in the alluvial fan deposits, Bluck postulates a series of fault-bounded basins along the NW margin of the main graben, within which each sequence thickens to the northwest (Figure 8.28). As the basement moves north-eastwards with the creation of each new extensional fault, a succession of new basins is formed, filling from

this same source. Each is an asymmetric half-graben, with a wedge-shaped conglomerate deposit thinning southeastwards away from the source. This type of sedimentary basin structure is typical of strike-slip extensional regions (see 6.1, and Reading, 1980).

The Southern Uplands (zone 5)

This terrane is bounded on its northwestern side by the Southern Uplands fault, and on its southeastern side by the hidden Solway suture (Figure 8.29). It has suffered strike-slip displacement in relation to the Midland Valley and the terranes to the northwest. The extent of displacement is not known but may be considerable, as the source areas for the Ordovician sediments in the terrane can not be found in ground presently adjacent to it. Moreover the products of the Ordovician uplift of the Grampian terrane to the north are not found in the Ordovician basin to the south (Bluck and Leake, 1986).

The terrane is approximately 75 km in width, and consists of a thick sequence of highly deformed Ordovician and Silurian strata, resting on a hidden basement thought by Watson and Dunning (1979) to be of Grenville or Lewisian type. A seismic discontinuity at a depth of 12 km beneath the central part of the terrane is interpreted as a major décollement, the Ettrick Valley thrust, which carries the folded and thrust cover over the weak Moffat Shales horizon (Weir, 1974). A major seismic discontinuity dipping northwestwards at an angle of 15–25° has also been detected in deep seismic profiles offshore (Beamish and Smythe, 1986) and is equated with the Solway suture. The deep basement of the Southern Uplands thus underlies this suture and presumably belongs to the southern plate.

The major component of the sedimentary pile is a thick sequence of greywackes of turbidite origin, attributed generally to a trench environment. Associated with the sediments is an older sequence of pelagic sediments with occasional basalts.

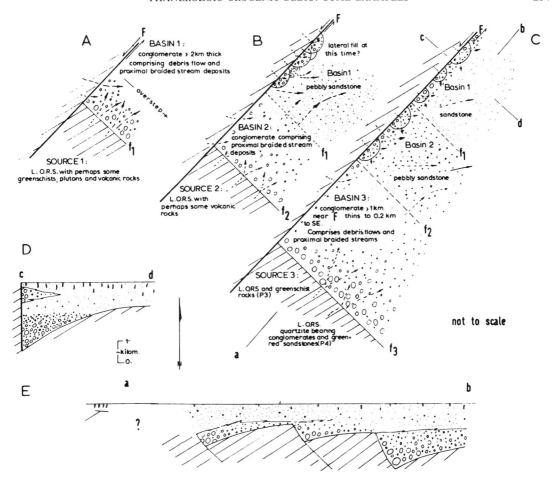

Figure 8.28 Diagrammatic interpretation of the Upper Old Red Sandstone (Upper Devonian) facies distribution in the southwest Midland Valley of Scotland in terms of a sinistral strike-slip model. A, B and C represent the progressive evolution of sedimentary basins as the basement to the Midland Valley moves northeastwards in relation to the block bounded by the Highland Boundary fault on the NW side of the Midland Valley. Successive positions of the SW boundary fault ($f1$, $f2$, $f3$) to the main basin occur progressively southwestwards, each giving rise to a separate source of coarse clastic material. D is profile $c-d$ across the basin and E is profile $a-b$ along the basin (see C). From Bluck (1980)

The structure is dominated by NE–SW-trending, SE-facing, monoclinal folds separated by tracts of unfolded, steeply-dipping beds. Mudstones exhibit generally steep slaty cleavage. The terrane contains ten or more distinct stratigraphic sequences which differ considerably across-strike over distances of a few km, but which can be followed along-strike over distances of 100 km or more. Considerable repetition of stratigraphic sequences takes place across the strike, much of which is attri-

butable to a series of steep reverse faults. Some of these are relatively minor, but others separate blocks with major stratigraphic differences, whose constituent strata were originally formed large distances apart. Within these major fault-bounded blocks, shown in Figure 8.29A, the overall sequence generally youngs towards the northwest, while progressively younger deposits are found towards the southeast.

Leggett *et al.* (1979) interpret the Southern Uplands as a Lower Palaeozoic accretionary

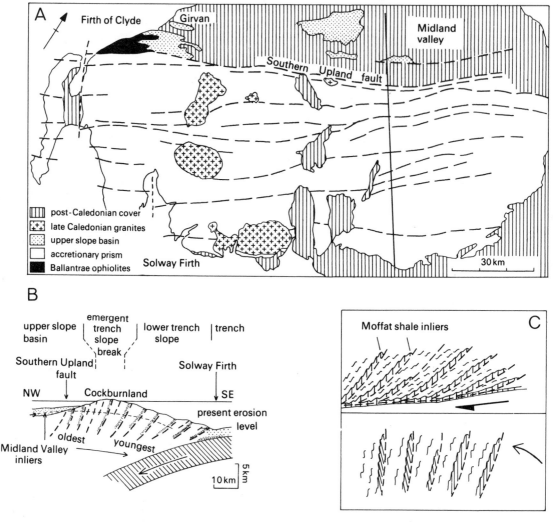

Figure 8.29 Structure of the Southern Uplands of Scotland. (*A*) Tectonic summary map of the Southern Uplands showing the positions of the main faults bounding the different blocks. (*B*) Diagrammatic profile across the area to illustrate the accretionary prism model. (*A*), (*B*), after Leggett *et al.* (1979). (*C*) Diagram to show how the emplacement of successive underthrust wedges, using the Moffat shales as a detachment horizon, can result in steepening and back-rotation of the earlier-formed thrust blocks. After Eales (1979).

prism, formed at a subduction zone on the northwest side of the Iapetus Ocean, along the Laurentian continental margin.

The fault-bounded blocks are interpreted as packets of sediment, offscraped from the downgoing plate, and stacked up and steepened in the fore-arc region to form an eventually emergent ridge at the trench–slope break (Figure 8.29*B*). This ridge corresponds to the conjectural 'Cockburnland' long recognized by stratigraphers as a Silurian landmass in the northern part of the terrane. This tectonic environment is reminiscent of the Barbados ridge complex discussed in 5.2. Accretion is considered to have commenced during Llandeilo time (mid-Ordovician) and continued until the end of the Wenlockian, a period of about 45 Ma.

The south side of the Solway suture (zones 6–8)

A major intracontinental suture has long been postulated within the Caledonides of the British Isles because of differences between the Cambro-Ordovician faunas of Girvan and the NW Highlands on the one hand, which exhibit North American affinities, and those of Wales on the other, which exhibit European affinities. The suture is therefore held to represent the site of closure of a vanished Lower Palaeozoic ocean, part of the Iapetus Ocean.

The suture can be detected by major differences in basement characteristics on either side of the Solway line, and can be traced through Ireland along the Navan–Shannon fault (Watson and Dunning, 1979).

South of the suture, four Caledonian tectonic units are recognized, the *Lake District zone*, the *Irish Sea block*, the *Welsh basin*, and the *Midlands platform*. The Lake District–Isle of Man–Leinster belt is generally regarded as the site of an early Ordovician volcanic arc on the northern margin of a southern continental plate (Figure 8.32) (see e.g. Moseley, 1977). South of this zone, the Irish Sea block exposes pre-Caledonian Cadomian basement in Anglesey, the Lleyn peninsula in N. Wales, and in SE Ireland. This basement is part of the Cadomian orogenic belt, subjected to deformation and metamorphism prior to the Lower Ordovician. The age of this orogeny in Anglesey, and the stratigraphic relationship between the Irish Sea block and the zones on either side, is still controversial. All three zones are suspect terranes with possible strike-slip displacements between them.

The *Welsh basin* has long been regarded as a classical geosyncline, following Jones (1938). It contains a thick sequence (*c.*10 km) of Lower Palaeozoic sediments, including a high proportion of turbidites. Volcanic rocks form an important component of the basin, particularly in the Ordovician of north and southwest Wales. Deformation is highly variable, ranging from tight folds associated with highly-strained, cleaved slates to rather gentle flexures with no associated cleavage. Rocks from deeper parts of the basin exhibit prehnite-pumpellyite metamorphic facies.

The Welsh basin has been widely interpreted as a back-arc extensional basin, formed on thinned continental crust, and related to a southward-dipping subduction zone (see Figure 8.32). Okada and Smith (1980) point out, however, that a better tectonic analogy for the Silurian is a fore-arc basin of the type presently forming landward of the Nankai trough off SW Japan. A similar basin was described west of the Barbados ridge, in the Lesser Antilles subduction zone of the Caribbean (see Figure 5.13). In the modern Japanese analogy, the Japanese islands correspond with the Midlands platform, and the Nankai trough subduction zone is represented by a postulated trench northwest of Anglesey. Okada and Smith believe that the subduction zone may have migrated southeastwards towards the Midlands platform, due to tectonic erosion, from its position in the Ordovician when the North Wales volcanic rocks were formed. The Irish Sea block, forming the trench–slope break in early Silurian times, became emergent in the later Silurian, thus isolating the basin. The main Caledonian deformation of the basin, like that of all the zones up to and including the Southern Uplands, is end-Silurian in age and is attributed to collision between the Midlands platform and the Laurentian continent.

Woodcock (1984) has demonstrated the importance of dextral strike-slip movements along major lineaments running parallel to the south-eastern margin of the Welsh basin. Along one of these features, the Pontesford lineament, Woodcock demonstrates patterns of branching faults in basement inliers and localized belts of compressional folds in the cover characteristic of a strike-slip regime (Figure 8.30).

The *Midlands platform* is not, as once thought, the northern margin of a large European continent in Lower Palaeozoic times, but a comparatively small region of undeformed Lower Palaeozoic strata resting on late Precambrian volcanic and sedimentary rocks that have been affected by the Cadomian orogeny.

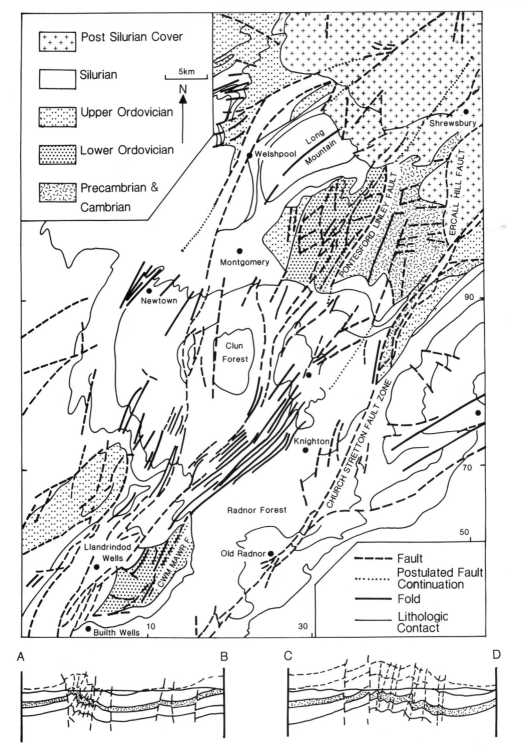

Figure 8.30 Map summarizing important Caledonian structures of mid-Wales and the Welsh Borders. Folds, continuous thick lines; faults, dashed lines. Note: (1) the major Cwm Mawr– Pontesford–Linley and Church Stretton fault zones; (2) the concentration of faults, and particularly folds along a relatively narrow zone between Llandrindod Wells and Clun; (3) occasional oblique relationships between fold axes and the major faults. *A–B* and *C–D* are cross-sections across the central part of the lineament, showing positive flower structures (see Figure 6.4). After Woodcock (1984).

The late Precambrian sequence rests in turn on an earlier Precambrian basement, of which only a few fragments are preserved, but whose geophysical characteristics differ considerably from those of the Laurentian basement north of the suture (Watson and Dunning, 1979). Similar rocks form the Avalon terrane that constitutes the southwestern margin of the Appalachian belt from Newfoundland to eastern Massachusetts (Figure 8.21). Our knowledge of the nature of this southern craton in Caledonian times is very limited, but it is clear that, in contrast to the two large continental masses of Laurentia and Baltica, the Midlands platform was part of a relatively minor block. The end-Silurian collision gave rise to relatively mild deformation with little shortening in the southern part of the British sector, quite unlike the complex thrust belts of Scandinavia and NW Scotland.

The Scandinavian Caledonides

The relationship between the Scandinavian and British Caledonides is shown in Figures 8.21 and 8.33. Although the Scandinavian belt appears to be a direct continuation along-strike of the Scottish Highlands, separated by only 500 km across the North Sea, there are considerable differences in structure. The Scandinavian Caledonides consist of an 1800 km-long belt, with a present outcrop width of up to 300 km, of nappes directed eastwards on to the Fennoscandian shield. The polarity of the belt is thus opposed to that of the NW-directed nappes of zones 1–3 in the British sector. The eastern margin of the belt curves through Denmark towards the Polish Caledonides and there is no direct connection with the southeastern part of the British Caledonides.

The allochthonous nappes of the Scandinavian Caledonides (Figure 8.31) contain stratigraphic successions extending up to the early Silurian, and the main Caledonian orogeny (the *Scandian*) is dated as mid-Silurian to Early Devonian. There is evidence also, in certain districts, of an early-Caledonian deformation (the *Finnmarkian*) of Late Cambrian to early Ordovician age (540–490 Ma BP).

The nappe sequence, summarized by Roberts and Gee (1985), is divided into four separate complexes: the Lower, Middle, Upper and Uppermost Allochthons. The *Lower Allochthon* is composed mainly of a sequence of late Precambrian to early Palaeozoic sediments that have been involved in thin-skinned thrust deformation in a zone along the entire eastern margin of the belt, bounded in the east by a sole thrust. Further west, Precambrian crystalline basement is also involved in nappes of the Lower Allochthon. This sequence is overlain by the *Middle Allochthon*, which is composed mainly of highly-deformed Precambrian crystalline rocks, together with unfossiliferous, probably late Precambrian, psammites. The rocks of this unit are extensively mylonitized, exhibit greater and more penetrative deformation than the underlying unit, and are metamorphosed in middle- to upper-greenschist facies. The Middle Allochthon contains a suite of late Precambrian pre-tectonic basic dykes and an extensive syntectonic mafic/ultramafic to alkaline igneous complex of late Cambrian to early Ordovician age.

The *Upper Allochthon* contains deformed volcano-sedimentary successions derived from a variety of tectonic environments: island arc, fore-arc and back-arc basin, and oceanic. The nappes containing these sequences have been transported several hundred km eastwards. Two deformations can be recognized, an early Ordovician phase that has been attributed to the obduction of the ophiolite assemblages, and a later Siluro-Devonian event. Metamorphism during both phases reached upper amphibolite facies. This unit may thus correspond with the missing ophiolite nappe of the Scottish Highlands.

The *Uppermost Allochthon* consists of a series of nappes comprising migmatitic gneisses and high-grade metasediments, together with units of lower-grade supracrustal rocks. These rocks are intruded by synorogenic granitoid and gabbro plutons. Some of the gneisses in these nappes appear to represent pre-Caledonian basement with the Precambrian supracrustal cover. Several thrust slices contain ophiolite fragments and associated low-grade metasediments.

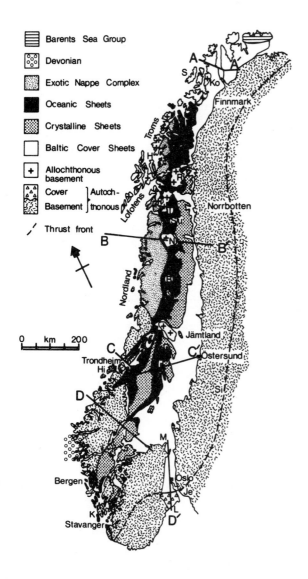

Figure 8.31 Structural summary of the Scandinavian Caledonides. (*A*) Map showing principal tectonic units: external units comprise the autochthonous basement and cover and the external crystalline nappes; internal units comprise the oceanic and exotic nappe complexes.

The later stages in the evolution of the belt were marked by regional uplift of the western part of the orogenic belt, leading to the accumulation of early to mid-Devonian molasse deposits in fault-controlled extensional basins.

Ramsay *et al.* (1985) estimate that a total shortening of *c.*400 km may have taken place across the belt in northern Norway, although the displacement on the lowermost nappe diminishes to the northeast. In the northern section a clear distinction can be made into two groups of nappes: an earlier *Finnmarkian* nappe complex and a later *Scandian* nappe complex, each containing distinctive sedimentary sequences — late Precambrian to Cambrian, and Ordovician to Silurian respectively.

The Finnmarkian orogeny commenced during the Upper Cambrian in the interior of the belt and progressed towards the craton, ending in early Ordovician times. The Scandian orogeny created a new group of nappes that in places overrode the Finnmarkian nappes, al-

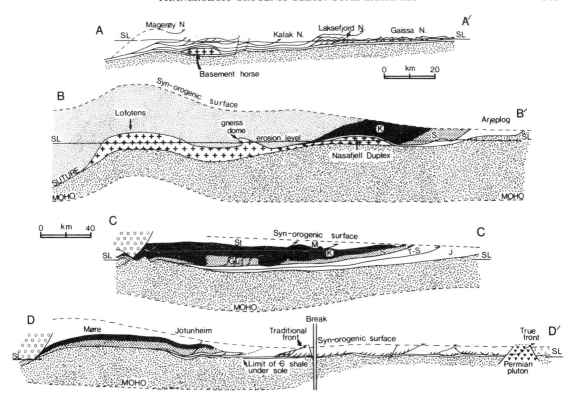

Figure 8.31 Structural summary of the Scandinavian Caledonides. (*B*) Schematic profiles across the lines marked in (*A*). Note that *A–A'* is at a different scale to the others. Vertical scale equals horizontal. From Hossack and Cooper (1986), after various sources.

ready deeply eroded. In addition, Finnmarkian thrusts were locally reactivated. Hossack and Cooper (1986) divide the nappe complex into two zones (Figure 8.31*A*): an external zone of thrust sheets that have been emplaced southeastwards onto the Fennoscandian craton, and an internal zone of exotic nappes with NE–SW stretching directions, parallel to the strike of the belt. These two sets of nappes must therefore have different emplacement histories. The external zone comprises the Lower, Middle and Upper Allochthons of Roberts and Gee, and the internal zone corresponds to their Uppermost Allochthon.

Hossack and Cooper claim that the pre-erosion thrust front lay much further to the east than the present outcrop, based on its position in the Oslo Graben (Figure 8.31*A*). Thus the width of the thrust belt in the cover

sheets of the Lower Allochthon is very much wider (up to nearly 300 km) than is apparent at outcrop. The width of this zone expands in the north to cover the whole exposed width of the belt (see section *AA'* in Figure 8.31*B*). The exposed thrust front, according to Hossack and Cooper, corresponds to the position of a series of frontal ramps where the sole thrust cuts down from the Cambrian black shales into the late Precambrian sequence.

The crystalline basement nappes of the Middle Allochthon, like the cover of the Lower Allochthon, are thought to have been derived from the Fennoscandian shield. A restored section of a profile in the south through Oslo (section *DD'* in Figure 8.31*B*) indicates that the sole thrust ramps down through crystalline basement beneath the Jotun nappe, about 475 km west of the present thrust front, but east of the autochthonous basement outcrop of SW Norway.

The 'oceanic' nappes of the Upper Allochthon cover a wide surface area in central and northern Norway (Figure 8.31A) and include the well-known Seve and Koli nappes. Sediments within these nappes contain faunas with Baltic affinities but a higher oceanic nappe contains material with North American affinities (Gee, 1975). The geochemistry of the volcanic rocks is consistent with an ocean-floor origin (Furnes et al., 1982). These nappes are therefore considered to represent obducted slices of oceanic crust originating on both sides of the Iapetus Ocean. Since the major deformation of the ophiolites of the Upper Allochthon is Finnmarkian in age, the authors date the obduction of the oceanic material as Lower to Middle Ordovician.

The exotic nappes of the internal zone, or Uppermost Allochthon, form large outcrops along the coastal belt of central and northern Norway (Figure 8.31A). They overlie the Lofoten basement complex, which Hossack and Cooper believe to be allochthonous and part of the Middle Allochthon (see Figure 8.31B, section BB'). Hossack and Cooper point out that these internal nappes must have been derived either from a micro-continent within the Iapetus Ocean separating Baltica from the Laurentian continent, or from the latter continent itself, since they overlie the oceanic material of the Upper Allochthon.

The highly deformed rocks of these nappes exhibit NE–SW-oriented sheath folds and elongation lineations indicating either emplacement parallel to the strike of the orogen, or possibly oblique southward emplacement in a transpressional regime. The main deformation of the Uppermost, Upper and Middle Allochthons is regarded as a Finnmarkian event, rather similar to the Grampian orogeny in Scotland.

The Scandian or end-Caledonian orogeny produced the first deformation in the thrust sheets of the Lower Allochthon, but can also be recognized in the higher nappes, suggesting that their emplacement was in part a Scandian event. Some reactivation of the earlier thrusts is indicated by a metamorphic overprint of c.420 Ma BP found in certain of the upper nappes. The dates of the Scandian event are diachronous across the orogen from c.450 Ma BP in the central part of the belt to c.420 Ma at the thrust front, indicating a movement rate of c.2.8 cm/year.

Hossack and Cooper propose a plate tectonic model for the evolution of the Scandinavian Caledonides that explains the Finnmarkian event in terms of the obduction of a slab of oceanic crust containing an island arc; and the Scandian or end-Caledonian event as a collision of Laurentia with Baltica during which the latter underthrust the former (see Figure 8.34B).

Plate tectonic interpretation of the North Atlantic Caledonides

The first plate tectonic interpretation of this region was made by Dewey (1969). He describes a model (Figure 8.32) involving a NW-dipping subduction zone below the Grampian Highlands in the British sector, and two SE-dipping subduction zones, one below the Irish Sea block in Cambrian to Lower Ordovician times, and a later one below the Lake District in Upper Ordovician times, with continent–continent collision taking place during the late Silurian. Many subsequent refinements and alternatives have been suggested, but the tectonic framework suggested by Dewey is still the basis of most modern views.

The evolution of the belt may be said to commence with the break-up of a late Precambrian continent that is documented by palaeomagnetic evidence (Piper, 1985) and by the diversification of faunas in the early Cambrian. Differences in early Cambrian shelly faunas between Laurentia and Baltica are well known. McMenamin (1982) presents evidence that this faunal separation commenced in the late Precambrian with certain benthic 'Ediacaran' soft-bodied faunas in the period 650–600 Ma BP. Evidence for widespread rifting and intracontinental extension preceding this

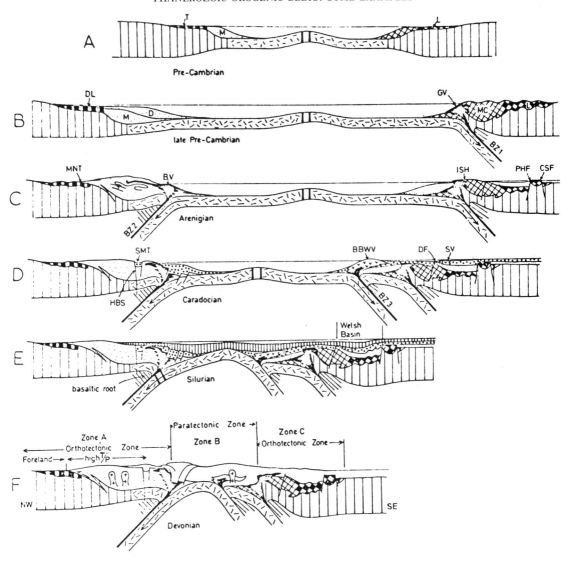

Figure 8.32 Schematic sections illustrating the Dewey (1969) model for the tectonic evolution of the British Caledonides. *DL*, Durness limestone; *M*, Moine complex; *D*, Dalradian sediments; *GV*, Girvan; *MNC*, Moine thrust; *BV*, Ballantrae volcanics; *MC*, Mona complex; *ISH*, Irish Sea block; *PHF*, Pontesford Hill fault; *CSF*, Church Stretton fault; *HBS*, Highland Border Series; *SMT*, South Mayo trough; *BBWV*, Borrowdale volcanics; *DF*, Dinorwic fault; *SV*, Strokestown volcanics; *BZ1, 2, 3*, successive subduction zones. From Dewey (1969)

continental separation is provided by the development of fault-controlled Torridonian and Grampian–Dalradian basins in Scotland around 800 Ma ago.

The Iapetus Ocean is widely regarded as having opened initially in the early Cambrian (Anderton, 1980) at a time of dramatic change in the nature of the Dalradian basin. After the deposition of the Port Askaig tillite, a product of the Vendian glaciation of *c.*640 Ma BP, rapid subsidence took place along growth faults, producing turbidite-filled basins, succeeded by the extrusion of basic volcanics and by further subsidence and turbidite sedimentation. By

Lower Cambrian times, both faunal and palaeo-magnetic evidence indicate an appreciable separation of Laurentian, Baltic and southern continental masses, with an intervening Iapetus Ocean. The nature of the southern continental mass is unclear. The British part, termed the Midlands platform, is usually regarded as the lateral equivalent of the Avalonian terrane in the Appalachians. The faunas of this continental mass show differences from both Laurentia and Baltica, and Avalonia probably represents a large island or series of islands separated from Baltica by oceanic crust.

In a revised model, Dewey (1982) demonstrates the importance of oblique plate convergence during Ordovician and Silurian times, a possibility explored earlier by Phillips *et al.* (1976). Dewey's model involves a large dextral strike-slip component of relative movement along the British sector of the belt during the life of the subduction zones. This relative motion changed after end-Silurian collision to a sinistral strike-slip regime which produced around 1000 km of relative motion along the Appalachian–British sector of the belt.

Dewey and Shackleton (1984) present a further revised tectonic model (Figures 8.33, 8.34) in which the early Caledonian event, termed the *Grampian* in Scotland and Ireland, the *Finnmarkian* in Scandinavia, the *Humberian* in Newfoundland, and the *Taconic* in the

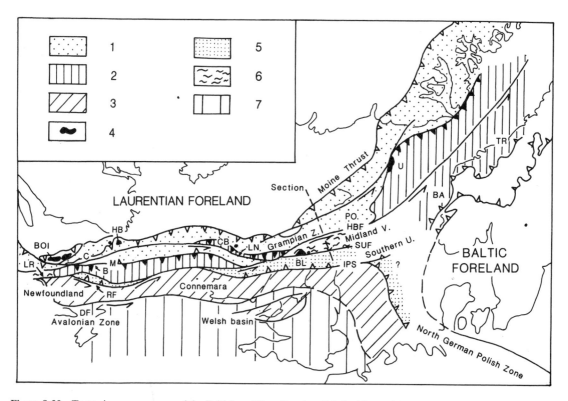

Figure 8.33 Tectonic summary map of the British and Scandinavian Caledonides and Northern Appalachians, restored to the pre-Mesozoic fit of Bullard *et al.* (1965): 1, infraCambrian to early Ordovician sequences in Grampian zone; 2, Ordovician arc terranes; 3, oceanic and arc terranes south of the suture; 4, ophiolite complexes; 5, Ordovician-Silurian accretionary prisms; 6, Grenville basement within the orogen; 7, Avalonian-Cadomian terrane on the south side of the orogen. *BOI*, Bay of Islands; *LR*, Long Range; *HB*, Hare Bay ophiolite; *M*, Mings Bight ophiolite; *B*, Bett's Cove ophiolite; *RF*, Reach fault; *DF*, Dover fault; *CB*, Clew Bay ophiolite; *LN*, Loch Nafooey fault zone; *BL*, Ballantrae ophiolite; *PO*, Portsoy; *SUF*, Southern Uplands fault; *IPS*, Iapetus suture; *U*, Unst ophiolite; *BA*, Bergen arcs; *TR*, Trondheim. After Dewey and Shackleton (1984).

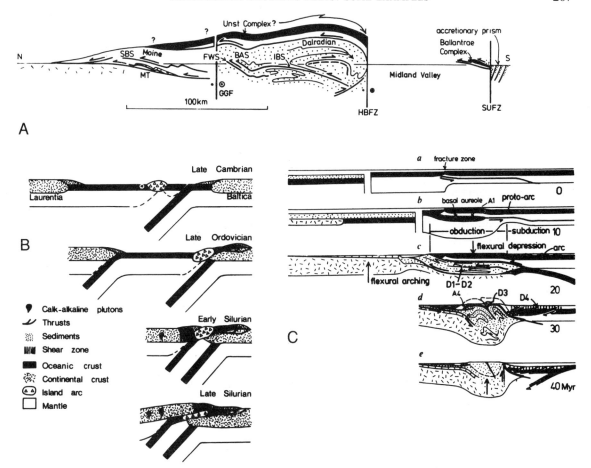

Figure 8.34 (*A*) Schematic N–S structural profile along line indicated in Figure 8.33. *MT*, Moine thrust; *SBS*, Sgurr Beag slide; *FWS*, Fort William slide; *BAS*, Ballachulish slide; *IBS*, Iltay boundary slide; *GGF*, Great Glen fault; *HBFZ*, Highland Boundary fault zone; *SUFZ*, Southern Uplands fault zone. From Dewey and Suhackleton (1984). (*B*) Sequential cartoons to illustrate a tectonic interpretation of the Scandinavian Caledonides. From Hossack and Cooper (1986). (*C*) Approximately true-scale schematic sections illustrating a model for the evolution of the Grampian orogeny. An ophiolite is overthrust onto oceanic lithosphere at a fracture zone and thereafter progressively obducted onto the continental rise and shelf, producing the D1–2 deformations in the Grampian Highlands (sections *a–c*). Section (*d*) shows shortening and thickening of the sedimentary pile, leading eventually to a reversal of subduction polarity (*e*). From Dewey and Shackleton (1984)

Appalachians of New York and New England, is attributed to the obduction of a giant ophiolite nappe, up to 15 km thick, generated at a ridge-transform intersection in the Ordovician Iapetus Ocean. This event appears to have been diachronous, taking place in pre-Llandeilian times in Britain, before the end of the Llandeilo period in Newfoundland, and in mid-Caradoc time in New York and New England. In Newfoundland, progressive dia-

chronous deformation can be established, commencing at about 495 Ma BP in the southeast, and ending at 455 Ma BP in the northwest.

The obducted ophiolite complexes were formed in early Ordovician time and only a short period elapsed before their obduction. The earlier stages of this process would have been submarine (Figure 8.34C). Emergence would only have taken place when the stacked crust reached thicknesses of *c*.30 km. During the same period, a volcanic arc, resting on an oceanic foundation, lay to the southeast of the obducted portion of the ophiolite, and is represented in numerous places along the belt, including the Midland Valley terrane of Britain and the Notre Dame terrane in Newfoundland. The collision of this arc terrane with Laurentia may have ended the obduction. Dewey and Shackleton suggest that the continued convergent motion caused firstly (i) the northwestwards obduction, then (ii) south-eastwards subduction, forming the volcanic arc, then (iii) collision of the arc with Laurentia, followed by (iv) northwestwards subduction in the mid-Ordovician to produce the Southern Uplands accretionary prism. The latter, Llandeilian, event is held to explain the late S-facing structures in the Grampian Highlands referred to earlier, and is correlated with the uplift of that terrane.

After this diachronous orogenic phase, during which convergent motion was oblique with a dextral strike-slip component, the direction of plate convergence appears to have changed. Evidence for large sinistral displacements have already been discussed, for example the 160 km sinistral displacement of Grampian metamorphic zones along the Great Glen Fault. Dewey and Shackleton suggest that a total sinistral displacement of 1800 km occurred between the Caradocian and the early Devonian.

Oceanic closure and continental collision appear to have been diachronous. Collision of Laurentia and Baltica took place during the Silurian period, and was followed by end-Silurian to mid-Devonian collision of Avalonia with the consolidated Laurentian–Baltic continent. The authors suggest that it was the latter collision event which induced the sinistral strike-slip motion between Laurentia and Baltica, and between Laurentia and Avalonia. Of the 1800 km strike-slip displacement between Laurentia and Avalonia, they estimate that 500 km may have been taken up by convergence across the north German–Polish branch of the Caledonides and the remaining 1300 km in the displacement of Scandinavia relative to North America (see Figure 8.33).

9 Orogeny in the Precambrian

As we have seen in the previous chapter, it becomes progressively more difficult to interpret the structure of orogenic belts as we look further back in time, beyond the period when accurate continental reconstructions can be made from oceanic palaeomagnetic data. This difficulty is magnified in the Precambrian because of the inherent inaccuracies of continental palaeomagnetic reconstructions, and is compounded by other problems. The familiar stratigraphic certainties conveyed by accurate palaeontological dating in the Phanerozoic are absent throughout much of the Precambrian. The widely-used radiometric dating methods cannot yet give the same precision in subdividing and correlating stratigraphic sequences. The nature of the rocks is different: much of the Precambrian outcrop consists of uplifted deep-crustal material, often highly-deformed and metamorphosed, whose original geometrical relationships to adjoining outcrops and regions cannot be accurately reconstructed.

Probably the most important problem, and certainly the most stimulating one, in interpreting Precambrian rocks is the uncertainty over the extent to which present-day processes can be regarded as useful analogues in the Precambrian. Although it is a fundamental tenet of geology that 'the present is the key to the past', we find difficulty in unlocking some of the older doors!

Precambrian chronology

In this chapter we shall discuss examples of Precambrian orogenesis from the four main periods of Precambrian time: the *Archaean*, and the *Early*, *Middle*, and *Late Proterozoic* (Table 9.1). The basis of the stratigraphic subdivision of the Precambrian is chronological: arbitrary dates are chosen to represent the boundaries between the subdivisions, which may correspond to major stratigraphic breaks in some areas but not in others. The size of the time divisions is much larger than in the

Table 9.1 Chronological subdivision of the Precambrian.

Eon	Era	Canadian scheme	Ma
Proterozoic	Late	Hadrynian	
			1000
	Middle	Helikian	
			1600
	Early	Aphebian	
			2500
Archaean	Late		
			2900
	Middle		
			3400
	Early		
			4000 ?

Phanerozoic. Thus the major subdivisions, Proterozoic and Archaean, have the status of *eons*, like the Phanerozoic, and the boundary between them is at 2500 Ma BP. The subdivisions Late, Middle and Early Proterozoic are in widespread use, although the time planes marking the boundaries between them are not yet generally agreed. Table 9.1 shows the system used in the Canadian shield as an example. Further stratigraphic subdivisions are in use in several different countries (see Harland *et al.*, 1982) but are not internationally agreed.

Major orogenies in the Phanerozoic occur at intervals of the order of 200 Ma, and last for periods of the order of 100 Ma. In the Precambrian, a series of 'orogenic cycles' have been recognized (Sutton, 1963) that are much longer than the Phanerozoic orogenies. The Archaean alone extends at least from 3800 Ma BP, the age of the oldest dated continental rocks, to 2500 Ma — a period longer than the whole of Phanerozoic time. Within this period, it is not generally possible to recognize separate widespread orogenies such as the Caledonian or the Hercynian.

Nor can we define orogenic belts in the Archaean. Although local mobile belts have been recognized, such as the Limpopo belt of southern Africa, bordered by more stable zones, these are not comparable in scale or in inferred process to the Phanerozoic orogens.

9.1 Plate movements in the Precambrian

The question of what kind of plate tectonic process operated during the Precambrian (and particularly during the Archaean), is one that has caused considerable debate among Precambrian geologists since the plate tectonic theory was established. Opinion has ranged widely. At one extreme is the view held by Baer (1977) and others that 'modern' plate processes did not commence until the late Precambrian. This view is based on the argument that in warmer, thinner, early Precambrian lithosphere, eclogite could not form in the mafic oceanic crust, making the oceanic lithosphere too buoyant to subduct. Heat loss and tectonic movements were concentrated in the softer continental lithosphere. At the other extreme is the position taken by Burke *et al.* (1976) and Tarney and Windley (1977) for example, who believe that the early Precambrian plate tectonic processes were essentially similar to those operating at present, and differ only in their rate, in the size of the plates, and in other relatively minor respects.

The key to the problem lies in understanding Precambrian heat production and heat loss. This problem is addressed by Bickle (1980) who, following McKenzie and Weiss (1975), points out that higher heat production during the earlier history of the Earth would lead to more rapid movement of thinner plates, whose motions would be governed by smaller-scale convective cells than those inferred for the present.

Since the thickness of the oceanic lithosphere is governed by the supply of heat to the base of the lithosphere, an increase in heat supply will result in a thinner lithosphere and less efficient operation of the plate creation–subduction processes. However this effect may be offset by a faster rate of plate creation, and by the subduction of younger, hotter oceanic plate. Bickle points out that about 45% of the present heat loss through the Earth's surface arises from the plate creation–subduction process, and that if this process ceased, average thermal gradients in the continents would be nearly doubled. Our knowledge of Archaean thermal gradients is imperfect, but it is clear from pressure–temperature estimates in Archaean lower-crustal rocks that thermal gradients in the Archaean continental crust were not greatly different from today's values, despite the much higher rate of heat production. Pressure estimates from lower-crustal granulite-facies gneisses, such as the Archaean Scourie gneisses of NW Scotland, indicate crustal thicknesses of at least 30 km and possibly 50 km, at the time of formation (see e.g. Cartwright and Barnicoat, 1987). A marked rise in geothermal gradient would have resulted in the wholesale melting of the rocks at such depths.

Bickle concludes that the prolific eruption of volcanic rocks appears to be the only mechanism capable of transporting sufficient heat, and that a plate tectonic process is necessary for recycling these volcanic rocks. Bickle's thermal model suggests a sixfold increase in the areal rate of plate production in the late Archaean, compared with the present. A corresponding increase in the overall rate of subduction is required to keep the system in balance.

Many authors have noted the high proportion of tonalitic plutonic intrusions in the Archaean that are interpreted as melting products of subducted oceanic crust, the deep-seated counterparts of volcanic-arc andesites. It is generally believed by Precambrian geologists that a large proportion of the present continental crust was created during the late Archaean (see e.g. Dewey and Windley, 1981). A detailed study by Taylor *et al.* (1980) of the Pb-isotope composition of the Archaean craton of South Greenland (see below) indicates two ages of derivation for the mantle-derived Pb in the igneous rocks: one at around 3000–2800 Ma BP and the other at c.3700 Ma. They found no evidence in regions of younger Archaean crust of the material with older Pb, suggesting that the greater part of the craton was formed, rather than reworked, in the late Archaean. Rb–Sr and Sm–Nd isotope studies of Early Proterozoic terrains in Canada indicate that the bulk of the crust there was also formed in the late Archaean period (Mc-Culloch and Wasserburg, 1978).

The opinion of many Precambrian workers seems to be in favour of a convective plate-tectonic process in the early Precambrian, operating at a much faster rate than at present, and that this process was responsible for the creation of a substantial proportion of the continental crust during the approximately 500 Ma period of the late Archaean.

The way in which this plate tectonic process operated must have been rather different from the model accepted for the Phanerozoic. One of the most significant features of early Precambrian mobile belts is the absence of ophiolites and of low-temperature, high-pressure metamorphic rocks, the accepted indicators of subduction in mobile belts from the Late Proterozoic onwards. This was one of the factors that led Baer (1977) to reject subduction in the early Precambrian. Baer argues that in a thinner, warmer lithosphere, eclogite could not form, preventing the density inversion necessary to power the subduction process from taking place. A possible solution to the Archaean subduction problem is suggested by Arndt (1983) and by Nisbet and Fowler (1983), who point out that the Archaean oceanic crust may have been largely komatiitic in composition, corresponding to that of the ultramafic komatiitic basalts found in Archaean greenstone belts. An oceanic crust of this composition, moving over an asthenosphere that was hotter, and consequently less dense and less viscous than at present, would move laterally more easily. When cooled, such a lithosphere could be denser than the asthenosphere and be readily subductible.

Proterozoic plate movements

Apparent polar wander paths have been constructed for the Proterozoic of all the major Precambrian shield regions. These are regarded as evidence of continental plate movements during this period. The polar wander curves exhibit a number of 'hairpin' bends interpreted as major collision orogenies resulting in abrupt changes in plate motion. Irving and McGlynn (1981) discuss the Precambrian palaeomagnetic record from North America (Laurentia) and conclude that the data are not sufficiently sensitive to detect relative movements within the Laurentian domain during the Archaean and Early Proterozoic, but that large latitudinal movements relative to the pole occurred with an average rate of 5–6 cm/year. Three loops and hairpins in the Early to Mid-Proterozoic apparent polar wander path for Laurentia (Figure 9.1) are identified with the Hudsonian orogeny at c.1750 Ma BP, the Keweenawan or early Grenville extensional

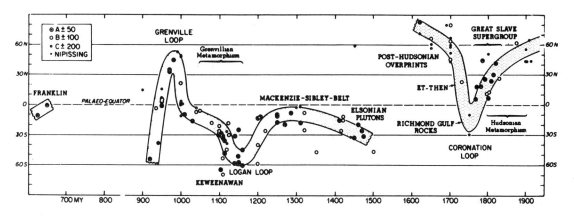

Figure 9.1 Plot showing the variation in magnetic palaeolatitude with time for the Laurentian shield. Note the abrupt changes ('hairpins') at 1750, 1150 and 1000 Ma BP. Calculated for Winnipeg (50°N, 97°S). Age uncertainty of individual points are in Ma. From Irving and McGlynn (1986)

event at *c*.1150 Ma BP, and the main end-Grenville convergent deformation and uplift at *c*.950 Ma BP (see 9.3).

Piper (1982) believes that the polar wander paths for the various continental blocks during the Proterozoic are sufficiently similar to be interpreted in terms of a Proterozoic supercontinent (Figure 9.12). However, it should be emphasized that considerable relative motion may be concealed by these paths, and that there is still some controversy over this proposal.

Nevertheless, we may conclude that the palaeomagnetic data for the Proterozoic support continental plate movements similar to those of the Phanerozoic. There is considerable controversy about the possible existence of collisional sutures in Precambrian mobile belts. We shall discuss some of the evidence in the examples to follow. On the one hand, there is the belief that most Proterozoic mobile belts are intracontinental with respect to a major Proterozoic supercontinent, and that plate boundary processes, if they exist, occur only around the margins of the supercontinent. Others believe that cryptic sutures are present in many supposedly intracontinental belts but are palaeomagnetically undetectable. The evidence hinges at present on the interpretation of magmatic rocks: whether or not they can be interpreted as subduction-derived. We shall discuss examples of intracontinental belts where such rocks are absent; in such cases a collisional model is difficult to substantiate. The impression gained is that intraplate continental mobility was more widespread during the Proterozoic than during the Phanerozoic, and that this difference is to be explained by an increase in continental lithosphere strength in late Precambrian times.

In their study of the evolution of the continental crust, Dewey and Windley (1981) interpret the Precambrian record as follows: the Archaean is a period of rapid crustal growth achieved by the amalgamation of volcanic arcs to form about 85% of the present continental mass by 2500 Ma BP. During the Early Proterozoic, large cratons were formed

by the amalgamation of Archaean fragments, and gradually stabilized, thickened and differentiated. Internal movement belts resulted from marginal collisions. By Late Proterozoic time, modern accretionary margins were in existence.

9.2 Late Proterozoic Pan-African belts

Areas of Late Proterozoic to early Palaeozoic tectono-thermal activity form a large part of the surface area of Gondwanaland (Figure 9.2), particularly in Africa. These tectonically active zones cannot be termed orogenic belts, since in many instances they do not form clearly defined linear features. Nor is the term 'orogenic' necessarily applicable, since it implies an analogy with Phanerozoic mountain belts whose tectonic origins are clearly established. Many Precambrian geologists prefer the term *mobile belt* or *mobile zone* for such regions, and we shall follow this practice here.

The Pan-African system of mobile crustal zones comprises areas of tectonic, magmatic and metamorphic activity in the age range 1000–450 Ma. These mobile zones separate cratons that were tectonically stable during this period, such as the West African, Congo and Kalahari cratons of Africa, Guyana–Brazil in South America, and Peninsular India.

Palaeomagnetic evidence (McWilliams, 1981) shows that Gondwanaland existed as a supercontinent only from latest Precambrian or early Palaeozoic time until its break-up in the Mesozoic. The formation of Gondwanaland is therefore a Pan-African event. Prior to this event, the palaeomagnetic record suggests that two supercontinents existed with markedly different apparent polar-wander paths. One comprises the Precambrian shield areas of Africa and South America, and may be termed West Gondwana; the other, East Gondwana, comprises the Precambrian shields of India, Australia and Antarctica. The two supercontinents are presumed to have collided along the Pan-African Mozambique belt that extends along the east coast of Africa (Figure 9.2). Other Pan-African mobile belts may be inter-

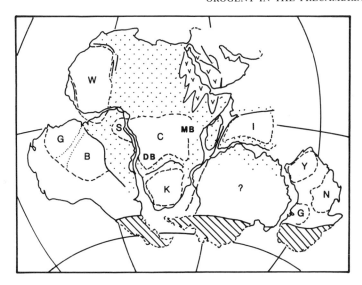

Figure 9.2 Tectonic summary map of Gondwanaland reconstructed according to Smith and Hallam (1970), showing the extent of Pan-African mobile regions and the Late Proterozoic cratons. *A*, Arabian craton; *W*, West African; *G*, Guyanan; *B*, Brazilian; *S*, Sao Francisco; *C*, Congo; *I*, Indian; *Y*, Yilgarn; *N*, Kimberley etc.; *G*, SE Australian. *MB*, Mozambique belt; *DB*, Damaran belt. The area with 'V' ornament is the Hejaz accretionary arc province. The Palaeozoic orogenic belt is shown in ruled ornament. After McWilliams (1981).

preted as intracratonic or intracontinental, such as the Damara–Zambesi belt between the Congo and Kalahari cratons. In this case there is no palaeomagnetic evidence for ocean closure. A third type of Pan-African mobile zone is represented by the large region of tectono-thermal activity covering much of northeastern Africa from Morocco to Ethiopia and the neighbouring Arabian shield. This zone is marginal to the West Gondwana continent, and could be analogous, to some extent, to the Cordilleran orogenic belt of the western Americas. We shall discuss three different Pan-African domains to illustrate these contrasting tectonic settings: the Arabian–Nubian shield, the Mozambique belt, and the Damara belt.

Pan-African history of the Arabian–Nubian shield

The Pan-African mobile zone of northeast Africa and Arabia is exposed in the adjoining shield areas of Nubia and Saudi Arabia. After closing the Cenozoic Red Sea rift, the zone is about 1100 km in width, and extends from the northern end of the Red Sea to Ethiopia, where it is interrupted by younger rocks of the African Rift system. To the south, it is continuous with the Mozambique belt.

The tectonic interpretation of the region is discussed by Gass (1981), who points out that tectonothermal activity extended from *c*.1200 Ma until *c*.450 Ma BP, from Mid-Proterozoic to Lower Palaeozoic times. This period of mobility thus spans a longer time than the whole of the Phanerozoic, although it is comparable with the life of the Cordilleran orogenic belt of North America, for example. The Pan-African as originally defined by Kennedy (1964) was restricted to the period 650–450 Ma BP. Gass recognizes three major divisions termed *Lower*, *Middle* and *Upper Pan-African* respectively, although there is considerable overlap in the radiometric age ranges of rocks assigned to these divisions (Figure 9.3*B*).

The oldest rocks of the Lower Pan-African are metasedimentary quartzites thought to represent a passive-margin sedimentary wedge flanking the Archaean Nile craton (Figure 9.3*A*). No evidence of continental basement older than 1200 Ma BP has been found within this Pan-African zone. The quartzites are overlain by a thick sequence (> 12 km) of basalts and basaltic andesites with oceanic affinities, intercalated with greywackes, carbonates and cherts. These supracrustal rocks were invaded by basic plutons dated at *c*.900 Ma BP.

In the Middle Pan-African, the bulk of the rocks of the region were formed; 50–60% of

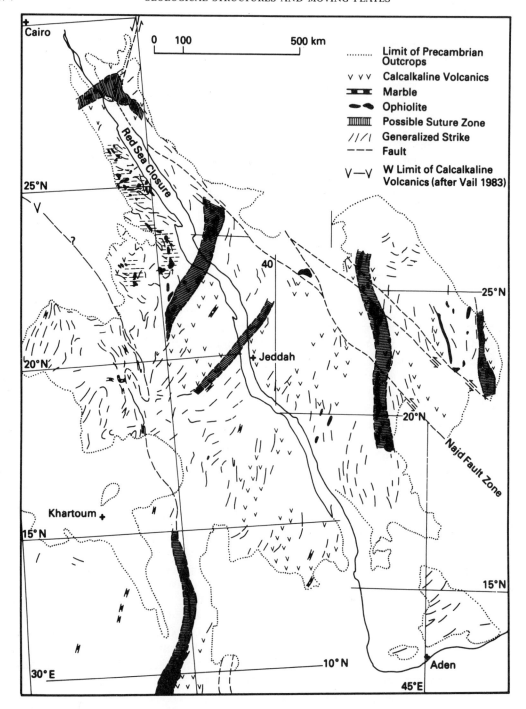

Figure 9.3 (*A*) Outline tectonic map of NE.Africa and the related part of Arabia, allowing for the estimated displacement on the Najd fault. Note the positions of postulated suture zones. From Shackleton (1986)

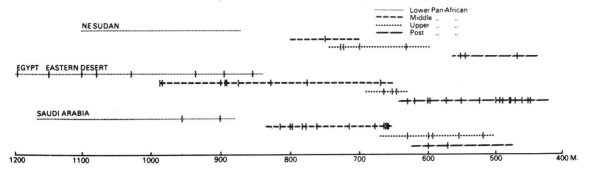

Figure 9.3 (*B*) Schematic representation of the age ranges of magmatic activity in NE Africa and Arabia, divided into Lower, Middle and Upper Pan-African and post-Pan-African periods. Note that activity is effectively continuous. From Gass (1981)

these comprise plutonic rocks of gabbroic to granitic composition. These invade supracrustal volcanic rocks of andesitic to rhyolitic character. Stromatolitic limestones and cherts are widespread, but form a relatively minor component by volume of the supracrustal assemblage. The overall picture obtained from a very complex outcrop pattern is of a series of emergent volcanic arcs depositing material in shallow subsiding basins, some of which accumulated more than 10 km of material. The volcanic rocks appear to have become more siliceous with time. This supracrustal assemblage was deformed and invaded by a series of syntectonic and post-tectonic dioritic to granodioritic plutons. The latter range in age from *c*.820 to *c*.660 Ma. The plutons often form broad linear zones around 50 km wide, aligned in a N–S to NE–SW direction. These zones are interpreted as the roots of volcanic island arcs. Between the zones lie belts of mainly supracrustal material in which a number of ophiolite masses have been identified (Figure 9.3). Belts of ophiolite fragments are interpreted as pieces of oceanic lithosphere preserved along collision sutures (Shackleton, 1986). The relationship of some of these linear ophiolite zones to adjoining dated rocks suggests at least three periods of emplacement, and therefore collision, at *c*.1000 Ma, 800 Ma and 600 Ma BP respectively.

The base of the *Upper Pan-African* is marked by a regional unconformity at the base of a series of unmetamorphosed and undeformed

silicic volcanics and volcaniclastic sediments. The basal units comprise fluviatile to shallow-water clastic sediments with intercalated stromatolitic limestones. This supracrustal assemblage is intruded by calc-alkaline granites and granodiorites with ages in the range 650–610 Ma. The geochemical characteristics of the Upper Pan-African igneous rocks are also indicative of a volcanic arc, but the lack of metamorphism and deformation, except for block faulting, suggests that this arc was developed on continental crust.

The end of the Pan-African, according to Gass, is signalled by the cessation of calc-alkaline magmatism (i.e. of subduction) and the incoming of peralkaline magmatism commencing about 600 Ma BP.

The above sequence of events is interpreted as evidence of the gradual and progressive cratonization of northeast Africa by the successive building and subsequent accretion of oceanic volcanic arcs. Evidence for intervening oceanic material is present in the form of ophiolite zones, which may represent either cryptic sutures or the remnants of obducted sheets (Shackleton, 1986). According to Shackleton, in the Eastern Desert of Egypt, a huge sheet of ophiolitic melange overlain by calc-alkaline volcanic rocks has been thrust over continental shelf deposits. The thrust has a gentle regional dip to the southeast, but is refolded so as to form a number of separate outcrops (Figure 9.4) over a distance of 350 km. A NW–SE elongation lineation

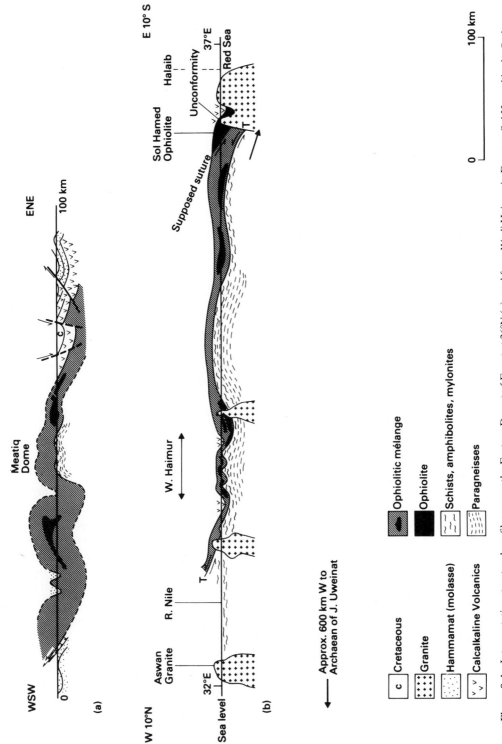

Figure 9.4 Interpretative structural profiles across the Eastern Desert of Egypt at 26°N (*a*), and from Wadi Haimur in Egypt to Sol Hamed in the Sudan at around 20°N (*b*), to show how the ophiolite outcrops may be linked to form the regional gently dipping sheets interpreted as suture zones (see the three northwestern sheets of Figure 9.3A). From Shackleton (1985)

indicates northwestwards translation with a minimum displacement of 100 km. The sheet is considered to represent a collisional suture trending NE–SW across the Red Sea and dipping south-eastwards.

Present-day volcanic island arcs rarely exceed 150 km in width with active magmatic cores less than 50 km across. If the Pan-African arc system were of similar dimensions, more than ten could be accommodated within the Nubian-Arabian belt during the 300 Ma or so of subduction-related magmatism, representing numerous subduction zones probably widely separated in time and space.

The presence of older continental crust on the eastern side of the Pan-African belt, in Saudi Arabia (Stacey and Hedge, 1984) suggests that arc accretion may have been finally terminated by continent–continent collision, possibly representing the northward continuation of the Mozambique belt now to be described.

The Mozambique belt

The Pan-African mobile zone of the Arabian-Nubian shield is directly along-strike of the Mozambique belt, which, according to Shackleton (1977, 1978), is a direct continuation of it. However, as Shackleton points out, the low-grade metamorphic assemblages of island-arc derivation in the north are very different from the high-grade metasediments and granitoid gneisses of the Mozambique belt in East Africa. This belt contains large areas of older continental basement yielding Archaean to Mid-Proterozoic ages and variably affected by Pan-African deformation and metamorphism with ages ranging from c.950 to 550 Ma. In the central part of the Mozambique belt in Kenya and Tanzania, only Pan-African ages have been obtained, which suggests that older continental material here may have been thoroughly reworked and its geochronological signature obliterated.

The western front of the belt in the E. African sector lies along the margin of the Tanzanian craton, where E–W Archaean struc-tures are sharply truncated along a zone of mylonites several km wide (Figure 9.5). East of the mylonite belt, there is a transition from brittle to ductile deformation corresponding to an increase in metamorphic grade. The eastern limit of this zone of strong Pan-African deformation and metamorphism is thought to lie between mainland Africa and Madagascar, which contains extensive Archaean and Mid-Proterozoic domains. On the eastern side of Madagascar, prior to the Mesozoic break-up of Gondwanaland, lay the early Precambrian shield of Peninsular India, which exhibits a quite different palaeomagnetic polar-wander path from Africa and is therefore considered to have collided with Africa during the Pan-African period of activity.

The rock assemblage in the Mozambique belt in the E. African sector consists of granitoid gneisses with widespread and abundant intercalations of metasediments, including quartzites, marbles and graphitic pelites. The quartzites are more abundant in the west, whereas marbles are commoner in the east. Along the western front, quartzites rest unconformably on the Archaean basement of the Tanzanian craton. Shackleton considers that the metasedimentary assemblage as a whole probably represents a continental shelf sequence with an original eastwards facies change from near-shore sands to deeper-water limestones, resting on older granitoid basement now transformed to gneiss. It is a familiar problem in Precambrian research that the detailed stratigraphic data necessary to substantiate this interpretation cannot be obtained.

There are three zones within the belt that Shackleton identifies as possible ophiolite horizons, one of which can be traced for more than 150 km. Metavolcanic rocks associated with one of these ophiolite bodies yields a Pan-African age of c.660 Ma.

The western half of the E. African sector of the belt is dominated by E- or NE-dipping planar structures (mylonite fabrics, foliations and lithological layering). This structural package is interpreted by Shackleton as an imbricate stack of thrust slices of Archaean base-

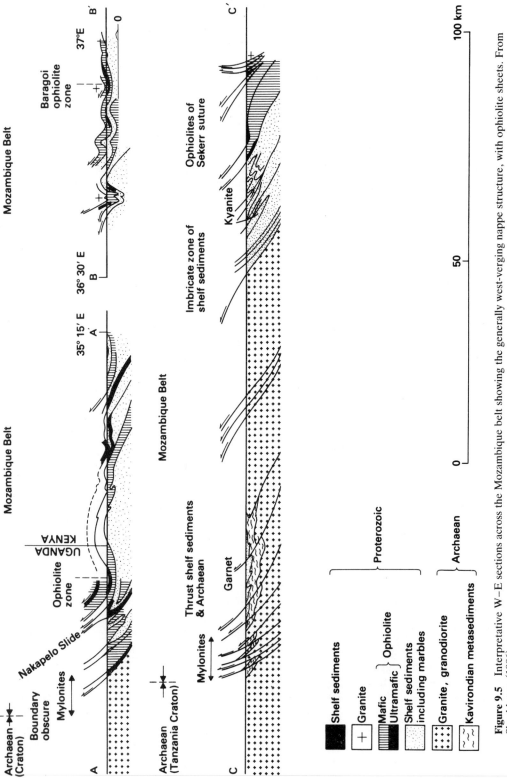

Figure 9.5 Interpretative W–E sections across the Mozambique belt showing the generally west-verging nappe structure, with ophiolite sheets. From Shackleton (1986)

ment and subordinate Proterozoic sedimentary cover, with the ophiolite sheet representing a high structural level. If the ophiolite belts represent the same sheet, overlying the platform cover, this sheet must root in a suture lying on the eastern side of the belt (Figure 9.5). Elongation lineations in this zone are consistently NW–SE in trend. However, in the central part of the sector, the lineations are N–S. Shackleton and Ries (1984) interpret this change in trend as the result of a later strike-slip relative movement between the two continents after collision. They point in support to the lack of correspondence between the E–W-trending Archaean greenstone belts of the Tanzanian craton, on the west side of the belt, and the N–S-trending high-grade gneisses of the Madagascar Archaean.

The Damaran belt in Namibia

The Pan-African mobile belt system in SW Africa consists of two branches (Figure 9.2): one parallel to the coast on the west side of the Congo and Kalahari cratons, the *Gabon-Cape belt*, and the other trending almost at right angles to it and separating the two cratons. The latter belt is known as the *Damaran belt* and links with the Zambesi belt further east. The structure of the southwestern end of the Damaran belt, representing the zone of intersection of the two Pan-African belts, has been studied by Coward (1981). The Damaran belt is often described as an example of an intra-cratonic mobile belt (see e.g. Shackleton, 1976). There is no evidence of relative movement between the Congo and Kalahari cratons, within the limits of resolution of the palaeomagnetic data. The West Gondwana craton (Figure 9.2) appears to have behaved as a coherent unit both before and after the Pan-African events. However, quite large relative movements (of perhaps up to 1000 km) could not be detected palaeomagnetically, and the evidence leaves open the possibility either of quite large intracontinental displacements, or of the opening and subsequent closure of a small ocean basin.

A map of the Damara belt (Figure 9.6A) is described by Hawkesworth *et al.* (1986). Basement granites of the Congo and Kalahari cratons form respectively the northwestern and southeastern margins of the belt, which is about 350 km across. Small basement inliers occur within the belt as well. A large proportion of the outcrop is made up of metasediments divided into two separate sequences. The older sequence, the Nosib Group, comprises coarse clastic deposits overlain by shelf carbonates up to 4 km in thickness. Potassic lavas and associated alkaline plutons found along the northern margin are thought to be coeval with this Group. The overlying sequence consists of carbonates, quartzites, pelites and, locally, graphitic schists that reach a thickness of 16 km in the north. Carbonates are dominant in the northwest, but biotite-schists representing clastic deposits are more abundant in the southwest. The latter contain a layer of mafic volcanic rock, the *Matchless belt*, traced for over 350 km, although only a few m thick.

The centre of the belt is marked by a zone of syntectonic to post-tectonic granitoid plutons ranging from diorites to granites and including highly potassic varieties. The Damaran sediments were deposited between 1000 and 750 Ma BP, and most of the Damaran intrusions were emplaced in the period 570–450 Ma BP during the later part of the Pan-African time-span. Widespread regional metamorphism occurred at the same time. According to Coward (1981), the major phase of deformation in the belt, which produced the NE–SW structural trend, predates the earliest granites dated at 560–550 Ma. Later deformation in the central zone produced block-faulting and arching. However, along the southwestern margin, deformation associated with SE-directed thrusting continued until about 500 Ma BP. A later suite of post-tectonic granites was emplaced around 500–450 Ma BP.

Geochemical studies of the metasediments suggest that they contain very little new material, but that the rocks of the belt, both sedimentary and igneous, mainly consist of

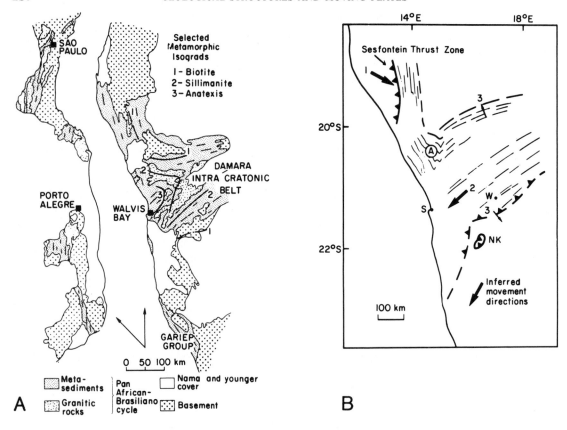

Figure 9.6 Structure of the Damara belt. (*A*) Tectonic sketch indicating the inferred movement directions relating to the three main phases of deformation. *NK*, Naukluft nappes; *S*, Swakopmund; *W*, Windhoek; *A*, area of interference structures between *F1* and *F2*; toothed lines, thrust zones. From Coward (1983), with permission. (*B*) Simplified map of the Damara–Ribeira belt, showing the distribution of basement, granitic intrusions and metasedimentary cover, and the arrangement of metamorphic isograds in the region of intersection between the Damaran and coastal belts. From Hawkesworth *et al.* (1986), with permission.

reworked basement material. A possible exception is the mafic Matchless belt, which has a mid-ocean-ridge basalt chemistry. However, the granitoid intrusions display intraplate rather than subduction-related magmatism. The balance of the geochemical evidence is therefore in favour of an intracontinental origin for the belt.

The structural study by Coward (1981) identifies three main regional movement phases from the structural history (Figure 9.6*B*): the first involved sinistral convergence along the northern arm of the Gabon–Cape belt, pro-ducing overthrusting towards the southeast along the Sesfontein thrust zone. This phase, according to Coward, is not represented within the Damara belt itself. The second phase produced refolding on N–S axes in the northern arm, but in the Damara belt is associated with recumbent west-verging structures with an ENE–WSW elongation lineation. These structures are interpreted by Coward as a low-angle shear zone with a movement sense sub-parallel to the trend of the belt. The most intense deformation occurred along the central zone, where the synkinematic granitoid intrusions are concentrated. These structures in turn are deformed by more upright folds, trending NE–SW to ENE–WSW, that verge north-westwards along the northern margin of the belt, and southeastwards along the southern margin. The latter margin is a major thrust zone carrying strongly-deformed Damaran metasediments onto the Kalahari craton.

Coward interprets the structural pattern in terms of an early NW–SE oblique convergence along the northern arm, followed by ortho-gonal convergence across the northern belt, coupled with differential strike-slip movements in the Damaran belt along a low-angle shear zone parallel to the trend of the belt. The third phase represents orthogonal convergence across the Damaran belt. However, Hawkes-worth *et al.* (1986) cite evidence indicating that the Damaran belt may be older than the Gabon–Cape belt.

There seems to be no general agreement concerning the plate-tectonic interpretation of these two belts. It has been suggested that the Matchless mafic belt may represent a colli-sional suture. However there is no evidence for subduction-related magmatism, and the other evidence is suggestive more of an extensional rift environment, producing a high-tempera-ture belt. Subsequent transpressional and con-vergent movements along this thinned and thermally weakened zone could have been purely intracontinental, reflecting differential movements between the neighbouring cratons.

Uncertainties of this kind are common in the interpretation of Precambrian mobile belts and illustrate the problems caused by imperfections in the quality of the evidence available to us.

9.3 The Mid-Proterozoic Grenville–Sveconorwegian system

During the Middle Proterozoic period (1800–1000 Ma BP), an extensive network of mobile belts was formed in both Laurasia and Gond-wanaland. The 'Grenville orogenic cycle' was recognized by Sutton (1963) as one of the major subdivisions in the tectonic history of

the Earth, and Dearnley (1966) believed that the Grenville mobile belts heralded the start of a third major orogenic cycle, fundamentally different from the preceding two, that was to extend into Phanerozoic time to include all subsequent orogenies. His belief was founded on the observation that the Grenville mobile belts were relatively narrow, regular, and linear, cutting across all previous belts, and were more analogous to the Phanerozoic belts than to the mobile zones of the preceding Hudsonian cycle and of the Archaean.

Later work, and the application of plate tectonic principles, has tended to reduce the significance of the Grenville 'revolution' and to emphasize similarities between pre-Grenville and post-Grenville belts. Thus there are Early Proterozoic belts with strong similarities to some of the Pan-African zones described above.

The Mid-Proterozoic mobile belt system of the North Atlantic region comprises three branches: one is the Grenville belt, *sensu stricto*, along the south-eastern margin of the Canadian shield; the second is the Sveconor-wegian belt of SW Sweden and southern Norway; and the third is the Carolinidian belt of E. Greenland (Figure 9.7). After restoring the Mesozoic extension across the North Atlantic, these three belts form a continuous Y-shaped system intersecting in the region of the British Isles. The northern or Carolinidian arm is poorly known and is interrupted by the Caledonian belt. We shall discuss only the Grenville and Sveconorwegian belts, and their possible interrelationships.

The Grenville Province

The exposed part of the North American Grenville belt, in Canada and in the Adiron-dack Mountains of New York State in the northeast USA, is known as the *Grenville Province* (Figure 9.8). In Canada, this belt extends for 1500 km along strike, and is over 600 km in width. The date of the Grenville 'orogeny' has traditionally been regarded as

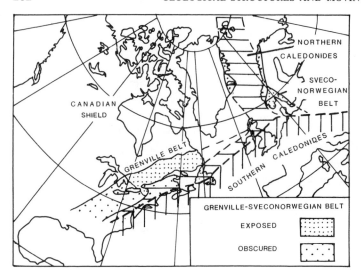

Figure 9.7 Location of the Mid-Proterozoic Grenville–Sveconorwegian orogenic belt system of the North Atlantic region (dotted ornament), together with the later northern and southern Caledonide belts.

c.1000 Ma BP, and the Grenville Province was affected by a major tectono-thermal event around this time. Most K–Ar ages range from 1100 Ma to 800 Ma, signifying slow uplift and cooling over a long period. However older igneous and tectonic events also affect the Province: in particular, a regional suite of intrusive plutons with dates ranging from c.1500 to c.1400 Ma. A general description of the Province and a tectonic interpretation are provided by Baer (1981) and Davidson (1985).

The major part of the Province appears to consist of Archaean and Early Proterozoic basement that can be traced into the Grenville belt from the adjoining craton. The distinctive sedimentary assemblage of the Labrador belt (see 9.4), deformed about 1800 Ma BP in the Hudsonian orogeny, can be recognized over a distance of 400 km from the margin of the Grenville belt into the interior, where it is progressively deformed and disrupted by Grenvillian deformation (Roach and Duffell, 1974). This deformation has produced overturned folds trending sub-parallel to the Grenville front. A second major phase of deformation produced N–S to NNW–SSE-trending folds in a wide zone in the interior of the belt.

The Archaean and Early Proterozoic basement of the Province is intruded by a major suite of plutons consisting of anorthosites and related rocks dated at 1500–1400 Ma. These bodies form a wide belt that extends from the Hopedale region of northern Labrador to the Adirondacks (Figure 9.8A). The belt thus cuts obliquely across the Grenville Province, extending 400 km north of the Grenville front at

Figure 9.8 (A) Simplified tectonic map of the Grenville belt. Note the distribution of pre-Grenville anorthosite-mangerite plutons and of Proterozoic supracrustal rocks. The Early Proterozoic sediments of the Labrador belt (see Figure 9.15) can be traced for a long distance into the Grenville belt. The principal tectonic zones are: I, the Foreland zone; II, the Grenville front zone, consisting dominantly of basement gneisses uplifted in relation to the foreland; III, the Central Gneiss zone, consisting of high-grade gneisses with interfolded Grenville Group metasediments, sub-divided into the Ontario (A) and Quebec (B) sectors, which differ mainly in the orientation of the structures; IV, the Central Metasedimentary zone, which contains the main outcrop of the Grenville Supergroup; V, the Central Granulite terrain, characterized by granulite-facies gneisses and anorthosites; VA is the Adirondack sector; VI, the Baie Comeau segment, which is similar to V but in amphibolite facies; VII, the Eastern Grenville zone, which contains similar rocks to VI but contains in addition low-grade supracrustal cover similar to that of the Nain province to the north. (B) Structural sketch sections across the Grenville belt (NW–SE) to illustrate the general relationship between basement gneisses (wiggly ornament), Archaean metasediments (greenstone belt facies) (close stipple), and Proterozoic supracrustal cover (dashed ornament). Sills and diapirs of the anorthosite-mangerite suite are also shown. After Wynne-Edwards (1972).

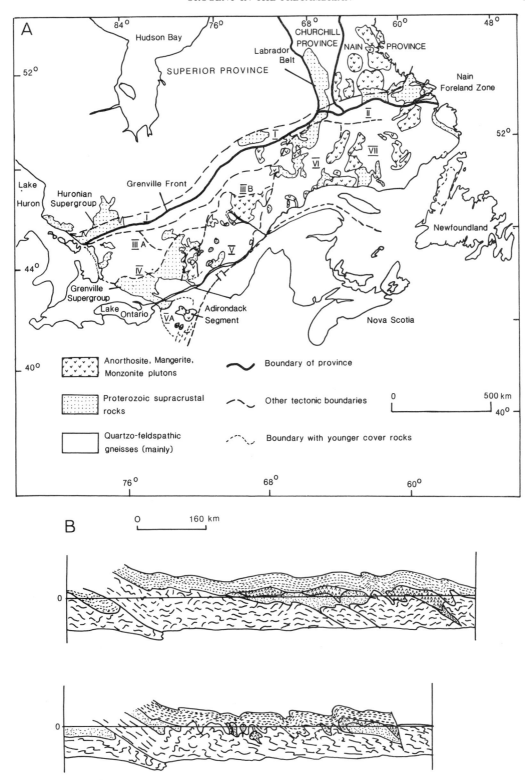

its northern end. The emplacement of these plutons is generally considered to pre-date the deposition of the sediments of the Grenville Group, although Baer believes that later mobilization of certain plutons occurred during the Grenvillian deformation. The western boundary of the anorthosite belt coincides with the Chibougamau–Gatineau lineament, which marks a major change in character of the basement gneisses. Baer suggests that a vertical displacement on this lineament pre-dates the deposition of the Grenville Group sediments that occur on both sides of it.

The metasediments comprise a sequence of carbonates, quartzites and pelites associated with both felsic and mafic volcanic rocks. The marbles alone are at least 15 km thick. Dates of 1300 Ma and 1250 Ma on members of the volcanic suite are held to give the depositional age for the Group. The northwestern margin of the metasediment outcrop is marked by a line of alkaline plutons dated at 1280 Ma, which is interpreted by Baer as evidence of a continental rift graben, in which the thick sedimentary pile was deposited. Regional extension is also indicated by the formation of graben at Seal Lake (at $c.1300$ Ma BP), along the Keweenawan rift (at $c.1130$ Ma), in the Gardar Province of S. Greenland (at 1245–1020 Ma), and by the occurrence of dyke swarms parallel to the Grenville front (at $c.1200$ Ma).

The greater part of the Grenville Province has been affected by a high-grade metamorphic event yielding upper-amphibolite to granulite-facies assemblages in sediments and basement rocks. This event is dated at $c.1150$ Ma BP in the southwestern part of the Province. Baer notes the concentration of Rb–Sr whole-rock dates of around 1100 Ma in a central zone of strong NE–SW deformation, and suggests that there may be two main Grenville tectonic events, one at $c.1100$ Ma BP, and the second at $c.950$ Ma, related to the final uplift of the Province.

Major thrust belts occur along several sectors of the Grenville front. In the north, in the Seal Lake region, thrusting is dated at $c.950$ Ma BP. In the Central Gneiss belt (zone II of

Figure 9.8), a series of sinuous shear zones separating distinct gneissose domains is described by Davidson (1985). A major SE-dipping shear zone also separates this domain from the adjoining Central Metasedimentary belt (zone III). The minor structures indicate a northwestwards sense of movement on these zones, tentatively dated at $c.1100$ Ma BP. The present crustal thickness of the Province varies from a maximum of 50 km along parts of the front to between 30–45 km elsewhere. These data suggest that tectonic thickening by thrusting in the marginal zone is still preserved, whereas the highly metamorphosed central zone corresponds to an uplifted segment of originally thickened crust, now much reduced in thickness.

According to Baer and Davidson, there is no good evidence in the exposed Grenville belt for oceanic closure in the form of obducted ophiolites or cryptic sutures; however, Windley (1987) believes that there is evidence for a collisional suture within the Central Metasedimentary belt. We shall consider the plate tectonic implications of the Grenville orogeny later, after discussing the Sveconorwegian.

The Sveconorwegian belt

The Mid-Proterozoic Sveconorwegian mobile belt occupies the western half of southern Sweden, and forms the Precambrian foreland to the Caledonides in southern Norway. Mid-Proterozoic dates in the Precambrian basement gneisses within the Caledonides enable the belt to be traced northwards to link presumably with the Carolinides belt of E. Greenland (Figure 9.7). This belt, and the associated tectonothermal events, has been correlated with the Grenville belt for many years. The maximum exposed width of the belt is about 600 km across south Norway and the northern part of the Swedish sector, and is divided into two parts by the Permian Oslo Graben. The main Sveconorwegian tectonothermal event (orogeny?) is generally assigned to the period 1200–900 Ma BP.

A general description of the Sveconorwegian

belt (Figure 9.9) is provided by Berthelsen (1980). The eastern boundary of the belt is marked by a major shear zone, termed the *Protogine Zone*, separating the *Eastern Segment* of the belt from the Svecokarelian (Early Proterozoic) craton to the east. This shear zone extends from Skåne in the south to the Caledonian front, and is marked by a wide belt of intense deformation, known in Sweden as the 'central Swedish schistosity zone'. South of Lake Vättern, and in the Värmland–Kopparberg sector north of Lake Vänern, the zone is a low-angle west-dipping thrust that, in the latter sector, carries *c*.1670 Ma-old late Svecokarelian granites eastwards on to post-Svecokarelian sedimentary cover on the eastern craton. Berthelsen estimates a minimum displacement of 50 km on the thrust in Värmland. This thrust cuts an earlier easterly-dipping shear zone from Lake Vänern northwards.

The rocks of the eastern segment consist predominantly of granitoid gneisses apparently derived mainly from Svecokarelian plutons, and known as the Pregothian gneisses. Occasional supracrustal remnants become more frequent towards the west, and minor amphibolite bodies of various ages are common. K–Ar cooling ages record regional uplift of the belt between 1050 and 950 Ma BP. Most of the deformation of the Pregothian gneisses in the Eastern Segment is probably Svecokarelian in age (Larson *et al.*, 1986).

The Eastern Segment is separated from the *Median Segment* by another major west-dipping shear zone termed the *Mylonite Zone*. North of Lake Vanern, this zone also cuts an earlier NE-dipping shear zone (Figure 9.10) in a similar manner to the Protogine zone to the east. The rocks of the Median Segment consist of Pregothian basement gneisses similar to

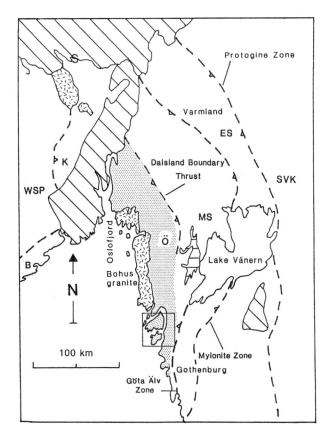

Figure 9.9 Principal tectonic sub-divisions of the central part of the Sveconorwegian mobile belt in SW Sweden and S. Norway. *SVK*, Svecokarelian (Early Proterozoic) craton; *ES*, Eastern segment; *MS*, Median segment; *Ö*, Östfold–Marstrand segment; *K*, Konsberg and *B*, Bamble, separated parts of the Kongsberg–Bamble segment; *WSP*, Western sub-province. The important Östfold–Marstrand segment is stippled; the box shows the Lysekil–Marstrand area of Figure 9.11, discussed in the text. Horizontal ruling, Dal formation; oblique ruling, younger cover; late Sveconorwegian granites shown in hachured ornament. After Park *et al.* (1987)

Figure 9.10 Simplified tectonic map of the Östfold, Median and Eastern segments of the Sveconorwegian belt in SE Norway and adjoining areas of Sweden. Note the pattern of linear structures (dots with bars) re-oriented near the major low-angle shear zones. The outcrop of the Dal formation is shown with dense shading. The post-tectonic Bohus–Iddefjord granite is also shown. After Berthelsen (1980).

those of the Eastern Segment but more grano-dioritic in composition, on average. These rocks have yielded an Rb–Sr age of c.1700 Ma. In the western part of the zone, supracrustal gneisses of the Åmål–Kroppefjäll Group, thought to be of Svecokarelian age, occur together with a younger sequence known as the Dal Group, deposited about 1050 Ma BP.

The Dal Group consists of weakly-metamorphosed sandstones, pelites and impure carbonates, with basic volcanic rocks, and rests unconformably on the Pregothian and Åmål gneisses (Figure 9.10). Both the Dal sequence and the underlying gneisses are involved in strong Sveconorwegian deformation, particu-

larly along a third major west-dipping shear zone known as the *Dalsland Boundary Thrust*. This thrust separates the Median from the *Östfold* or *Östfold-Marstrand Segment*, and displaces older gneissose basement of the Östfold Segment over Dal rocks. This major displacement seen in the north appears to be replaced southwards by a network of shear zones, one of which is the prominent *Göta Älv Zone* that runs through Gothenburg (Figure 9.9).

The Östfold Segment is dominated by supracrustal gneisses known as the Stora Le–Marstrand formation. These rocks comprise semipelitic to psammitic metasediments and mafic volcanic rocks, and are cut by a large number of intrusive bodies of various ages and compositions. The older of these yield Sveco-karelian ages of *c*.1700 Ma that provide a minimum age of deposition for the supracrustal sequence.

The structural history of the Norwegian part of this complex segment is described by Hage-skov (1980), and of the Swedish part by Park *et al*. (1987; see Table 9.2). The magmatic rocks are described by Samuelsson and Åhäll (1985) and Åhäll and Daly (1985). The older calc-alkaline granitoid plutons of groups A and B give Svecokarelian ages, and may be regarded as part of the basement to the Sveconorwegian belt. However, the group C intrusions dated at 1420–1220 Ma represent an important post-Svecokarelian magmatic suite, which has experienced only Sveconorwegian tectono-thermal activity. The suite includes a regional mafic dyke swarm that is recognized along the west coast from the Oslo district to western Orust, including the Koster islands west of Stromstad. The suite also includes alkaline augen granites of markedly different chemistry to that of the preceding calc-alkaline plutons.

Three main phases of Sveconorwegian de-formation are recognized by Park *et al*., the earliest of which is accompanied by amphi-bolite-facies metamorphism dated at *c*.1090 Ma, and the later by mainly greenschist-facies. The late Sveconorwegian structures are cut by the post-tectonic peraluminous Bohus-Iddefjord granite, dated at 891 Ma.

The Sveconorwegian structures in the Östfold-Marstrand belt (Figure 9.11) are re-lated by Park *et al*. to movements on several shear zones, including a major low-angle shear zone with an early NE–SW and a later NW–SE movement direction. The belt is bounded in the west by another major N–S shear zone, which is steep, with a sinistral, strike-slip sense of movement. This zone ex-tends off the west coast of Sweden to connect with the Oslo Fjord shear zone in southern Norway (Figure 9.9). The Östfold-Marstrand

Table 9.2 Simplified chronology of the Lysekil-Marstrand area (see Figure 9.9). From Park *et al*. (1987).

	Rocks and structures	Metamorphism	Age, Ma
SVECOKARELIAN	1. Deposition of Stora Le-Marstrand sediments and emplacement of contemporaneous mafic volcanics		*c*. 1700
	2. Emplacement of group A intrusions (mainly granitoid)		
	3. D1 deformation	M1 amphibolite-facies with (weak?) migmatitic veining	
	4. Emplacement of group B intrusions (mainly granitoid)		*c*. 1650
	5. D2 deformation	M2 amphibolite-facies with intense regional migmatitic veining	
	6. Emplacement of group C intrusions (bimodal)		1420–1220
SVECONORWEGIAN	7. D3 deformation	M3 amphibolite-facies	1090
		----- ? ------- ? ------- ? ------- ? -----------	
	8. D4 deformation	probably lower grade, and locally retrogressive	
	9. D5 deformation		
	10. Emplacement of group D Bohus granite		890

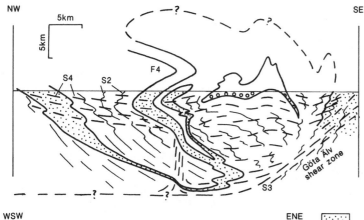

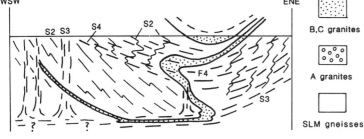

Figure 9.11 Interpretative structural profiles across the Lysekil–Marstrand area (see Figure 9.9). S2 and the emplacement of the A, B and C granites are pre-Sveconorwegian in age. The Sveconorwegian structures D3 and D4 are related to movements on major low-angle shear zones and steep transfer zones. From Park *et al.* (1987)

Segment is interpreted as a deeper-level, more ductile part of the mobile belt where the basement is more intensely reworked by the Sveconorwegian deformation than in the more easterly segments.

West of the Oslo Fjord shear zone lies the *Kongsberg–Bamble* segment of south Norway (Figure 9.9), interpreted by Berthelsen as a western equivalent to the Östfold segment. However, the precise relationship between the two segments is obscured by the younger rocks of the Oslo graben, and by an unknown strike-slip displacement along the Oslo Fjord shear zone. The Kongsberg-Bamble segment is separated from the Western Subprovince by yet another major shear zone, the *Kristiansand-Bang* shear zone (Hageskov, 1980). This zone dips to the east beneath the Kongsberg–Bamble segment. The Western Subprovince contains large numbers of plutonic igneous rocks including charnockites, monzonites and anorthosites, in addition to granites and granodiorites. Most of these intrusions appear to have been emplaced between 1200 and 850 Ma

BP, although older Svecokarelian ages have also been obtained (see review by Demaiffe and Michot, 1985).

In western Rogaland, two major Sveconorwegian deformations are recognized (Falkum and Pedersen, 1979) dated at *c.*1100 Ma and *c.*1000 Ma respectively. These produce large-scale west-verging nappes with a N–S trend. Magmatic episodes preceded, separated and followed these deformations. The regional metamorphism is characteristically high-temperature granulite- to amphibolite- facies in this western sub-province.

The Telemark supracrustal sequence in the eastern part of the sub-province, consisting predominantly of quartzites and metavolcanic rocks, probably deposited between 1225 and 1000 Ma BP, may represent the lateral equivalent of the Dal Formation of the median segment to the east.

Despite the fact that the Sveconorwegian mobile belt is comparatively well known, there are major difficulties in explaining it in terms of plate tectonic models. The wide zone of calc-alkaline plutons making up the basement of the eastern segments is generally considered to represent an easterly-dipping subduction zone, or series of zones (see Berthelsen, 1980), which is a westwards continuation of the Svecokarelian magmatic arc. There is no evidence of any pre-Svecokarelian basement within the Sveconorwegian belt, which is probably composed of a succession of Svecokarelian volcanic arcs situated along the Svecokarelian continental margin. The post-Svecokarelian history of the region commences with extensional rifting associated with the emplacement of dyke swarms and alkali-granites in the period 1420–1220 Ma BP in SW Sweden and SE Norway. During this period, the Telemark shelf sequence was deposited in SE Norway and, possibly later, the Dal deposits were formed in Sweden in an extensional basin.

The main Sveconorwegian deformation in the period 1100–1000 Ma BP represents, according to Berthelsen, a collision between the Western Subprovince and the eastern segments. The structural evidence from SW Sweden suggests that the convergent movements involved an earlier northeastwards movement on a low-angle shear zone and on a related steep NE–SW transfer zone, followed by later south-eastwards or eastwards movements on the zones further east (Park et al., 1987). In Norway, the southeastwards movements appear to be dominant (Figure 9.10).

Plate-tectonic interpretation of the Grenville-Sveconorwegian system

The palaeomagnetic evidence suggests that the major Early Proterozoic shield regions may have formed a single supercontinent at the commencement of Mid-Proterozoic time (Piper, 1982). On Piper's reconstruction (Figure 9.12A), the Grenville and Sveconorwegian belts are sub-parallel, and form the southern margin of the supercontinent. An important rifting phase is reflected in magmatism and in the formation of extensional sedimentary-volcanic basins both in the Grenville Province and in the Östfold-Marstrand seg-

ment of the Sveconorwegian belt. This phase may be related to a regional rifting that resulted in the break-up of the eastern end of the supercontinent (Figure 9.12B), and its subsequent rotation between 1200 and 1000 Ma BP (Patchett et al., 1978; Stearn and Piper, 1984). Subsequent closure of the intervening oceanic region would have resulted in collision along the Grenville sector, and convergence across the major shear-zone boundaries in the Sveconorwegian belt (Figure 9.12C).

The Western Subprovince in S. Norway, with its apparently mantle-derived Sveconorwegian magmatic rocks (see Demaiffe and Michot, 1985), represents the northern part of a hypothetical continental wedge that is interposed between the Grenville and Swedish sectors at the time of collision. The relationships between the basement rocks in the remainder of the wedge have been obliterated by subsequent orogenies. However, the shape of the wedge is determined by palaeomagnetic constraints assuming a continuous continental area between the two sectors at the time of collision, interpreted as the end-Grenvillian deformation. The geometry of the wedge necessitates oblique dextral convergence along the Grenville belt and sinistral convergence along the Sveconorwegian belt (Figure 9.12C). The late sinistral convergence in the Sveconorwegian sector may have followed an earlier phase of convergence with a more easterly or northeasterly direction. This event may be related to the initial collision with the Western Subprovince (Figure 9.12B). The c.1150 Ma-old deformation in the Grenville belt may also be attributable to an early phase of orthogonal convergence, possibly involving a western extension of the same continental block.

9.4 Early Proterozoic belts of the North Atlantic region

Mobile belts or zones of Early Proterozoic age form the largest proportion of the Precambrian surface area of the North Atlantic region, and surround several Archaean cratons, of which the largest is the Superior Province of the

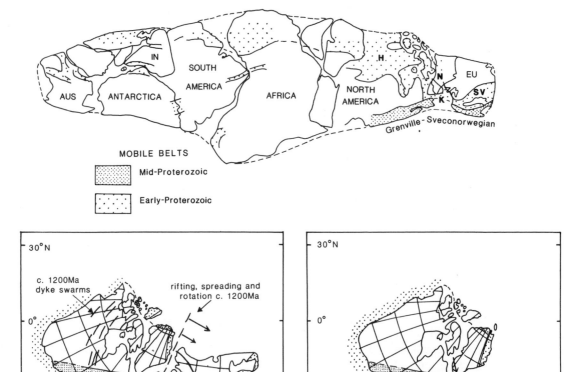

Figure 9.12 (A) Palaeomagnetic reconstruction of the continents in the Proterozoic (the Proterozoic 'super-continent') based on Piper (1976). *AUS*, Australia; *IN*, India; *EU*, Europe. Note the position and extent of Early Proterozoic mobile belts in North America and Europe (*H*, Hudsonian; *N*, Nagssugtoqidian; *K*, Ketilidian; *SV*, Svecokarelian) and the position of the Grenville–Sveconorwegian belt along the margin of the super-continent. (*B*, *C*) Schematic model based on the palaeomagnetic restorations (see Stearn and Piper, 1984) of North America-Europe before and after the 1000 Ma-old end-Grenville event, interpreted as a collisional orogeny. Note that the commencement of rotation of Europe relative to North America is related to the regional extensional event at *c.*1200 Ma, and that early Sveconorwegian movements (and perhaps Grenvillian also) are attributed to collision of a microplate at *c.*1100 Ma, prior to the main collision event. Note also the difference in convergence direction both between the two events, and, in the later event, between the sinistral component along the Sveconorwegian suture and the dextral component along the Grenville suture.

Canadian shield. Other important Archaean cratons are the Slave craton in northwest Canada, the North Atlantic craton of S. Greenland, and the Kola craton of northern Finland and the USSR. The mobile zones that surround these cratons are very much wider than the Grenville or Sveconorwegian belts or, for example, the Caledonide belts of the Phanerozoic, but are not dissimilar in some respects to certain of the Pan-African zones, or to the Cenozoic belt of Central Asia. The Svecokarelian of Scandinavia and the Hudsonian of the Canadian shield both occupy zones about 1000 km across. These mobile zones differ fundamentally in character from one part to another, both along and across strike. Generalizations made about Early Proterozoic tectonic regimes, particularly in contrast with Archaean and Mid-Proterozoic regimes, are often based on one or two particular examples, whereas the network as a whole should ideally be viewed in its entirety.

We shall discuss briefly two parts of this complex network: the Labrador belt, which forms part of the Hudsonian regime of Canada; and the Lewisian-Nagssugtoqidian belt of Scotland and S. Greenland.

The Labrador belt

This belt forms part of the circum-Superior belt of the Canadian shield (Figure 9.13) summarized by Baragar and Scoates (1981). They believe that this belt evolved from an annular rift surrounding the Superior craton (Figure

9.14). The southern sector developed into an ocean that subsequently closed, whereas the U-shaped northern part of the system formed intra-continental rift basins by crustal thinning and stretching. This northern belt was subsequently deformed by shortening, which re-

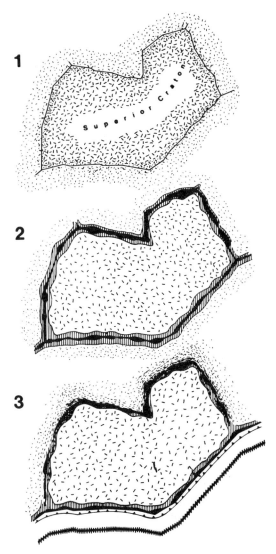

Figure 9.14 Interpretation of the circum-Superior belt (see Figure 9.13) in terms of an initial annular extensional zone created by southward movement of a continental plate (2) followed by collision and compression (or transpression, as appropriate) as plate movement was reversed. The events relating to the SE side of the craton are speculative because of overprinting in the Grenville belt. From Baragar and Scoates (1981)

Figure 9.13 Arrangement of Early Proterozoic mobile belts and Archaean cratons in North America–Greenland. The circum-Superior belt consists of the Southern province in the south and the Churchill province in the north and northeast. Within the latter province are the narrower supracrustal belts of the Belcher islands (*BB*), Cape Smith (*CSB*) and Labrador (*LB*) belts. The Labrador sector of the Churchill province (see Figure 9.15) is bounded on its northeast side by the North Atlantic craton (*NAC*) — see Figure 9.23.

sulted from the convergence accompanying oceanic closure along the southern sector. A simple opening and closing model along these lines would satisfy the palaeomagnetic data, which preclude any large relative displacements between the Slave, Superior and North Atlantic cratons. The simple geometry of the system demands that both divergent and convergent movements should be oblique (transtensional then transpressional) along the northeastern and western sectors respectively, but more orthogonal on the northwestern sector.

The Labrador belt, forming the northeastern sector of this system, is similar in many respects to the Cape Smith and Belcher belts on the north-western side of the Superior craton (Figure 9.13). It consists of a western supracrustal zone (Figure 9.15), often termed the Labrador 'geosyncline' or 'trough', about 100 km across, with a much wider zone of modified Archaean basement rocks on its northeastern side. The belt is bounded to the southwest by the Superior craton, and to the northeast by the East Nain Province (part of the N. Atlantic craton — see Figure 9.13). The Labrador trough has been intensively studied, and is described by Dimroth (1981). To the southeast, the belt is truncated by the Grenville belt (see above) and to the northwest it bends round into the Cape Smith belt.

The supracrustal rocks of the Labrador trough consist of a thick sequence of sediments and volcanic rocks. The earliest deposits rest on Archaean basement of the margins of the supracrustal belt, and consist of about 1500 m of coarse arkosic red beds. These are overlain by shelf deposits of orthoquartzite, dolomite and iron-formation, reaching a maximum thickness in the west of about 1500 m also. The shelf deposits are partly eroded and unconformably overlain in the central part of the trough by locally-derived conglomeratic mass-flow deposits and greywackes. These are succeeded by voluminous basaltic and andesitic lavas more than 5 km thick. Most of the clastic sediments appear to have been derived from the foreland, and the greywackes are compositionally dis-

tinct from the typical synorogenic flysch of the Alps and other Phanerozoic orogenic belts.

The western margin of the belt is defined by a zone of imbricate thrust sheets dipping to the east. Those detach on a low-angle décollement plane as in the classical thin-skinned thrust model. East of the thrust belt is a zone of tight to isoclinal folds overturned to the west. Fold axes generally trend NNW and are associated with a strong foliation. Several generations of folds with associated foliations are superimposed on the large-scale overfolds in the more complex eastern zone. According to Dimroth, the structural pattern is strongly influenced by early syn-sedimentary growth faults defining a series of blocks, some of which have subsequently become overthrust.

The overall structure is illustrated in Figure 9.16. It is clearly asymmetric with a westerly vergence. The metamorphic grade increases from pumpellyite-prehnite in the west to amphibolite in the east. The metamorphism is believed to be caused by deep burial under a moderate geothermal gradient. Metamorphism was syntectonic in the west but syn- to post-tectonic in the east.

The age of the trough deposits is not accurately known, but must post-date the 2700 Ma-old Archaean basement. The main deformation and metamorphism have been dated at 1800–1600 Ma, mostly by the K–Ar method; however the data are poorly constrained and their significance unclear. The neighbouring Cape Smith volcanic rocks have yielded an Rb–Sr date of 2300 Ma, which probably represents the date of formation of the supracrustal sequence in both areas.

Dimroth emphasizes that the mafic volcanic sequence is not an ophiolite complex but bears more similarity to plateau basalts of submarine origin. Moreover, none of the units is allochthonous: the thrusts are not continuous over long distances, and basement inliers can be recognized in the interior of the belt. The shortening of about 100 km estimated from the exposed structure is attributed by Dimroth to A-subduction, in which an eastern lithosphere slab is detached from the crust and dips below

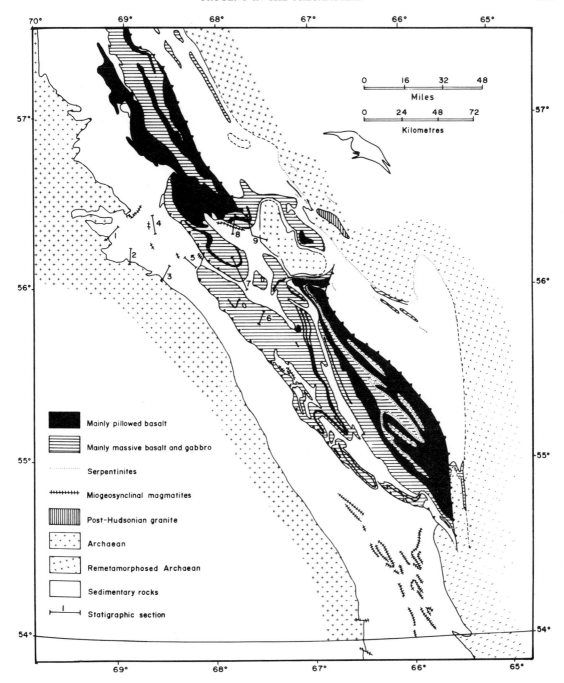

Figure 9.15 Simplified geological map of the central part of the Labrador belt. From Dimroth (1981)

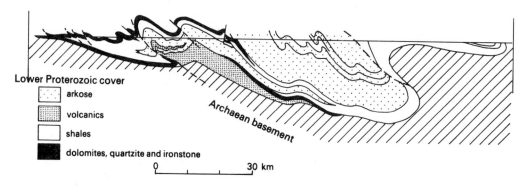

Lower Proterozoic cover

 ::: arkose

 ▦ volcanics

 □ shales

 ■ dolomites, quartzite and ironstone

0 _____ 30 km

Figure 9.16 Diagrammatic tectonic profile across the Labrador belt (see Figure 9.15) to illustrate the asymmetric arrangement of SW-verging thrusts and folds. After Dimroth (1981).

the western craton. He thus visualizes the crustal shortening being transferred to the base of the crust by the thrusting and then taken up by overlapping of the mantle lithosphere. Several questions are posed by this interpretation: the origin of the wide belt of reworked basement and Hudsonian granites in the east is not accounted for, nor does the model explain the early stretching phase in which the basin was formed. There seems no reason to suppose that the origin of this belt is any different from that of other intra-continental belts, such as the Damara belt already described, where early divergence and crustal thinning is followed by convergent shortening. How the mantle part of the lithosphere accommodates to such movements is as yet speculative. If the model suggested by Baragar and Scoates is correct, we might expect the convergent deformation in the belt to be achieved by dextral transpression.

The Lewisian-Nagssugtoqidian system: the western Nagssugtoqidian

The Nagssugtoqidian mobile belt of S. Greenland marks the northern boundary of the North Atlantic craton (Figure 9.17A). On the west coast, the belt is over 300 km in width. A major sinistral shear zone separates it from the Rinkian mobile belt, of probably similar age, to the north, but the relationship between the two belts is not yet understood. On the east

coast, the belt is 240 km across and is bounded to the north by an Archaean block. A general description of the belt is given by Escher et al. and Bridgwater (1976), and in a series of papers edited by Korstgård (1979). The belt consists predominantly of re-worked Archaean basement gneisses with interlayered and infolded belts of supracrustal metasediments and metavolcanic rocks (Figure 9.17B).

The supracrustal rocks are divided into an earlier Archaean group, consisting of pelites, impure marbles, and quartzites, and are associated with amphibolites of probable igneous origin. Younger supracrustal rocks that postdate the Archaean metamorphism occur only in isolated outcrops, mainly in the northern part of the belt. These consist of pelitic schists, quartzites, marbles and amphibolites. The quartzites preserve current bedding, and the metasediments as a whole exhibit a lower degree of metamorphism than the basement gneisses. Only a few small granitic intrusions occur, which post-date the main Nagssugtoqidian deformation. Pegmatites, on the other hand, are abundant. In the southern part of the belt, a regional swarm of metadolerite dykes, the Kangâmiut dykes, intrudes the Archaean craton; these dykes are progressively deformed and metamorphosed within the Nagssugtoqidian belt.

The Nagssugtoqidian deformation has produced several broad, NE–SW, steep shear zones separated by areas of weaker deforma-

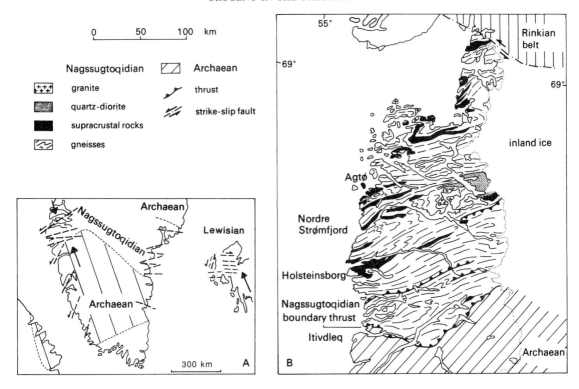

Figure 9.17 The Early Proterozoic Nagssugtoqidian mobile belt of Greenland. (*A*) Regional summary sketch (pre-Mesozoic restoration) of the Nagssugtoqidian-Lewisian belt situated on the northern margin of the Archaean North Atlantic craton. The coeval Ketilidian belt lies on the south side of the craton. Arrows denote inferred movement directions of the craton in relation to the mobile belt inferred from shear zones. After Watterson (1978). (*B*) Tectonic summary map of the Nagssugtoqidian belt of SW Greenland showing structural trends and distribution of main rock units. Note the thrusts at and near the boundary with the Archaean craton in the south.

tion. Figure 9.18 is a map of the Nordre Strømfjord shear zone, south of Agtø, which is a 16 km-wide zone with a sinistral strike-slip sense of movement. Near the southern margin of the belt, between Itivdleq and Holsteinsborg, two main phases of Nagssugtoqidian deformation can be distinguished: the earlier, termed *Nag 1* by the Greenland geologists, produces dextral strike-slip movements on the steep E–W Itivdleq shear zone, and pre-dates the emplacement of Kangâmiut dykes. The later *Nag 2* phase is a regionally more important and pervasive deformation that results in overthrusting to the southeast. Detailed studies of the structure of this critical marginal region of the belt have been made by Watterson and his colleagues (eg. see Grocott, 1979). The Nagssugtoqidian metamorphism is high-

grade throughout the belt; a central granulite-facies zone is bounded to the north and south by regions of amphibolite facies.

The Lewisian complex

After removing the effects of the North Atlantic opening, the eastern Nagssugtoqidian and the Lewisian complex of NW Scotland lie along-strike, and only about 400 km apart (Figure 9.22*A*). Similarities between the two belts have been noted by Myers (1987). However as the Lewisian complex is probably the most intensively studied piece of Precambrian crust in the world, we shall concentrate our attention now on that region. A general description of the Lewisian complex is provided by Park and Tarney (1987), and a structural

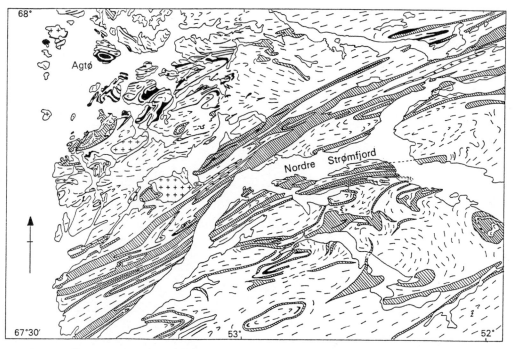

Figure 9.18 Simplified structural map of the Nordre Strømfjord shear zone in the central part of the belt (see Figure 9.17B). This sinistral shear zone is defined by various rock units and structural trends that become aligned within a NE–SW belt of intense deformation. Black, amphibolites; ruled ornament, metasediments; crosses, granitic and charnockitic intrusions; blank areas, granitic to tonalitic gneisses. After Olesen *et al.* (1979).

interpretation by Coward and Park (1987). The complex forms a well-exposed strip along the northwest coast of the Scottish mainland and in the islands of the Outer Hebrides (Figure 9.19). The exposed width of the belt is about 260 km but, unlike the eastern Nagssugtoqidian, the margins of the belt are not seen. The bulk of the complex is formed of Archaean tonalitic to granodioritic gneisses that are preserved in a relatively unmodified state in the central mainland region and in several small enclaves elsewhere. The remainder of the Lewisian complex has experienced intense Early Proterozoic deformation and high-grade metamorphism.

In addition to the Archaean basement gneisses, the complex includes two narrow belts of Early Proterozoic supracrustal rocks and associated intrusions, at South Harris in the Outer Hebrides and at Gairloch and Loch Maree on the mainland. The supracrustal assemblage, known on the mainland as the *Loch Maree Group*, consists of mafic volcanic rocks with associated narrow bands of siliceous schist, banded-iron-formation, graphite schist and marble, overlain by a thick sequence of metagreywackes. These rocks were deposited probably around 2000 Ma BP. The other important Proterozoic addition to the complex is the well-known Scourie dyke swarm. These mafic dykes are dated at *c*.2400 Ma BP, but some members of the swarm may be as young as 1900 Ma in age. They occupy the same key stratigraphic position in Early Proterozoic chronology as the Kangâmiut dykes in Greenland, separating the earlier *Inverian* deformation from the later *Laxfordian* phases.

As in W. Greenland, the Early Proterozoic deformation appears to be related to the development of major steep shear zones, the effect of which, over much of the Lewisian, has been obscured by the younger deformations.

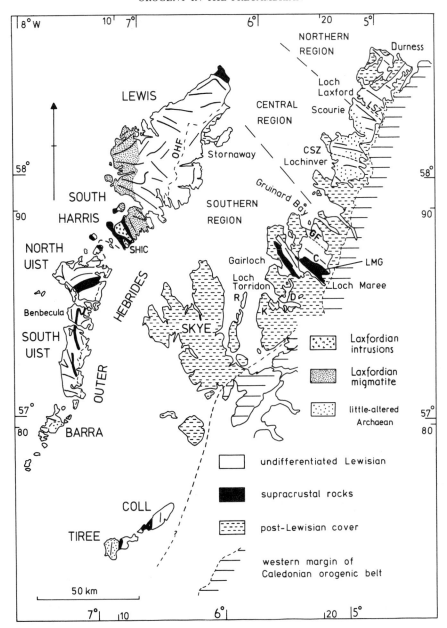

Figure 9.19 Location map illustrating some important features of the Lewisian complex of NW Scotland. *OHF*, Outer Hebrides fault; *SHIC*, South Harris igneous complex; *LSZ*, Laxford shear zone; *CSZ*, Canisp shear zone; *GF*, Gruinard 'front'; *LMG*, Loch Maree Group; *C*, Carnmore; *D*, Diabaig; *K*, Kenmore; *R*, Rona. From Park and Tarney (1987).

The major Inverian shear zones probably occupied the whole of the northern and southern regions of the mainland Lewisian, and most of the Outer Hebrides. Their initiation is thought to be associated with the uplift of the central Archaean block that is dated at *c*.2500 Ma by the K–Ar method. The Inverian deformation took place under amphibolite-facies conditions, resulting in extensive retrogression of Archaean granulite-facies assemblages.

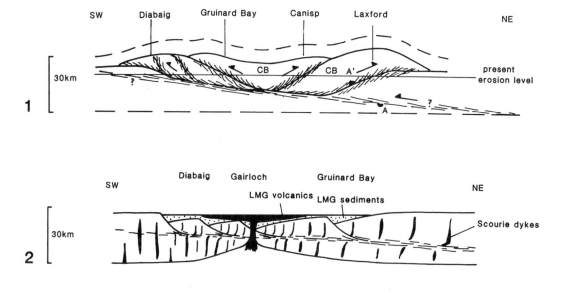

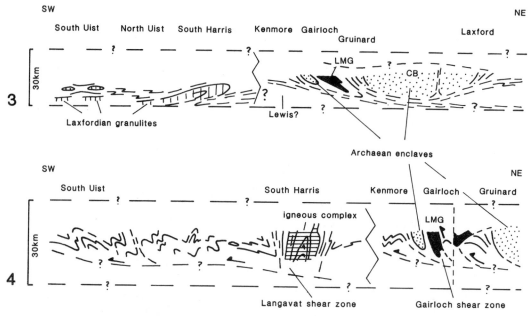

Figure 9.20 Sequence of cartoon profiles illustrating the Early Proterozoic tectonic evolution of the Lewisian complex. (1) The Inverian event, interpreted as resulting from the initiation of a network of shear zones: note that deep-crustal granulites from *A* are transferred to *A'*. (2) The emplacement of the Scourie dykes and the Loch Maree Group (*LMG*) in an extensional (transtensional) environment. (3) The Laxfordian D1–D2 events, attributed to movements approximately perpendicular to the line of section on the shear zone network.established in (1). Note that the relationship between the Outer Hebrides and mainland parts of the composite profile are speculative. (4) The Laxfordian D3 event, producing upright folds and shear zones by movements oblique to the line of section and with a component of compression along it. CB, central block (central region of Figure 9.19). From Park and Tarney (1987)

Minor Inverian shear zones also cut the otherwise unmodified Archaean central block. According to Coward and Park, the major zones dip beneath the central block, detaching on a low-angle shear zone that underlies the block (Figure 9.20 (1)). The Inverian shear zones of the mainland exhibit a mainly dip-slip sense of movement but with a small dextral strike-slip component.

The Scourie dyke swarm was emplaced over a wide area from the north coast to as far south as Barra in the Outer Hebrides. In areas where they are less deformed, the dykes exhibit evidence of dilatational emplacement in a dextral shear regime. An extensional regime is also indicated by the emplacement of the supracrustal rocks of the Loch Maree Group (Figure 9.20 (2)). Although deposited possibly 400 Ma after the earliest dated Scourie dykes, it is likely that later members of the dyke swarm may have been associated with the emplacement of the Group, and supracrustal rocks at Gairloch are cut by dykes with typical Scourie dyke chemistry. The dyke swarm and the supracrustal basin may represent a long-continued extensional or transtensional regime.

Although four phases of Laxfordian (post-dyke) deformation can be recognized over much of the Lewisian outcrop, the sequence can be simplified to two events of major regional significance. The earlier (D1/D2) is associated with amphibolite-facies metamorphism, and is more pervasive and intense than the second (D3), which results in a refolding of the D1–2 fabric under mainly greenschist-facies conditions. The D3 phase is responsible for the widespread, upright, NW–SE folds which are a prominent feature of the Lewisian outcrop pattern. Minor pegmatite and granite sheets with an age of $c.1800$ Ma post-date D1–2 and pre-date D3. The latter event is provisionally dated at $c.1600$ Ma from widespread resetting of K–Ar ages in the range 1600–1400 Ma.

The major D1–2 shear zones are oriented NW–SE, approximately parallel to the Inverian structures and to the Scourie dykes. They vary in attitude: steep zones occur within the central block; inclined zones at Laxford in the north and Torridon in the south dip beneath the central block and detach on a low-angle shear zone that comes to the surface in the north and south, according to the model of Coward and Park (Figure 9.20 (3)). The whole of the Outer Hebrides is interpreted as part of this originally gently-inclined zone. The shear zones are regarded as an inter-connected network of displacement planes separating less deformed or undeformed blocks whose relative movements can be established by studying the sense of movement on the shear zones. The inclined D1 zones show typically moderate-plunging lineations indicating both dextral strike-slip and normal dip-slip components, whereas the sub-horizontal 'flats' display NW–SE movement directions (Figure 9.21). The D1–2 deformation pattern overall is probably indicative of a continued component of dextral strike-slip movement. During the D3 deformation, two major steep NW–SE shear zones were formed, at Langavat in South Harris, and at Gairloch. Both exhibit a dextral strike-slip component of movement and are associated with strong compressional shortening across the belt, forming the prominent NW–SE upright F3 folds (Figure 9.20 (4)). The D3 regime is interpreted as dextral transpressional overall.

These changes in tectonic regime are summarized in Figure 9.22, which shows how they might be interpreted by changes in convergence direction across the Lewisian-Nagssugtoqidian belt. The first phase (Inverian-Nag 1) may relate to overall N–S or NNW–SSE convergence. This arrangement was originally suggested by Watterson (1978) to explain the conjugate pattern of Early Proterozoic shear zones. It relates the coeval sinistral displacements on NE–SW zones in W. Greenland and the dextral/overthrust displacements on the NW–SE zones in Scotland. The early Laxfordian-Nag 2 regime is marked by a change to dextral strike-slip in Scotland and overthrust in W. Greenland.

There is no evidence either in the Lewisian or in the Nagssugtoqidian complexes of colli-

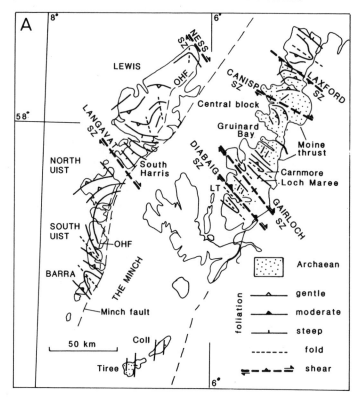

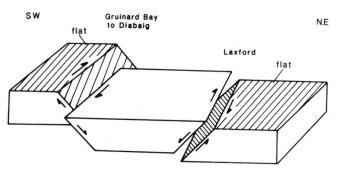

Figure 9.21 Block diagram illustrating the effects of oblique-slip extensional movements on the inclined shear zones at Laxford and between Gruinard Bay and Diabaig, combined with NW–SE strike-parallel movements on the shear-zone flats to the northeast and southwest. From Coward and Park (1987).

sion sutures or of former ocean basins. Nor is there any indication of calc-alkaline magmatism that might betray the former presence of a subduction zone. The belt appears to consist almost entirely of pre-existing Archaean basement that has been subjected to essentially intraplate tectonic movements. These movements have not only caused intense deformation, but have resulted in considerable crustal heating and the local emplacement of magmas.

It is instructive to compare this intraplate belt with the coeval Ketilidian belt to the south (in S. Greenland) and with the Svecokarelian belt to the east (see Figure 9.12A). These belts display abundant calc-alkaline volcanic and plutonic magmatism, and are widely thought to represent an Early Proterozoic destructive continental margin. It is tempting, following Watterson (1978) to ascribe the intraplate deformation of the belts we have just examined to processes occurring at that margin, about 1000 km to the south.

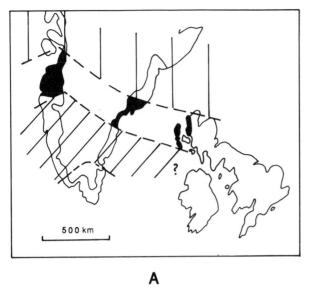

A

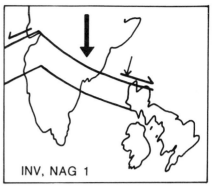

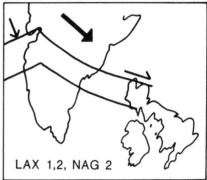

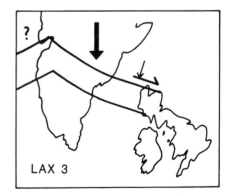

B

Figure 9.22 (*A*) Reconstruction of the Early Proterozoic belts of Greenland and Scotland after removing the effects of the North Atlantic opening. The restoration is based on the removal of oceanic crust and on the assumption of an average 50% thinning of continental crust on the continental shelves. The Nagssugtoqidian–Lewisian belt (black) appears to lie between two more stable Archaean 'plates' to the north and south (ruled ornament). The Ketilidian belt lies on the south side of the Archaean craton of S. Greenland. (*B*) Sequence of cartoon diagrams illustrating an interpretation of the kinematic history of the belt, based on a change in movement direction of the northern plate with respect to the southern. Dominantly convergent movement during the Inverian and Nag. 1 period changes to dominantly strike-slip in the Lewisian, but convergent in the western Nagssugtoqidian, during Laxfordian D1–2 Nag. 2, and back to dominantly convergent in Laxfordian D3 times in the Lewisian. From Coward and Park (1987).

9.5 The Archaean: a different kind of orogeny?

Rocks of Archaean age form a number of stable cratons within the Proterozoic shield regions of all the main continental masses (see e.g. Figure 9.13). In addition, a large proportion of the Proterozoic shields consists of reworked Archaean crust, as we have seen. Archaean regions are traditionally divided into two quite different types: the *granite-greenstone terrains* and the *high-grade gneiss terrains*. The granite-greenstone terrains consist of *greenstone belts* surrounded and cut by granitoid plutons, and metamorphosed typically in greenschist or lower facies. The high-grade gneiss terrains consist predominantly of granulite- to amphibolite-facies gneisses of varying types but including a high proportion of broadly granitic composition.

The high-grade gneiss terrains are the product of tectonothermal activity of a similar nature to that associated with younger Precambrian mobile belts, although the Archaean terrains exhibit certain special characteristics. The granite-greenstone terrains are unique to the Archaean: there are no precise analogues in the younger stratigraphic record. Certain Archaean cratons consist entirely of one or other of these two types of terrain, while in others the two are found in association. We shall discuss two examples in detail: the high-grade gneiss terrain of the *North Atlantic craton*, and the *Superior Province* of the Canadian shield. The latter is basically a granite-greenstone terrain but is crossed by several belts of high-grade gneiss, and is bordered by regions of high-grade gneiss terrain on its north-western and north-eastern sides. *The North Atlantic craton.* The Archaean high-grade gneiss terrain of S. Greenland and the

adjoining part of Labrador is known as the North Atlantic craton (Bridgwater *et al.*, 1973). This craton (Figure 9.23), about 700 km across from north to south, and over 500 km from west to east, was a region of continuous mobility during Archaean times, and consists almost solely of gneisses in upper amphibolite to granulite facies. Between 80 and 90% of these gneisses are broadly granitic in composition, predominantly tonalitic to granodioritic. Within the gneisses are relatively narrow bands and inclusions of metasedimentary gneisses such as quartzites, pelitic and semipelitic schists, marbles and banded-iron-formation, and of meta-igneous amphibolites and anorthosites. In the past, the origin of the granitoid gneisses has been hotly debated, and the opinion was widely held that many of the gneisses represented granitized sediments of broadly semipelitic composition. However, modern geochemical studies have demonstrated that the bulk of the gneisses are deformed and metamorphosed calc-alkaline rocks probably of plutonic origin (see e.g. Weaver and Tarney, 1987). The sedimentary assemblage is suggestive of an epicontinental shelf environment, and contrasts markedly with the greenstone-belt assemblage. Another important component of the terrain is the anorthosite-leucogabbro complex described by Windley (1973). This complex consists of an associa-

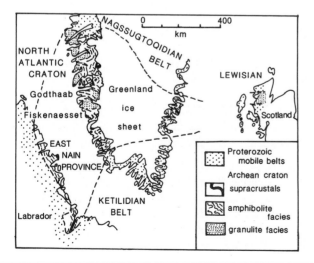

Figure 9.23 Summary tectonic map of the North Atlantic craton, with the surrounding Early Proterozoic belts. After Bridgwater *et al.* (1976).

tion of anorthosite, leucogabbro, and minor gabbro, and exhibits prominent igneous layering. The assemblage is exposed over a large area (Figure 9.24) due to complex folding, but is considered to represent a single sheet.

The rocks of the craton have undergone intense and repeated deformation and metamorphism over a period of more than 100 Ma.

Bridgwater *et al.* (1976) propose a sequence of 15 separate events for the Archaean of S. Greenland, summarized in Table 9.3. The sequence may be divided into two main crust-forming cycles. The earlier cycle (events 1–3, Table 9.3) culminated in the emplacement of granites at *c*.3750 Ma BP. The later cycle includes a number of separate intrusive events,

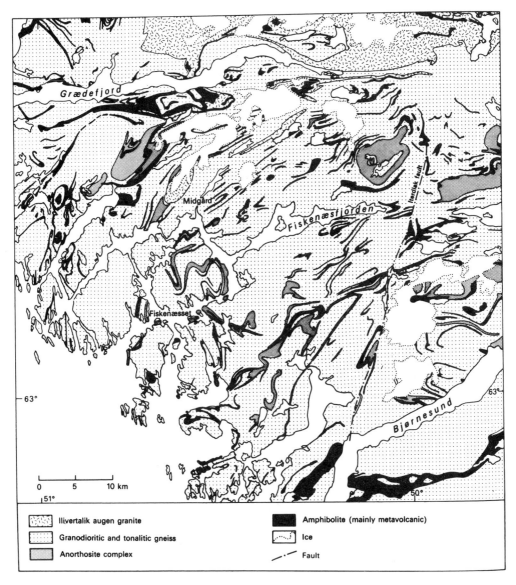

Figure 9.24 Simplified geological map of the Fiskenaesset region, in the central part of the Archaean craton on the west coast of Greenland. From Bridgwater *et al.* (1976)

Table 9.3 Simplified sequence of Archaean events in S. Greenland. After Escher *et al.* (1976).

Ma		
?	1.	Formation of early crust (source for Isua sediments)
	2.	Deposition of Isua sediments and volcanic rocks
3750	3.	Intrusion of Amitsoq granitic rocks
	4.	Deformation and metamorphism
	5.	Emplacement of Ameralik basic dyke swarm
	6.	Deposition of Malene sediments and volcanic rock
	7.	Emplacement of stratiform gabbro-anorthosites
	8.	Intense deformation
3040	9.	Emplacement of ultrabasic bodies and calc-alkaline granitic sheets (Nuk gneisses)
	10.	Intense deformation
3000–2800	11.	Emplacement of granites and other igneous bodies
3000–2700	12.	High-grade metamorphism
	13.	Deposition of Tartoq Group supracrustal rocks
*c.*2700	14.	Localized deformation in shear zones
*c.*2600	15.	Emplacement of K-granites and regional pegmatites

of which the most important is the emplacement of the Nuk granitic suite and the accompanying deformation and metamorphism in the period 3040–2700 Ma BP. There is very little evidence as to the history of the region in the intervening period of about 700 Ma.

The Isua supracrustal assemblage of the earlier cycle bears some similarity to that of the greenstone belts described below. It consists of a mafic and ultramafic volcanic suite with associated metasediments including carbonates, banded-iron-formation, quartzites and metagreywackes. The supracrustal belt is only about 2 km wide at its maximum, but extends for over 30 km in an arcuate outcrop. Because of the limited outcrop and the effects of later events, it is not possible to draw definite conclusions about the tectonic environment of these very early rocks, except that they indicate, in a general way, a similarity in all essential respects to the much later (*c.*3300 Ma old) greenstone belts of Africa, Australia and elsewhere.

As pointed out earlier, the Pb-isotopic evidence indicates that the early crust-forming event is limited to a relatively small area and that the bulk of the continental crust of the craton was added during the younger cycle. The Malene supracrustal assemblage of this younger cycle represents a sequence that is much more typical of the high-grade gneiss terrains in general. It consists of basic metavolcanic rocks, including well-preserved pillow lavas, together with semipelitic to pelitic gneisses, pure and impure marbles, and thin quartzites. This assemblage is very widely distributed throughout the craton, and has been interpreted by Bridgwater and Fyfe (1974) as indicating small marine basins overlying thin continental crust. However, other authors (Burke *et al.*, 1976; Windley and Smith, 1976) view the high-grade terrains, including the North Atlantic craton, as the product of Andean-type active continental margins created by the subduction of oceanic lithosphere. The voluminous calc-alkaline tonalitic magmas in their view are generated by subduction. It has been pointed out that the smaller thickness and lower relative density of Archaean oceanic lithosphere may produce much shallower angles of subduction (see e.g. Dewey, 1977) that would have important implications for the width of the mobile belt, the pattern of magma emplacement, and the style of deformation.

A dominant feature of the North Atlantic craton and of other high-grade terrains is a high-strain structure, produced by very intense deformation, which is expressed in complex interleaving of basement, supracrustal cover, and various intrusive igneous sheets on a regional scale. This structure appears to have been initially sub-horizontal, although subsequently refolded by more upright folds. The deformation producing this structure in S. Greenland embraces events 8 and 9 of Table 9.3. Figure 9.24 shows the outcrop pattern of the Fiskenaesset anorthosite sheet. Although refolded by later structures, it is still traceable over an area of around 3600 km, indicating that the high-strain structure is effectively horizontal over areas of that size.

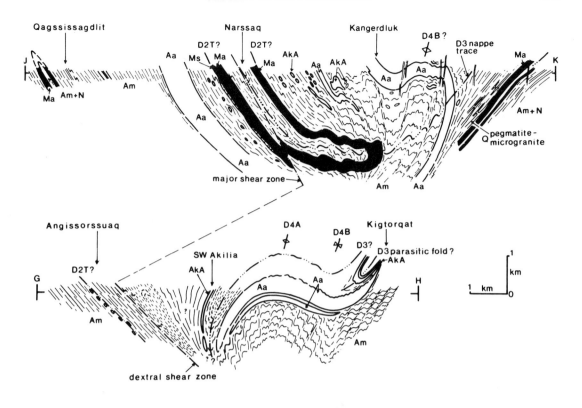

Figure 9.25 WNW–ESE structural profiles across the northwest Buksefjorden area in the central west-coast Archaean. The section illustrates the refolding of the early recumbent fold-thrust structure (D1–3) by later upright folds (D4). *Ma*, Malene amphibolite; *Am*, Amitsoq gneiss; *N*, Nuq gneiss; *Aa*, Amitsoq augen gneiss; *Ms*, Malene supracrustal gneisses; *AkA*, Akilia association; *Q*, pegmatite and microgranite; D2T, D2 thrust. From Chadwick and Nutman (1979)

Chadwick and Nutman (1979) describe in detail the structural evolution of the Buksefjorden area in the central part of the west coast outcrop of the craton. They divide the structural sequence into four main events, of which the first is related to the early Amitsoq crust-forming cycle. Their D2 event is the first regional high-strain deformation (corresponding to event 8 of Table 9.3) and resulted in the isoclinal folding and thrusting of Amitsoq basement and Malene cover. The D3 structures are large recumbent nappes, refolding the D2 structures and also deforming the Nûk granitic sheets. D4 produced the upright folds that dominate the outcrop pattern. High-grade metamorphism accompanied the D3 deformation and continued after D4.

No modern analogue has been established for the regional, sub-horizontal, high-strain structure of the D2–D3 type. According to the uniformitarian view expressed by Windley (1981) and others, the high-strain horizontal structures would be found at low levels in a typical modern volcanic arc complex. However, this analogy has not yet been satisfactorily demonstrated.

Park (1981, 1982) discusses a number of possible mechanisms that could explain this type of structure: *subduction, gravity spreading, mantle decoupling, thinned-crust collision,* and *tectonic underplating (A-subduction)* (Figure 9.26). Major differences in the subduction angle, the thickness and strength of the lithosphere, and the ease of detachment of

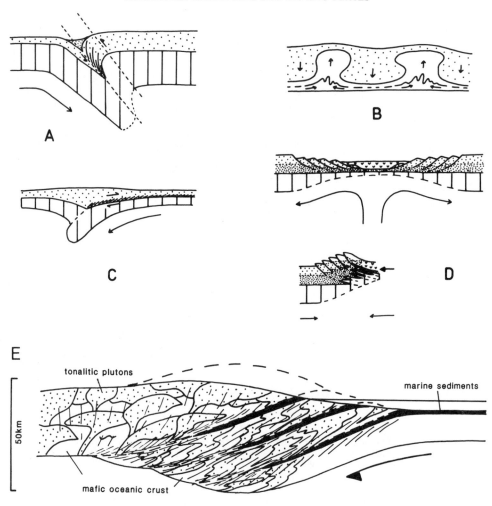

Figure 9.26 Possible mechanisms for explaining the high-strain structures of Archaean high-grade terrains. (*A*) Subduction: more likely to produce steep to moderately inclined high-strain zones at deep levels in the crust. (*B*) Gravity spreading associated with solid diapirism: requires sub-horizontal mass flow in source area for domes (blank) which would produce radial or convergent constrictional strain patterns (not observed). (*C*) Mantle decoupling: horizontal shear zone forms at the base of the crust (see Figure 2.29) due to differential movements in either convergent or divergent regimes (this mechanism is similar to that required in Ampferer subduction). (*D*) Compression of thinned crust: extensional strain in the extensional phase is re-inforced by thrust-sense shear zones during subsequent convergence. (*E*) Tectonic underplating: this process is associated with Ampferer subduction, and requires the tectonic separation of crust from mantle lithosphere. Repeated slicing of the crust juxtaposes supracrustal and deep-crustal material as found in SW Greenland. (*A*)–(*D*) from Park (1982); (*E*) from Park and Tarney (1987).

continental from oceanic crust, or from mantle lithosphere, can be expected in the Archaean. Any tectonic model for the Archaean at present will be highly speculative, and no generally acceptable model yet exists. If we assume that subduction operated at a faster rate during the Archaean than at present,

the 3000–2700 Ma crust-building cycle of the North Atlantic craton must presumably involve the welding together of a whole series of volcanic arcs. The structural pattern should therefore reflect a succession of extensional, translational and collisional events associated with the gradual accumulation and tectono-magmatic thickening of the Archaean crust. The high-strain horizontal structures probably formed in low-angle shear zones representing large horizontal translations at deep crustal levels at some stage in this process.

Greenstone belts of the Superior Province

The Superior Province (Figure 9.27) is one of the best examples of a granite-greenstone terrain (see e.g. Goodwin, 1981). It is the largest of the Archaean cratons, being about 450 × 300 km in extent, and is separated from the neighbouring North Atlantic craton by the 300 km-wide Labrador belt (see Figure 9.13). The Province contains two main components: crystalline granitoid rocks varying from undeformed igneous plutons to highly-deformed gneisses, and the greenstone belts, which are outcrops of supracrustal sequences consisting of volcanic rocks of various kinds, predominantly mafic, together with metasediments. The proportion of greenstone belt to 'granite' varies throughout the province, being much higher in the south than in the north (Figure 9.27). High-grade gneisses occur in several narrow belts, and also in the north-western and north-eastern parts of the province. Since the greenstone belt outcrops are usually delimited by the discordant margins of younger plutons, the former extent, shape and relationships of the belts is impossible to reconstruct. It is thought that many of the greenstone outcrops represent the fragmented relics of much larger greenstone basins. However, some of the large greenstone belts, such as the Abitibi belt, contain more localized basins that existed for part of their evolutionary history.

The supracrustal assemblages of the greenstone belts consist of roughly equal proportions of sediments and volcanic material deposited around 2700 Ma BP. There is typically a lower sequence dominated by mafic lavas, and an upper dominated by coarse clastic sediments, especially greywackes. Frequently, an unconformity separates the two sequences. In the Abitibi belt in the southeast, a number of separate volcanic piles can be recognized, that grade laterally into sediments. The lavas are commonly tholeiitic basalts and andesites, but rhyolites, rhyo-dacites and ultramafic types also occur. Of lesser importance volumetrically are sediment types such as quartzites, banded-iron-formation, sulphide-rich black shales, and minor carbonates. These sediments are usually found in association with the earlier volcanic sequences, whereas the coarse clastic deposits are more common in the upper part of the succession. Estimated thicknesses shown in Figure 9.27 range from 20 000–59 000 feet (6–18 km). While some of these figures may be overestimates, due to unrecognized structural repetition, there can be no doubt that the greenstone sequences are commonly of the order of 5–10 km in thickness. Although the original extent of the basins is impossible to estimate, it is thought that, in general, greenstone sequences were formed in relatively shallow marine basins of probably regional extent. Through time, vertical movements thought to be due to differential loading of the granitic basement, caused a restriction of the basins to approximately their present size, and gave rise to the younger clastic-dominated sequences (see e.g. Bickle and Eriksson, 1982).

The granitoid element of the terrain consists partly of gneissose basement that pre-dates the formation of the greenstone belts, and partly of younger post-greenstone plutons. In practice, it is not possible at present to determine the origin of much of the granite outcrop because of poor geochronological control. Ages of c.2900–3000 Ma have been obtained from some of the older basement gneisses, in which fragments of an older series of greenstone belts occur.

Overall, the province exhibits an E–W structural trend, expressed in the elongate elliptical shape of many of the granitoid areas, by the orientation of the greenstone belts themselves, and also by the orientation of the high-grade gneiss belts. On the scale of the individual greenstone belts, the deformation pattern is much more variable. Figure 9.28A is a map showing the granite-greenstone relationships in the Kenora district, in the south-

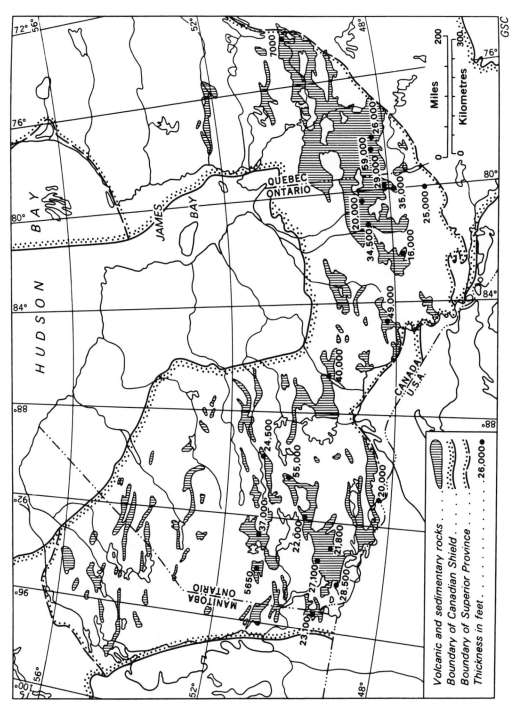

Figure 9.27 Distribution of greenstone belts (ruled ornament) in the Superior province of the Canadian shield. The blank areas are mainly granitoid rocks. The Abitibi belt is the large greenstone outcrop in the southeast, crossing from Ontario to Quebec. From Stockwell et al. (1970), with permission.

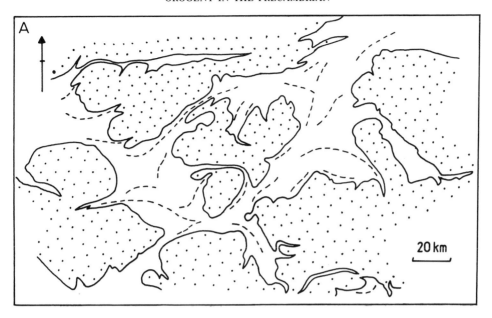

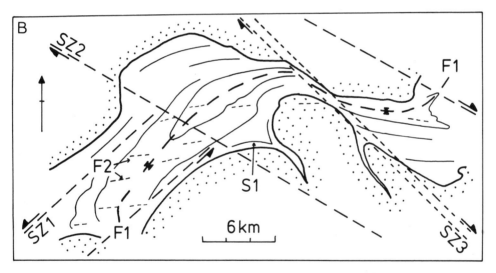

Figure 9.28   Tectonic patterns in greenstone belts (*A*) Granite-greenstone relationships in the Kenora area, SW Superior province. Granite, dotted; greenstone outcrop, blank; fold axes, dashed. Note the roughly equidimensional shape of the granites, and how they exhibit convex margins to the greenstones, which, together with the fold axes, 'wrap around' them. (*B*) Simplified structural map of the Bigstone Lake greenstone belt, northern Manitoba. The early foliation (S1) wraps around the granite margins whereas the later deformation appears to relate to large strike-slip shear zones (SZ1–3). (*A*), (*B*) from Park (1982).

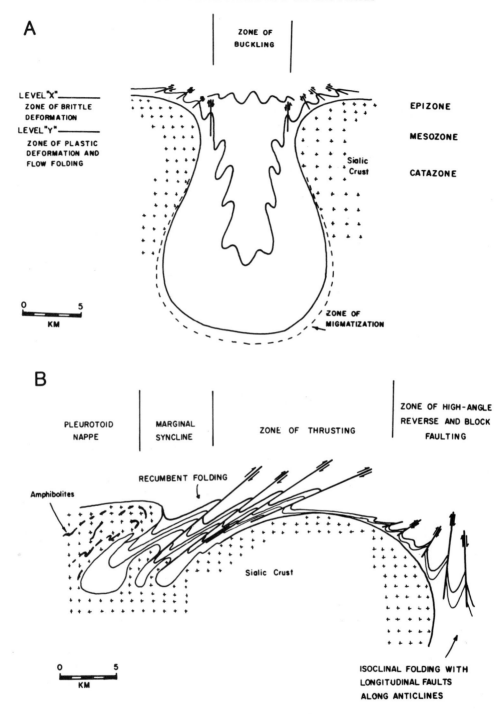

Figure 9.29   Structural models to show the patterns of structures associated with the solid diapirism model for greenstone belt evolution. (*A*) Structures expected near the centre of the subsiding greenstone belt. Attitude of faults and fold axial planes changes from steep to inclined, moving from the central 'sink' towards the margins. (*B*) Structures expected near the margins of a subsiding greenstone belt. A highly asymmetric pattern of folds and thrusts is predicted at the site of the rim syncline produced by marginal depression of the greenstones. (*A*), (*B*) from Gorman *et al.* (1978).

western part of the province. Here, the granite plutons are weakly deformed or undeformed, and more or less equidimensional, and their original relationships to the greenstone outcrops is clear. The rocks of the greenstone sequence are deformed in such a way that the fold axes and foliations 'wrap around' the margins of the plutons. This relationship, obscured in many belts further north because of more intense late Archaean deformation, is typical of greenstone belts of many other granite-greenstone terrains.

In a study of the Bigstone Lake greenstone belt in northern Manitoba, Park and Ermanovics (1978) demonstrated that the deformation sequence in the greenstone belt consisted of two main phases. The first is associated with regional high strains, penetrative fabrics, and isoclinal folds which locally are seen to wrap around the margins of the bordering plutons (Figure 9.28B). The second produces local refolding and crenulation schistosities, and is associated with major steep ductile shear zones that cut across the belt and affect both greenstone and granite outcrops.

The type of structural pattern associated with the first deformation here, and in the Kenora outcrops, is most easily explained by the diapiric behaviour of a solid granitoid basement loaded by a denser greenstone basin. The gravitational effect of the density inversion is to induce upward and outward flow of the granitoid substratum. The greenstone material becomes trapped in synclinal keels whose shapes are controlled by the diapir margins. This process has been discussed by Gorman *et al.* (1978) and Schwerdtner *et al.* (1979) and modelled numerically by Mareschal and West (1980). Figure 9.29 illustrates the structural patterns expected in central and marginal areas of greenstone belts according to the Gorman *et al.* model. The symmetrical upright or mildly-fanned D1 structure of the Bigstone Lake belt fits profile A of the model. Tight overfolds and thrusts found as early structures in other greenstone belts (see e.g. Coward, 1976, figure 4) fit profile B of the model.

The later steep ductile shear zones that affect the Bigstone Lake belt are part of a regional set of such structures that dominate the later Archaean deformation of the Superior Province. Park (1981) shows that, in the

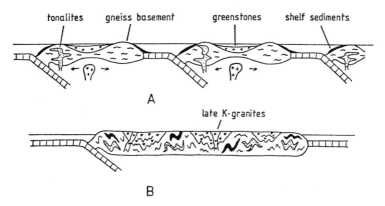

Figure 9.30 Uniformitarian plate-tectonic interpretation of high-grade gneiss and greenstone terrains. High-grade terrains are interpreted as Andean-type convergent margins and greenstone belts as back-arc basins (*A*). Collision welds microcontinents together and causes greenstone deformation (*B*) Contrast Figure 9.26. After Windley (1977).

western part of the province, these shear zones form a conjugate set of NE–SW sinistral and NW–SE dextral zones that indicate a late Archaean N–S compression across the province, and suggests that the craton was by that time strong enough to transmit a constant regional stress. This tectonic regime is very similar to the one described for the Early Proterozoic shear zones of the S. Greenland-Scotland region (see Figure 9.17*A*). Both sets of structures probably formed around the time of the Early Proterozoic-Archaean boundary between 2600 and 2500 Ma BP.

The origin and tectonic significance of the greenstone basins is still speculative. Figure 9.30 shows the Tarney-Windley model in which the greenstone belts are interpreted as back-arc extensional basins related to subduction. In an alternative model suggested by Drury (1977), the basins are visualized as resulting from lateral compression of a thin, weak lithosphere exerted by a shallow-subducting slab. Park (1982) interprets the basins as intraplate extensional structures analogous to the initial stages of present-day continental rift zones.

None of these explanations seems entirely satisfactory. The subduction-collision model requires a number of unrecognized sutures; the compressional model fails to explain the initially non-linear pattern of the greenstone deformation; and the intraplate model ignores the bimodal nature of the volcanic suites, which argues for a volcanic arc analogue.

Opinion since these papers were published appears to favour some kind of extensional basin model, involving thinned continental crust in the Superior examples, but oceanic crust in some of the older examples, closely associated with volcanic island arcs derived by subduction. Present-day back-arc basins are the closest modern analogues, but many important differences are apparent between such basins and the granite-greenstone terrains.

The granite-greenstone terrains represent one of the best demonstrations in the geological record, that uniformitarian principles and present-day plate tectonic analogues may not always be appropriate in dealing with the distant past.

References

Åhäll, K.-I. and Daly, J.S. (1985) Late Presveconorwegian magmatism in the Östfold-Marstrand belt in Bohuslän, SW Sweden. In Tobi, A.C. and Touret, J.L.R. (eds.), *The Deep Proterozoic Crust in the North Atlantic Provinces*, D. Reidel, Dordrecht, 359–367.

Ahorner, L. (1975) Present-day stress field and seismotectonic block movements along major fault zones in Central Europe. *Tectonophysics* **29**, 233–249.

Aleinikov, A.L., Bellavin, O.V., Bulashevich, Yu.P., Tavrin, I.F., Maksimov, E.M., Rudkevich, M.Ya., Nalivkin, V.D., Shablinskaya, N.V. and Surkov, V.S. (1980) Dynamics of the Russian and West Siberian platforms. In Bally, A.W., Bender, P.L., McGetchin, T.R. and Walcott, R.I., *Dynamics of Plate Interiors*, American Geophysical Union, Geological Society of America, Geodynamics Series, vol.1, 53–71.

Allegré, C.J. and others (1984) Structure and evolution of the Himalaya–Tibet orogenic belt. *Nature (London)* **307**, 17–36.

Allmendinger, R.W., Sharp, J.W., Von Tish, D., Serpa, L., Kaufman, S. and Oliver, J. (1983) Cenozoic and Mesozoic structure of the eastern Basin and Range province, Utah, from COCORP seismic-reflection data. *Geology* **11**, 532–536.

Anderson, D.L. (1971) The San Andreas fault. *Scientific American* **225**, 87–102.

Anderson, D.L. and Bass, J.D. (1986) Transition region of the Earth's upper mantle. *Nature (London)* **320**, 321–327.

Anderson, E.M. (1951) *The Dynamics of Faulting*. Oliver and Boyd, Edinburgh.

Anderson, R.E. (1971) Thin-skin distension in Tertiary rocks of southeastern Nevada. *Bull. geol. Soc. Am.* **82**, 43–58.

Anderton, R. (1980) Did Iapetus start to open during the Cambrian? *Nature (London)* **286**, 706–708.

Angelier, J. (1979) Determination of the mean directions of stresses from a given fault population. *Tectonophysics* **56**, 17–26.

Angelier, J. (1985) Extension and rifting: the Ziet region, Gulf of Suez. *J. struct. Geol.* **7**, 605–612.

Angelier, J. and Colletta, B. (1983) Tension fractures and extensional tectonics. *Nature (London)* **301**, 49–51.

Angelier, J., Lyberis, N., Le Pichon, X., Barrier, E. and Huchon, P. (1982) The tectonic development of the Sea of Crete: a synthesis. *Tectonophysics* **86**, 159–196.

Argand, E. (1916) Sur l'arc des Alpes Occidentales. *Eclog. geol. Helv.* **14**, 145–191.

Argand, E. (1924) La tectonique de l'Asie. *Repts. 13th Int. geol. Congr. Brussels*, vol.1, 170–372.

Arndt, N.T. (1983) Role of a thin, komatiite-rich oceanic crust in the Archaean plate-tectonic process. *Geology* **11**, 372–375.

Arthaud, F. and Matte, P. (1977) Late Paleozoic strike-slip faulting in southern Europe and northern Africa: result of a right-lateral shear zone between the Appalachians and the Urals. *Bull. geol. Soc. Am.* **88**, 1305–1320.

Artyushkov, E.V., Shlesinger, A.E., and Yanshin, A.L. (1980) The origin of vertical crustal movements within the lithospheric plates. In Bally, A.W., Bender, P.L.,

McGetchin, T.R. and Walcott, R.I. (eds.), *Dynamics of Plate Interiors*, American Geophysical Union, Geological Society of America, Geodynamics Series, vol.1, 37–51.

Atwater, T. (1970) Implications of plate tectonics for the Cenozoic tectonic evolution of western North America. *Bull. geol. Soc. Am.* **81**, 3513–3536.

Atwater, T. and Molnar, P. (1973) Relative motion of the Pacific and North American plates deduced from sea-floor spreading in the Atlantic, Indian and South Pacific oceans. *Stanford Univ. Publ. geol. Sci.* **13**, 136–148.

Avé Lallemant, H.G. (1978) Experimental deformation of diopside and websterite. *Tectonophysics* **48**, 1–27.

Baer, A.J. (1977) Speculations on the evolution of the lithosphere. *Precambrian Res.* **5**, 249–260.

Baer, A.J. (1981) A Grenvillian model of Proterozoic plate tectonics. In Kröner, A. (ed.), *Precambrian Plate Tectonics*, Elsevier, Amsterdam, 353–386.

Bailey, E.B. (1922) The structure of the Southwest Highlands of Scotland. *Q. J. geol. Soc.* **78**, 82–127.

Baker, B.H. and Wohlenberg, J. (1971) Structure and evolution of the Kenya rift valley. *Nature (London)* **229**, 538–542.

Baker, B.H., Mohr, P.A. and Williams, L.A.J. (1972) Geology of the eastern rift system of Africa. *Spec. Pap. geol. Soc. Am.* **136**.

Ballard, R.D. and van Andel, Tj.H. (1977) Morphology and tectonics of the inner rift valley at lat. 36°50′N on the mid-Atlantic ridge. *Bull. geol. Soci. Am.* **88**, 507–530.

Bally, A.W. (1980) Basins and subsidence — a summary. In Bally, A.W., Bender, P.L., McGetchin, T.R. and Walcott, R.I. (eds.), *Dynamics of Plate Interiors*, American Geophysical Union, Geological Society of America, Geodynamics Series, vol.1, 5–20.

Bally, A.W. (1982) Musings over sedimentary basin evolution. *Phil. Trans. R. Soc. London* **A305**, 325–338.

Bally, A.W., Gordy, P.L. and Stewart, G.A. (1966) Structure, seismic data and orogenic evolution of the southern Canadian Rockies. *Bull. Can. Pet. geol.* **14**, 337–381.

Bamford, D., Nunn, K., Prodehl, C. and Jacob, B. (1977) LISPB-III. Upper crustal structure of northern Britain. *J. geol. Soc. London* **133**, 481–488.

Bamford, D. (1979) Seismic constraints on the deep geology of the Caledonides of northern Britain. In Harris, A.L., Holland, C.H. and Leake, B.E. (eds.), *The Caledonides of the British Isles: reviewed, Spec. Publ. geol. Soc. London* 8, 93–96.

Baragar, W.R.A. and Scoates, R.J.F. (1981) The circum-Superior belt: a Proterozoic plate margin. In Kröner, A. (ed.), *Precambrian Plate Tectonics*, Elsevier, Amsterdam, 297–330.

Barazangi, M. and Isacks, B.L. (1976) Spatial distribution of earthquakes and subduction of the Nazca plate beneath South America. *Geology* **4**, 686–692.

Barazangi, M. and NI, J. (1982) Velocities and propagation characteristics of Pn and Sn beneath the Himalayan arc and Tibetan plateau: possible evidence for under-thrusting of Indian continental lithosphere beneath

Tibet. *Geology* **10**, 179–185.

Barber, A.J. and May, F. (1976) The history of the western Lewisian in the Glenelg inlier, Lochalsh, northern Highlands, Scotland. *Scott. J. Geol.* **12**, 35–50.

Bayly, B. (1982) Geometry of subducted plates and island arcs viewed as a buckling problem. *Geology* **10**, 629–632.

Beach, A. (1986) A deep seismic reflection profile across the northern North Sea. *Nature (London)* **323**, 53–55.

Beamish, D. and Smythe, D.K. (1986) Geophysical images of the deep crust: the Iapetus suture. *J. geol. Soc. London* **143**, 489–497.

Beck, M.E. (1983) On the mechanism of tectonic transport in zones of oblique subduction. *Tectonophysics* **93**, 1–11.

Beloussov, V.V. (1962) *Basic Problems in Geotectonics*. McGraw-Hill, New York.

Beloussov, V.V. (1968) *The Crust and Upper Mantle of the Oceans*. M. Nauka (in Russian).

Berthelsen, A. (1980) Towards a palinspastic tectonic analysis of the Baltic shield. *Int. geol. Congr.* Paris, colloq. C3, 5–21.

Bickle, M.J. (1978) Heat loss from the Earth: a constraint on Archaean tectonics from the relation between geothermal gradients and the rate of plate production. *Earth planet. Sci. Lett.* **40**, 301–315.

Bickle, M.J. and Eriksson, K.A. (1982) Evolution and subsidence of early Precambrian sedimentary basins. *Phil. Trans. R. Soc. London* **A305**, 225–247.

Biju-Duval, B., Le Quellec, P., Mascle, A., Renard, V. and Valery, P. (1982) Multibeam bathymetric survey and high resolution seismic investigations on the Barbados ridge complex (eastern Caribbean): a key to the knowledge and interpretation of an accretionary wedge. *Tectonophysics* **86**, 275–304.

Birch, F. (1964) Megageological considerations in rock mechanics. In Judd, W.R. (ed.), *State of Stress in the Earth's Crust*, Elsevier, New York, 55–80.

Blake, M.C., Campbell, R.H., Diblee, T.W., Howell, D.G., Nilsen, I.H., Normark, W.R., Vedder, J.C. and Silver, E.A. (1978) Neogene basin formation in relation to plate-tectonic evolution of San Andreas fault system, California, *Bull. Am. Assoc. petrol. Geol.* 62, 344–372.

Bluck, B.J. (1980) Evolution of a strike-slip fault-controlled basin, Upper Old Red Sandstone, Scotland. *Spec. Publ. Int. Assoc. Sedimentol.* **4**, 63–78.

Bluck, B.J. and Leake, B.E. (1986) Late Ordovician to early Silurian amalgamation of the Dalradian and adjacent Ordovician rocks in the British Isles. *Geology* **14**, 917–919.

Blundell, D.J., Hurich, C.A. and Smithson, S.B. (1985) A model for the Moist seismic reflection profile, N. Scotland. *J. geol. Soc. London* **142**, 245–258.

Bodine, J.H., Steckler, M.S. and Watts, A.B. (1981) Observations of flexure and the rheology of the oceanic lithosphere. *J. geophys. Res.* **86**, 3695–3707.

Bohannon, R.G. and Howell, D.G. (1982) Kinematic evolution of the junction of the San Andreas, Garlock, and Big Pine faults, California. *Geology* **10**, 358–363.

Bonatti, E. (1978) Vertical tectonism in oceanic fracture zones. *Earth planet. Sci. Lett.* **37**, 369–379.

Bott, M.P. (1971) *The Interior of the Earth*. Edward Arnold, London.

Bott, M.P. (1976) Formation of sedimentary basins of graben type by extension of the continental crust *Tectonophysics* **36**, 77–86.

Bott, M.P. (1980) Mechanisms of subsidence at passive continental margins. In Bally, A.W., Bender, P.L., McGetchin, T.R. and Walcott, R.I. (eds.), *Dynamics of Plate Interiors*, American Geophysical Union, Geological Society of America, Geodynamics Series, vol.1, 27–35.

Bott, M.P. (1982) *The Interior of the Earth: Structure, Constitution and Evolution*. Edward Arnold, London.

Bott, M.P. and Kusznir, N.J. (1979) Stress distributions associated with compensated plateau uplift structures with applications to the continental splitting mechanism. *Geophys. J. R. astron. Soc.* **56**, 451–459.

Bott, M.P. and Kusznir, N.J. (1984) Origins of tectonic stress in the lithosphere. *Tectonophysics* **105**, 1–14.

Boyer, S.E. and Elliott, D. (1982) Thrust systems. *Bull. Am. Assoc. petrol. geol.* **66**, 1196–1230.

Brace, W.F. (1964) Brittle fracture of rocks. In Judd, W.R. (ed.), *State of Stress in the Earth's Crust*, Elsevier, New York, 111–174.

Brace, W.F. and Kohlstedt, D.L. (1980) Limits on lithospheric stress imposed by laboratory experiments. *J. geophys. Res.* **85**, (B11) 6248–6252.

Brewer, J.A. and Smythe, D.K. (1984) MOIST and the continuity of crustal reflector geometry along the Caledonian-Appalachian orogen. *J. geol. Soc. London* **141**, 105–120.

Brewer, J.A., Cook, F.A., Brown, L.D., Oliver, J.E., Kaufman, S. and Albaugh, D.S. (1981) COCORP seismic reflection profiling across thrust faults. In McClay, K.R. and Price, N.J. (eds.), *Thrust and Nappe Tectonics, Spec. Publ. geol. Soc. London* **9**, 501–511.

Bridgwater, D. (1976) Nagssugtoqidian mobile belt in East Greenland. In Escher, A. and Watt, W.S. (eds.), *Geology of Greenland*, Geological Survey of Greenland, 96–103.

Bridgwater, D. and Fyfe, W.S. (1974) The pre-3Ga crust: fact — fiction — fantasy. *Geoscience Canada* **1** (3), 7–11.

Bridgwater, D., Keto, L., McGregor, V.R. and Myers, J.S. (1976) Archaean gneiss complex of Greenland. In Escher, A. and Watt, W.S. (eds.), *Geology of Greenland*, Geological Survey of Greenland, 18–75.

Bridgwater, D., Watson, J. and Windley, B.F. (1973) The Archaean craton of the North Atlantic region. *Phil. Trans. R. Soc. London* **A273**, 493–512.

Bronner, G., Roussel, J. and Trompette, R. (1980) Genesis and geodynamic evolution of the Taoudeni cratonic basin (upper Precambrian and Paleozoic), western Africa. In Bally, A.W., Bender, P.L., McGetchin, T.R. and Walcott, R.I. (eds.), *Dynamics of Plate Interiors*, American Geophysical Union, Geological Society of America, Geodynamics Series vol.1, 81–90.

Brown, R.O., Forgotson, J.M. and Forgotson. J.M., Jr. (1980) Predicting the orientation of hydraulically created fractures in the Cotton Valley Formation of East Texas. *55th Ann. Fall Conf. of the Society of Professional Engineers of AIME*, Dallas, Paper SPE 9269.

Brown, L.D. and Reilinger, R.E. (1980) Releveling data in North America: implications for vertical motions of plate interiors. In Bally, A.W., Bender, P.L., McGetchin, T.R. and Walcott, R.I. (eds.) *Dynamics of*

Plate Interiors, American Geophysical Union, Geological Society of America, Geodynamics Series, vol.1, 131–144.

Browne, S.E. and Fairhead, J.D. (1983) Gravity studies of the Central African rift system: a model of continental disruption. 1. The Ngaoundere and Abu Gabra rifts. *Tectonophysics* **94**, 187–204.

Bullard, E.C., Everett, J.E. and Smith, A.G. (1965) The fit of the continents around the Atlantic. *Phil. Trans. R. Soc. London* **A258**, 41–51.

Burke, K. and Dewey, J.F. (1973) Plume-generated triple junctions: key indicators in applying plate tectonics to old rocks. *J. Geol.* **81**, 406–433.

Burke, K., Dewey, J.F. and Kidd, W.S.F. (1976) Dominance of horizontal movements, arc and microcontinental collisions during the later permobile regime. In Windley, B.F. (ed.), *The Early History of the Earth*, John Wiley, London, 113–129.

Butler, R.W.H. (1982) The terminology of structures in thrust belts. *J. struct. Geol.* **4**, 239–245.

Butler, R.W.H. (1983) Balanced cross-sections and their implications for the deep structure of the northwest Alps. *J. struct. Geol.* **5**, 125–137.

Butler, R.W.H., Matthews, S.J. and Parish, M. (1986) The NW external Alpine thrust belt and its implications for the geometry of the western Alpine orogen. In Coward, M.P. and Ries, A.C. (eds.), *Collision Tectonics, Spec. Publ. geol. Soc. London* **19**, 245–260.

Byerlee, J.D. (1968) Brittle-ductile transition in rocks. *J. geophys. Res.* **73**, 4741–4750.

van Calsteren, P.W. and Den Tex, E. (1978) An early Palaeozoic continental rift in Galicia (W Spain). In Ramberg, I.B. and Neumann, E.R. (eds.), *Tectonics and Geophysics of Continental Rifts*, Reidel, Dordrecht, 125–132.

Cande, S.C. & Leslie, R.B. (1986) Late Cenozoic tectonics of the southern Chile trench. *J. geophys. Res.* **91**, **(B1)**, 471–496.

Carter, D.J., Audley-Charles, M.G. and Barber, A.J. (1976) Stratigraphical analysis of island arc-continental margin collision in eastern Indonesia. *J. geol. Soc. London* **132**, 179–198.

Cartwright, I. and Barnicoat, A.C. (1987) Petrology of Scourian supracrustal rocks and orthogneisses from Stoer, NW Scotland: implications for the geological evolution of the Lewisian complex. In Park, R.G. and Tarney, J. (eds.), *Evolution of the Lewisian Complex and Comparable Precambrian High Grade Terrains, Spec. Publ. geol. Soc. London* **27**, 93–107.

Chadwick, B. and Nutman, A.P. (1979) Archaean structural evolution in the northwest of the Buksefjorden region, southern West Greenland. *Precambrian Res.* **9**, 199–226.

Charlton, T.R. (1986) A plate tectonic model of the eastern Indonesia collision zone. *Nature (London)* **319**, 394–396.

Chase, C.G. (1978*a*), Plate kinematics: the Americas, East Africa, and the rest of the world. *Earth planet. Sci. Lett.* **37**, 355–368.

Chase, C.G. (1978*b*) Extension behind island arcs and motions relative to hot spots. *J. geophys. Res.* **83**, B11, 5385–5387.

Christie, P.A. and Sclater, J.C. (1980) An extensional origin for the Buchan and Witchground graben in the North Sea. *Nature (London)* **283**, 729–732.

Coleman, R.G. (1971) Plate tectonic emplacement of upper mantle peridotites along continental edges. *J. geophys. Res.* **76**, 1212–1222.

Collette, B.J. (1974) Thermal contraction joints in a spreading sea floor as origin of fracture zones. *Nature (London)* **251**, 299–300.

Condie, K.C. (1982) *Plate Tectonics and Continental Drift.* Pergamon Press, Oxford.

Coney, P.J. (1973) Non-collision tectogenesis in western North America. In Tarling, D.H. and Runcorn, S.K. (eds.), *Implications of Continental Drift to the Earth Sciences*, vol.2, Academic Press, London, 713–730.

Coney, P.J. and Reynolds, S.J. (1977) Cordilleran Benioff zones. *Nature (London)* **270**, 403–406.

Coney, P.J. Jones, D.L. and Monger, J.W.H. (1980) Cordilleran suspect terranes. *Nature (London)* **288**, 329–333.

Cook, F.A., Brown, L.D., Kaufman, S., Oliver, J.E. and Petersen, T.A. (1981) COCORP seismic profiling of the Appalachian orogen beneath the coastal plain of Georgia. *Bull. geol. Soc. Am.* **92**, 738–748.

Cook, N.G.W. (1965) The failure of rock. *Int. J. Rock Mech. Ming Sci.* **2**, 389–403.

Cook, N.G.W., Hoek, E., Pretorius, J.P.G., Oertlepp, W.D. and Salamon, M.D.G. (1966) Rock mechanics applied to the study of rockbursts. *J. S. A. Inst. Ming Metall.* **66**, 435–528.

Cooper, M.A., Collins, D., Ford, M., Murphy, F.X. and Trayner, P.M. (1984) Structural style, shortening estimates and the thrust front of the Irish Variscides. In Hutton, D.H.W. and Sanderson, D.J. (eds.), *Variscan Tectonics of the North Atlantic Region, Spec. Publ. geol. Soc. London* **14**, 167–175.

Coward, M.P. (1976) Archaean deformation patterns in southern Africa. *Phil. Trans. R. Soc. London* **A283**, 313–331.

Coward, M.P. (1980) The Caledonian thrust and shear zones of NW Scotland. *J. struct. Geol.* **2**, 11–17.

Coward, M.P. (1981) The junction between Pan African mobile belts in Namibia: its structural history. *Tectonophysics* **76**, 59–73.

Coward, M.P. (1983) Thrust tectonics, thin skinned or thick skinned, and the continuation of thrusts to deep in the crust. *J. struct. Geol.* **5**, 113–123.

Coward, M.P. and Butler, R.W.H. (1985) Thrust tectonics and the deep structure of the Pakistan Himalaya. *Geology* **13**, 417–420.

Coward, M.P. and Park, R.G. (1987) The role of mid-crustal shear zones in the Early Proterozoic evolution of the Lewisian. In Park, R.G. and Tarney, J. (eds.), *Evolution of the Lewisian Complex and Comparable Precambrian High-Grade Terrains, Spec. Publ. geol. Soc. London* **27**, 127–138.

Coward, M.P. and Smallwood, S. (1984) An interpretation of the Variscan tectonics of SW Britain. In Hutton, D.H.W. and Sanderson, D.J. (eds.), *Variscan Tectonics of the North Atlantic Region, Spec. Publ. geol. Soc. London* **14**, 89–102.

Coward, M.P., Jan, M.Q., Rex.D., Tarney, J., Thirlwall, M. and Windley, B.F. (1982) Geo-tectonic framework of the Himalaya of N Pakistan. *J. geol. Soc. London* **139**, 299–308.

Coward, M.P., Kim, J.H. and Parke, J. (1980) A correlation of Lewisian structures across the lower thrusts of the Moine thrust zone, Northwest Scotland. *Proc. Geol. Assoc. London* **91**, 327–337.

Coward, M.P., Knipe, R.J. and Butler, R.W.H. (1983) In 'Discussion on a model for the deep structure of the Moine thrust zone,' *J. geol. Soc. London* **140**, 519.

Cox, J.W. (1970) The high resolution dipmeter reveals dip-related borehole and formation characteristics. *11th Ann. Logging Symp.* Society of Professional Well Log Analysts, Los Angeles.

Cross, T.A. and Pilger, R.H. (1982) Controls of sub-duction geometry, location of magmatic arcs, and tectonics of arc and back-arc regions. *Bull. geological soc. Am.* **93**, 545–562.

Crowell, J.C. (1979) The San Andreas fault system through time. *J. geol. Soc. London* **136**, 293–302.

Dahlstrom, C.D.A. (1970) Structural geology in the western margin of the Canadian Rocky Mountains. *Bull. Can. Assoc. petrol. Geol.* **18**, 332–406.

Darracot, B.W., Fairhead, J.D., Girdler, R.W. and Hall, S.A. (1973) The East African rift system. In Tarling, D.H. and Runcorn, S.K. *Implications of Continental drift to the Earth Sciences*, Academic Press, London and New York, 757–766.

Davidson, A. (1985) Tectonic framework of the Grenville Province in Ontario and western Quebec, Canada. In Tobi, A.C. and Touret, J.L.R. (eds.), *The Deep Proterozoic Crust in the North Atlantic Provinces*, D. Reidel, Dordrecht, 133–149.

Davis, G. and Coney, P.J. (1979) Geologic development of the Cordilleran metamorphic core complexes. *Geology* **8**, 120–124.

Davis, G.H., Gardulski, A.F. and Anderson, T.H. (1981) Structural and structural-petrological characteristics of some metamorphic core complex terranes in southern Arizona and northern Sonora. In Ortlieb, L. and Roldan, Q. (eds.), *Geology of Northwestern New Mexico and Southern Arizona, Field Guides and Papers*, Univ. Nat. Auton de Mexico, Inst. de Geologia, Hermosillo, Sonora, Mexico, 323–366.

de Almeida, F.F.M. and Black, R. (1967) Comparaison Structurale entre le nord-est du Brésil et l'Ouest africain. *Symp. on Continental Drift*, Montevideo.

de Bremaecker, J.-C., Huchon, P. and Le Pichon, X. (1982) The deformation of Aegea: a finite element study. *Tectonophysics* **86**, 197–211.

de Charpal, O., Montadert, L., Guennoc, P. and Roberts, E.G. (1978) Rifting, crustal attenuation and subsidence in the Bay of Biscay. *Nature (London)* **275**, 706–710.

Dearnley, R. (1966) Orogenic fold belts and a hypothesis of Earth evolution. *Phys. Chemistry Earth (Oxford)* **7**, 1–114.

Debelmas, J., Escher, A. and Trumpy, R. (1983) Profiles through the western Alps. In Rast, N. and Delaney, F.M. (eds.), *Profiles of Orogenic Belts*, American Geophysical Union, Geological Society of America, Geodynamics Series, vol.10, 83–96.

Demaiffe, D. and Michot, J. (1985) Isotope geochronology of the Proterozoic crustal segment of southern Norway. In Tobi, A.C. and Touret, J.L.R. *The Deep Proterozoic Crust in the Atlantic Provinces*, D. Reidel, Dordrecht, 411–433.

Dewey, J.F. (1969) Evolution of the Appalachian/Caledonian orogen. *Nature (London)* **222**, 124–129.

Dewey, J.F. (1971) A model for the Lower Palaeozoic evolution of the southern margin of the early Caledonides of Scotland and Ireland. *Scott. J. Geol.* **7**, 219–240.

Dewey, J.F. (1975) Finite plate implications: some implications for the evolution of rock masses at plate margins. *Am. J. Sci.* **275A** (Rodgers vol.), 260–284.

Dewey, J.F. (1977) Ancient plate margins: some observations. *Tectonophysics* **33**, 397–385.

Dewey, J.F. (1980) Episodicity, sequence and style at convergent plate boundaries. *Spec. Pap. geol. Assoc. Can.* **20**, 553–574.

Dewey, J.F. (1982) Plate tectonics and the evolution of the British Isles. *J. geol. Soc. London* **139**, 371–412.

Dewey J.F. and Bird, J.M. (1970) Mountain belts and the new global tectonics. *J. geophys. Res.* **75**, 2625–2647.

Dewey, J.F. and Bird, J.M. (1971) Origin and emplacement of the ophiolite suite: Appalachian ophiolites in Newfoundland. *J. geophys. Res.* **76**, 3179–3206.

Dewey, J.F. and Burke, K.C.A. (1973) Tibetan, Variscan and Precambrian basement reactivation: products of continental collision. *J. Geol.* **81**, 683–692.

Dewey, J.F. and Burke, K.C. (1974) Hot spots and continental break-up: implications for collisional orogeny. *Geology* **2**, 57–60.

Dewey, J.F. and Shackleton, R.M. (1984) A model for the evolution of the Grampian tract in the early Caledonides and Appalachians. *Nature (London)* **312**, 115–121.

Dewey, J.F. and Windley, B.F. (1981) Growth and differentiation of the continental crust. *Phil. Trans. R. Soc. London* **A301**, 189–206.

Dewey, J.F., Pitman, W.C. III, Ryan, W.B.F. and Bonnin, J. (1973) Plate tectonics and the evolution of the Alpine system. *Bull. geol. Soc. Am.* **84**, 3137–3180.

Dietz, R.S. (1963) Collapsing continental rises: an actualistic concept of geosynclines and mountain building. *J. Geol.* **71**, 314–333.

Dimroth, E. (1981) Labrador geosyncline: type example of early Proterozoic cratonic reactivation. In Kroner, A. (ed.), *Precambrian Plate Tectonics*, Elsevier, Amsterdam, 331–352.

Donato, J.A. and Tully, M.C. (1981) A regional interpretation of North Sea gravity data. In Illing, L.V. and Hobson, G.D. (eds.), *Petroleum Geology of the Continental Shelf of Northwest Europe*, Heyden, London, 65–75.

Drury, S. (1977) Structures induced by granite diapirs in the Archaean greenstone belt at Yellowknife, Canada: implications for Archaean geotectonics. *J. Geol.* **85**, 345–358.

Dula, W.F. (1981) Correlation between deformation lamellae, microfractures, macrofractures, and in-situ stress measurements, White River uplift, Colorado. *Bull. geol. Soc. Am.* (part 1) **92**, 37–46.

Eales, M.H. (1979) Structure of the Southern Uplands of Scotland. In Harris, A.L., Holland, C.H. and Leake, B.E. (eds.), *The Caledonides of the British Isles: reviewed, Spec. Publ. geol. Soc. London* **8**, 269–273.

Eisbacher, G.H. and Bielenstein, H.U. (1971) Elastic strain recovery in Proterozoic rocks near Elliott Lake, Ontario. *J. geophys. Res.* **76**, 2012–2021.

Elder, J. (1976) *The Bowels of the Earth*. Oxford University Press, Oxford.

Elliott, D. and Johnson, M.R.W. (1980) Structural evolution in the northern part of the Moine Thrust Zone. *Trans. R. Soc. Edin. (Earth Sci.)* **71**, 69–96.

Elsasser, W.M. (1967) Convection and stress propagation in the upper mantle. In Runcorn, S.K. (ed.), *The Application of Modern Physics to the Earth and Planetary Interiors*, Wiley Interscience, London, 223–246.

Elsasser, W.M. (1971) Sea floor spreading as thermal convection. *J. geophys. Res.* **76**, 1101–1112.

England, P. and McKenzie, D.P. (1981) A thin viscous sheet model for continental deformation. *Geophys. J. R. astron. Soc.* **70**, 295–321.

Ernst, W.G. (1983) Phanerozoic continental accretion and the metamorphic evolution of northern and central California. *Tectonophysics* **100**, 287–320.

Escher, A., Sørensen, K. and Zeck, H.P. (1976) Nagssugtoqidian mobile belt in West Greenland. In Escher, A. and Watt, W.S. (eds.), *Geology of Greenland*, Geological Survey of Greenland, 76–96.

Fairhead, J.D. (1976) The structure of the lithosphere beneath the Eastern rift, East Africa, deduced from gravity studies. *Tectonophysics* **30**, 269–298.

Fairhead, J.D. and Girdler, R.W. (1971) The seismicity of Africa. *Geophys. J. R. astron. Soc.* **24**, 271–301.

Falkum, T. and Pedersen, S. (1979) Rb-Sr age determination on the intrusive Precambrian Homme granite and consequences for dating the last regional folding and metamorphism in the Flekkefjord region, SW Norway. *Nor. geol. Tidss.* **59**, 59–65.

Farrar, E. and Lowe, R.M. (1978) Age-length dependence of inclined seismic zones. *Nature (London)* **273**, 292–293.

Forsyth, D. and Uyeda, S. (1975) On the relative importance of the driving forces of plate motion. *Geophys. J. R. astron. Soc.* **43**, 163–200.

Francheteau, J. and Ballard, R.D. (1983) The East Pacific Rise near 21°N, 13°N and 20°S: inferences for alongstrike variability of axial processes of the mid-ocean ridge. *Earth planet. Sci. Lett.* **64**, 93–116.

Francheteau, J., Choukroune, P., Hekinian, R., Le Pichon, X. and Needham, H.D. (1976) Oceanic fracture zones do not provide deep sections in the crust. *Can. J. Earth Sci.* **13**, 1223–1235.

Franke, W. (1984) Late events in the tectonic history of the Saxothuringian zone. In Hutton, D.H.W. and Sanderson, D.J. (eds.) *Variscan Tectonics of the North Atlantic region*, *Spec. Publ. geol. Soc.* London **14**, 33–45.

Freund, R., Zak, I. and Garfunkel, Z. (1968) Age and rate of the sinistral movement along the Dead Sea rift. *Nature (London)* **220**, 253–255.

Froidevaux, C., Paquin, C. and Souriau, M. (1980) Tectonic stresses in France. *J. geophys. Res.* **85**, 6342–6346.

Furlong, K.P., Chapman, D.S. and Alfeld, P.W. (1982) Thermal modelling of the geometry of subduction with implications for the tectonics of the overriding plate. *J. geophys. Res.* **87**, 1786–1802.

Furnes, H., Thon, A., Nordås, J. and Garman, L.B. (1982) Geochemistry of Caledonian metabasalts from some Norwegian ophiolite fragments. *Contrib. Mineral. Petrol.* **79**, 295–307.

Gass, I.G. (1981) Pan-African (Upper Proterozoic) plate tectonics of the Arabian-Nubian shield. In Kröner, A. (ed.), *Precambrian Plate Tectonics*, Elsevier, Amsterdam, 387–405.

Gass, I.G., Chapman, D.S., Pollack, H.N. and Thorpe, R.S. (1978) Geological and geophysical parameters of mid-plate volcanism. *Phil. Trans. R. Soc. London* **A288**, 581–597.

Gee, D.G. (1975) A tectonic model for the central part of the Scandinavian Caledonides. *Am. J. Sci.* **275A**, 468–515.

Gephart, J.W. and Forsyth, D.D. (1985) On the state of stress in New England as determined from earthquake focal mechanisms. *Geology* **13**, 70–72.

Gibbs, A.D. (1984) Structural evolution of extensional basin margins. *J. geol. Soc. London* **141**, 609–620.

Gilully, J. (1971) Plate tectonics and magmatic evolution. *Bull. geol. Soc. Am.* **82**, 2387–2396.

Girdler, R.W. (1967) A review of terrestrial heat flow. In Runcorn, S.K. (ed.), *Mantles of the Earth and terrestrial Planets*, Wiley Interscience, New York.

Girdler, R.W. (1969) The Red Sea: a geophysical background. In Degens, E.T. and Ross, D.A. (eds.), *Hot Brines and Recent Heavy Metal Deposits in the Red Sea, a Geochemical and Geophysical Account*, Springer Verlag, New York, 38.

Girdler, R.W. and Darracott, B.W. (1972) African poles of rotation. *Comments on the Earth Sciences: Geophysics* **2** (5), 7–15.

Goetze, C. (1978) The mechanisms of creep in olivine. *Phil. Trans. R. Soc.* **A288**, 99–119.

Goodwin, A.M. (1981) Archaean plates and greenstone belts. In Kröner, A. (ed.), *Precambrian Plate Tectonics*, Elsevier, Amsterdam, 105–1135.

Gorman, B.E., Pearce, T.H. and Birkett, T.C. (1978) On the structure of Archaean greenstone belts. *Precambrian Res.* **6**, 23–41.

Gough, D.I. (1977) The geoid and single-cell mantle convection. *Earth planet. Sci. Lett.* **34**, 360–364.

Gough, D.I. (1984) Mantle upflow under North America and plate dynamics. *Nature (London)* **311**, 428–433.

Gough, D.I. and Bell, J.S. (1982) Stress orientation from borehole wall fractures with examples from Colorado, east Texas, and northern Canada. *Can. J. Earth Sci.* **19**, 1358–1370.

Graham, R.H. (1981) Gravity sliding in the Maritime Alps. In McClay, K.R. and Price, N.J. (eds.), *Thrust and Nappe Tectonics, Spec. Publ. geol. Soc. London* **9**, 335–352.

Grocott, J. (1979) Shape fabrics and superimposed simple shear strain in a Precambrian shear belt, W Greenland. *J. geol. Soc. London* **136**, 471–488.

Guillope, M. and Poirier, J.P. (1979) Dynamic recrystallisation during creep of single-crystalline halite; an experimental study. *J. geophys. R.* **84**, B10, 5557–5567.

Guo, S. (1980) Approaches to elevation and climatic changes of Qinghai-Xizang (Tibet) plateau from fossil angiosperms. *Symp. on Qinghai-Xizang (Tibet) plateau*, Beijing, China, abstracts, 13–14.

Gwinn, V.E. (1965) Thin-skinned tectonics in the Plateau and northwestern Valley and Ridge provinces of the central Appalachians. *Bull. geol. Soc. Am.* **75**, 863–900.

Hageskov, B. (1980) The Sveconorwegian structures of the Norwegian part of the Kongsberg — Bamble — Östfold segment. *Forh. geol. Fören. Stockholm* **102**, 150–155.

Haile, N.S. (1978) Paleomagnetic evidence for the rotation of Seram, Indonesia. *J. Phys. Earth* **26** (suppl.) S191–198.

Haimson, B.C. (1977) The hydrofracturing stress measuring method and recent field results. *Int. J. Rock Mech. Mining Sci. (Geomech, Abstr.)* **15**, 167–178.

Hamilton, W. (1975) Neogene extension of the western United States. *Geol. Soc. Am. Abstr. progr.* **7**, 1098.

Hancock, P.L., Dunne, W.M. and Tringham, M.E. (1983) Variscan deformation in southwest Wales. In Hancock, P.L. (ed.), *The Variscan Fold Belt in the British Isles*, Adam Hilger, Bristol, 108–129.

Harding, T.P. and Lowell, J.D. (1979) Structural styles, their plate tectonic habitats, and hydrocarbon traps in petroleum provinces. *Bull. Am. Assoc. petrol. Geol.* **63**, 1016–1058.

Harland, W.B. (1971) Tectonic transpression in Caledonian Spitzbergen. *Geol. Mag.* **108**, 27–42.

Harland, W.B., Cox, A.V., Llewellyn, P.G., Pickton, C.A.G., Smith, A.G. and Walters, R. (1982) *A Geologic Time Scale*. Cambridge University Press, Cambridge.

Harris, A.L., Bradbury, H.J. and McGonigal, M.H. (1976) The evolution and transport of the Tay nappe. *Scott. J. Geol.* **12**, 103–113.

Harris, A.L. and Pitcher, W.S. (1975) The Dalradian Supergroup. In Harris, A.L., Shackleton, R.M., Watson, J., Downie, C., Harland, W.B. and Moorbath, S. (eds.), *A Correlation of Precambrian Rocks in the British Isles*, *Spec. Publ. geol. Soc. London* **6**, 52–75.

Hatcher, R.D. Jr. (1981) Thrusts and nappes in the North American Appalachian orogen. In McClay, K.R. and Price, N.J. (eds.), *Thrust and Nappe Tectonics. Spec. Publ. geol. Soc. London* **9**, 491–499.

Hawkesworth, C.J., Menzies, M.A. and van Calsteren, P. (1986) Geochemical and tectonic evolution of the Damara belt, Namibia. In Coward, M.P. and Ries, A.C. (eds.), *Collision Tectonics, Spec. Publ. geol. Soc. London* **19**, 305–319.

Haxby, W.F. and Weissel, J.K. (1983) Evidence for small-scale mantle convection from Seasat altimeter data. *Trans. Am. geophys. Union* **64**, 676.

Heard, H.C. (1976) Comparison of the flow properties of rocks at crustal conditions. *Phil. Trans. R. Soc. London* **A283**, 173–186.

Heezen, B.C. (1962) The deep-sea floor. In Runcorn, S.K. (ed.), *Continental Drift*, Academic Press, New York and London, 235–288.

Hein, J.R. (1973) Deep-sea sediment source areas: implications of variable rates of movement between California and the Pacific plate. *Nature (Physical Sciences) (London)* **241**, 40–41.

Heirtzler, and van Andel (1977) Project FAMOUS: its origin, programs and setting. *Bull geol. Soc. Am.* **88**, 481–487.

Hermance, J.F. (1982) Magnetotelluric and geomagnetic deep-sounding studies in rifts and adjacent areas. In Pálmason, G. (ed.), *Continental and Oceanic Rifts*, American Geophysical Union, Geological Society of America, Geodynamics Series, vol.8, 169–192.

Herrin, E. (1969) Regional variations of P-wave velocity in the upper mantle beneath North America. In Hart, P.J. (ed.), *The Earth's Crust and Upper Mantle*, American Geophysical Union Geophysical Monogr. **13**, 242–246.

Herron, E.M. (1972) Sea-floor spreading and the Cenozoic history of the east-central Pacific. *Bull. geol. Soc. Am.* **83**, 1671–1692.

Hess, H.H. (1962) History of ocean basins. In Engel, A.E.J., James, H.L. and Leonard, B.F. (eds.) *Petrologic studies: a volume in honor of A.F. Buddington*, Geological Society of America, Boulder, Colorado, 599–620.

Hilde, T.W.C. (1983) Sediment subduction versus accretion around the Pacific. *Tectonophysics* **99**, 381–397.

Hirn, A., Lepine, J-C., Jobert, G., Sapin, M., Wittlinger, G., Xu, Z.X., Gao, E.Y., Wang, X.J., Teng, J.W., Xiong, S.B., Pandey, M.R. and Tater, J.M. (1981) Crustal structure and variability of the Himalayan border of Tibet. *Nature (London)* **307**, 23–25.

Holgate, N. (1969) Palaeozoic and Tertiary transcurrent movements on the Great Glen fault. *Scott. J. Geol.* **5**, 97–139.

Holmes, A. (1929) Radioactivity and earth movements. *Trans. geol. Soc. Glasgow* **18**, 559–606.

Holmes, A. (1978) *Principles of Physical Geology* (3rd edn.) Nelson, Sunbury, Middlesex, UK.

Hooker, V.E. and Duval, W.I. (1966) Stresses in rock outcrops near Atlanta, Georgia. *Rept. US Bureau of Mines* **6860**.

Hooker, V.E. and Johnson, C.F. (1967) *In situ* stresses along the Appalachian Piedmont. In *4th Symp. on Rock Mechanics*, US Department of Energy, Mines and Resources, Ottawa, 137–154.

Hooker, V.E. and Johnson, C.F. (1969) Near surface horizontal stresses, including the effects of rock anisotropy. *Rept. US Bureau of Mines* **7224**.

Hossack, J.R. and Cooper, M.A. (1986) Collision tectonics in the Scandinavian Caledonides. In Coward, M.P. and Ries, A.C. (eds.), *Collision Tectonics, Spec. Publ. geol. Soc. London* **19**, 287–304.

Hossack, J.R. (1978) The correction of stratigraphic sections for tectonic finite strain in the Bygdin area, Norway. *J. geol. Soc. London* **135**, 229–241.

House, L.S. and Jacob, K.H. (1982) Thermal stresses in subducting lithosphere can explain double seismic zones. *Nature (London)* **295**, 587–589.

Hsu, K.J. (1972) Alpine flysch in a Mediterranean setting. *Proc. 24th Int. geol. Congr.* Montreal, sect.6, 67–74.

Hsu, K.J. (1979) Thin-skinned plate tectonics during neo-Alpine orogenesis. *Am. J. Sci.* **279**, 353–366.

Hubbert, M.K. and Rubey, W.W. (1959) Role of fluid pressure in mechanics of overthrusting faulting. *Bull. geol. Soc. Am.* **70**, 115–166.

Huchon, P., Lyberis, N., Angelier, J., Le Pichon, X. and Renard, V. (1982) Tectonics of the Hellenic trench: a synthesis of SEABEAM and submersible observations. *Tectonophysics* **86**, 69–112.

Hunziker, J.C. (1986) The Alps: a case of multiple collision. In Coward, M.P. and Ries, A.C. (eds.), *Collision Tectonics, Spec. Publ. geol. Soc. London* **19**, 221–227.

Hussong, D.M. and Wippermann, L.K. (1981) Vertical movement and tectonic erosion of the continental wall of the Peru–Chile trench near 11°30′S latitude. In Kulm, L.D., Dymond, J., Dasch, E.J. and Hussong, D.M. (eds.) *Nazca Plate: Crustal Formation and Andean Convergence*, Geol. Soc. Am. Memoir 154, 519–524.

Hutton, D.H.W. and Sanderson, D.J. (eds.), (1984) *Variscan Tectonics. Spec. Publ. geol. Soc. London* **14**.

Hynes, A. and Mott, J. (1985) On the causes of back-arc spreading. *Geology* **13**, 387–389.

Illies, J.H. (1978) Two stages Rhinegraben rifting. In Neumann, E.-R. and Ramberg, I.B. (eds.) *Tectonics and Geophysics of Continental Rifts*, D. Reidel, Dordrecht, 63–71.

Illies, J.H. (ed.) (1981) *Mechanism of Graben Formation. Tectonophysics* **73**.

Illies, J.H. and Greiner, G. (1978) Rhinegraben and the Alpine system. *Bull. geol. Soc. Am.* **89**, 770–782.

Irving, E. and McGlynn, J.C. (1981) On the coherence, rotation and palaeolatitude of Laurentia in the Proterozoic. In Kröner, A. (ed.), *Precambrian Plate Tectonics*, Elsevier, Amsterdam, 561–598.

Isacks, B. and Molnar, P. (1969) Mantle earthquake mechanisms and the sinking of the lithosphere. *Nature (London)* **223**, 1121.

Isacks, B, Oliver, J, and Sykes, L.R. (1968) Seismology and the new global tectonics. *J. geophys. Res.* **73**, 5855–5899.

Jackson, J. and McKenzie, D.P. (1983) The geometrical evolution of normal fault systems. *J. struct. Geol.* **5**, 471–482.

James, D.E. (1972) Plate tectonic model for the evolution of the central Andes. *Bull. geol. Soc. Am.* **82**, 3325–3346.

Jarvis, G.T. and McKenzie, D.P. (1980) Sedimentary basin formation with finite extension rates. *Earth planet. Sci. Lett.* **48**, 42–52.

Jefferis, R.G. and Voigt, B. (1981) Fracture analysis near the mid-ocean plate boundary, Reykjavik-Hvalfjordur area, Iceland. *Tectonophysics* **76**, 171–236.

Jeffreys, H. (1970) *The Earth* (5th edn.). Cambridge University Press, Cambridge.

Jones, D.L., Irwin, W.P. and Ovenshine, A.T. (1972) Southeastern Alaska: a displaced continental fragment? *Prof. Pap. US geol. Surv.* **800B**, 211–217.

Jones, O.T. (1938) On the evolution of a geosyncline. *Q. J. geol. Soc.* **94**, 60–110.

Karig, D.E. (1971) Origin and development of marginal basins in the western Pacific. *J. geophys. Res.* **76**, 2542–2561.

Karig, D.E. (1974) Evolution of arc systems in the western Pacific. *Ann. Rev. Earth planet. Sci.* **2**, 51–75.

Karig, D.E., Caldwell, J.G., and Parmentier, E.M. (1976) Effects of accretion on the geometry of the descending lithosphere. *J. geophys. Res.* **81**, 6281–6291.

Kelley, S.P. and Powell, D. (1985) Relationships between marginal thrusting and movement on a major, internal shear zones in the Northern Highland Caledonides, Scotland. *J. struct. Geol.* **7**, 161–174.

Kennedy, W.Q. (1946) The Great Glen fault. *Q. J. geol. Soc. London* **102**, 41–72.

Kennedy, W.Q. (1964) The structural differentiation of Africa in the Pan-African (*c.* 500Ma) tectonic episode. *Ann. Rep. Univ. Leeds Res. Inst. African Geol.* **8**, 48–49.

Kenyon, P.M. and Turcotte, D.L. (1983) Convection in a two-layer mantle with a strongly temperature-dependent viscosity. *J. geophys. Res.* **88**, B8, 6403–6414.

Koch, P.S., Christie, J.M. and George, R.P. (1980) Flow law of 'wet' quartzite in the α-quartz field. *Eos* **61**, 376.

Kohlstedt, D.L. and Goetze, C. (1974) Low-stress high-temperature creep in olivine single crystals. *J. geophys. Res.* **79**, 2045–2051.

Korstgård, J.A. (ed.) (1979) *Nagssugtoqidian Geology. Rapp. Grønlands geol. Unders.* **89**.

Kröner, A. (1981) Precambrian plate tectonics. In Kröner, A. (ed.), *Precambrian Plate Tectonics*, Elsevier, Amsterdam, 57–90.

Kulm, L.D. Prince, R.A., French, W., Johnson, S. and Masias, A. (1981) Crustal structure and tectonics of the central Peru continental margin and trench. In Kulm, L.D., Dymond, J., Dasch, E.J. and Hussong, D.M. (eds.), *Nazca Plate: Crustal Formation and Andean Convergence, Geol. Soc. Am. Memoir* **154**, 445–468.

Kusznir, N.J. (1982) Lithosphere response to externally and internally derived stresses: a viscoelastic stress guide with amplification. *Geophys. J. R. astron. Soc.* **70**, 399–414.

Kusznir, N.J. and Bott, M.P. (1977) Stress concentration in the upper lithosphere caused by underlying viscoelastic creep. *Tectonophysics* **43**, 247–256.

Kusznir, N.J. and Karner, G.D. (1985) Flexural rigidity and its constraints on continental lithosphere temperature structure (abstr). *Geophys. J. R. astron. Soc.* **81**, 343.

Kusznir, N.J. and Park, R.G. (1982) Intraplate lithosphere strength and heat flow. *Nature (London)* **299**, 540–542.

Kusznir, N.J. and Park, R.G. (1984*a*) Intraplate lithosphere deformation and the strength of the lithosphere. *Geophys. J. R. astron. Soc.* **79**, 513–538.

Kusznir, N.J. and Park, R.G. (1984*b*) The strength of intraplate lithosphere. *Phys. Earth planet. Int.* **36**, 224–235.

Kusznir, N.J. and Park, R.G. (1986) The extensional strength of the continental lithosphere: its dependence on geothermal gradient, and crustal composition and thickness. In Coward, M.P., Dewey, J.F. and Hancock, P.L. (eds.), *Continental Extensional Tectonics, Spec. Publ. geol. Soc. London* **28**, 35–52.

Lachenbruch, A.H. and Sass, J.H. (1980) Heat flow and energetics of the San Andreas fault zone. *J. geophys. Res.* **85** (B11) 6185–6222.

Lambert, R.St.J. (1969) Isotopic studies relating to the Precambrian history of the Moinian of Scotland. *Proc. geol. Soc.* **1652**, 243–244.

Lambert, R.St.J. and McKerrow, W.S. (1976) The Grampian orogeny. *Scott. J. Geol.* **12**, 271–292.

Larson, S.A., Stigh, J. and Tullborg, E.-L. (1986) The deformation history of the eastern part of the southwest Swedish gneiss belt. *Precambrian Res.* **31**, 237–257.

Larson, R.L. and Pitman, W.C. (1972) Worldwide correlation of Mesozoic magnetic anomalies, and its implications. *Bull. geol. Soc. Am.* **83**, 3627–3644.

Laughton, A.S., Sclater, J.G. and McKenzie, D.P. (1973) The structure and evolution of the Indian Ocean. In Tarling, D.H. and Runcorn, S.K. (eds.), *Implications of Continental Drift to the Earth sciences*, Academic Press, London & New York.

Laughton, A.S., Whitmarsh, R.B. and Jones, M.T. (1970) The evolution of the Gulf of Aden. *Phil. Trans. R. Soc. London* **A267**, 227–266.

Lee, W.H.K. and Uyeda, S. (1965) Review of heat flow data. In Lee, W.H.K. (ed.), *Terrestrial Heat Flow, Geophys. Monogr. Washington* **8**.

Le Pichon, X. (1968) Sea-floor spreading and continental drift. *J. geophys. Res.* **73**, 3661–3697.

Le Pichon, X. and Angelier, J. (1979) The Hellenic arc and trench system: a key to the neotectonic evolution of the eastern Mediterranean area. *Tectonophysics* **60**, 1–42.

Le Pichon, X. and Huchon, P. (1984) Geoid, Pangea and convection. *Earth planet. Sci. Lett.* **67**, 123–135.

Le Pichon, X., Lyberis, N., Angelier, J. and Renard, V. (1981) Strain distribution over the east Mediterranean ridge: a synthesis incorporating new Sea-Beam data. *Tectonophysics* **86**, 243–274.

Leeder, M.R. (1982) Upper Palaeozoic basins of the British Isles: Caledonian inheritance versus Hercynian plate margin processes. *J. geol. Soc. London* **139**, 479–491.

Leggett, J.K., McKerrow, W.S. and Eales, M.H. (1979) The Southern Uplands of Scotland: a Lower Palaeozoic accretionary prism. *J. geol. Soc. London* **136**, 755–770.

Lichtman, G.S. and Eissen, J.–P. (1983) Time and space constraints on the evolution of medium-rate spreading centers. *Geology* **11**, 592–595.

Lin, J.-L., Fuller, M. and Zhang, W-y. (1985) Preliminary Phanerozoic polar wander paths for the North and South China blocks. *Nature (London)* **313**, 444–449.

Lockett, J.M. and Kusznir, N.J. (1982) Ductile shear zones: some aspects of constant slip velocity and constant shear stress models. *Geophys. J. R. astron. Soc.* **69**, 477–494.

Logatchev, N.A., Rogozhina, V.A., Solonenko, V.P. and Zorin, Y.A. (1978) Deep structure and evolution of the Baikal rift zone. In Neumann, E.R. and Ramberg, I.B. (eds.), *Tectonics and Geophysics of Continental Rifts*, D. Reidel, Dordrecht, 49–61.

Louden, K.E. and Forsyth, D.W. (1976) Thermal conduction across fracture zones and the gravitational edge effects. *J. geophys. Res.* **81**, 4869–4874.

Luyendyk, B.P. (1970) Dips of downgoing lithosphere plates beneath island arcs. *Bull. geol. Soc. Am.* **81**, 3411–3416.

Mareschal, J.-C. (1983) Mechanisms of uplift preceding rifting. *Tectonophysics* **94**, 51–66.

Mareschal, J.-C. and West, G.F. (1980) A model for Archaean tectonism. Part 2. Numerical models of vertical tectonism in greenstone belts. *Can. J. Earth Sci.* **17**, 60–71.

Mattauer, M. (1986) Intracontinental subduction, crust-mantle décollement and crustal-stacking wedge in the Himalayas and other collision belts. In Coward, M.P. and Ries, A. (eds.), *Collision Tectonics. Spec. Publ. geol. Soc. London* **19**, 37–50.

Matte, P. (1983) Two geotraverses across the Ibero-Armorican Variscan arc of western Europe. In Rast, N. and Delaney, F.M. (eds.), *Profiles of Orogenic Belts*, American Geophysical Union, Geological Society of America, Geodynamics Series, vol.10, 53–81.

Matte, Ph. and Burg, J.P. (1981) Sutures, thrusts and nappes in the Variscan Arc of western Europe: plate tectonic implications. In McClay, K.R. and Price, N.J. (eds.) *Thrust and Nappe Tectonics, Spec. Publ. geol. Soc. London* **9**, 353–358.

Mattskova, V.A. (1967) A revised velocity map of recent vertical crustal movements in the western half of the European USSR, and some remarks on the period of these movements. In Gerasimov, I.P. (ed.), *Recent Crustal Movements*, Israel Program for Scientific Translations, Jerusalem, 76–89.

McClay, K.R. and Coward, M.P. (1981) The Moine thrust zone: an overview. In McClay, K.R. and Price, N.J. (eds.), *Thrust and Nappe Tectonics, Spec. Publ. geol.*

Soc. London **9**, 241–260.

McCulloch, M.T. and Wasserburg, G.J. (1978) Sm-Nd and Rb-Sr chronology of continental crust. *Science* **200**, 1003–1011.

McElhinny, M.W., Embleton, B.J.J., Ma, X.H. and Zhang, Z.K. (1981) Fragmentation of Asia in the Permian. *Nature (London)* **293**, 212–216.

McGarr, A. (1980) Some constraints on levels of shear stress in the crust from observations and theory. *J. geophys. Res.* **85** (B11) 6231–6238.

McGarr, A. and Gay, N.C. (1978) State of stress in the Earth's crust. *Ann. Rev. Earth planet. Sci.* **6**, 405–436.

McGetchin, T.R., Burke, K.C., Thompson, G.A. and Young, R.A. (1980) Mode and mechanisms of plateau uplifts. In Bally, A.W., Bender, P.L., McGetchin, T.R. and Walcott, R.I. (eds.), *Dynamics of Plate Interiors*, American Geophysical Union, Geological Society of America, Geodynamics Series, vol.1, 99–110.

McKenzie, D.P. (1967) The viscosity of the mantle. *Geol. J. R. astron. Soc.* **14**, 297–305.

McKenzie, D.P. (1969) Speculations on the consequences and causes of plate motions. *Geol. J. R. astron. Soc.* **18**, 1–32.

McKenzie, D.P. (1972) Active tectonics of the Mediterranean region. *Geophys. J. R. astron. Soc.* **30**, 109–185.

McKenzie, D.P. (1978a) Some remarks on the development of sedimentary basins. *Earth planet. Sci. Lett.* **40**, 25–32.

McKenzie, D.P. (1978b) Active tectonics of the Alpine-Himalayan belt: the Aegean Sea and surrounding regions. *Geophys. J. R. astron. Soc.* **55**, 217–254.

McKenzie, D.P. (1983) The Earth's mantle. In *The Dynamic Earth, a Scientific American book*, Freeman, New York, 25–38.

McKenzie, D.P. and Morgan, W.J. (1969) Evolution of triple junctions. *Nature (London)* **224**, 125–133.

McKenzie, D.P. and Parker, R.L. (1967) The North Pacific: an example of tectonics on a sphere. *Nature (London)* **216**, 1276–1279.

McKenzie, D.P. and Sclater, J.G. (1971) The evolution of the Indian Ocean since the late Cretaceous. *Geol. J. R. astron. Soc.* **24**, 437–528.

McKenzie, D.P. and Weiss, N. (1975) Speculations on the thermal and tectonic history of the Earth. *Geol. J. R. astron. Soc.* **42**, 131–174.

McMenamin, M.A.S. (1982) A case for two late Proterozoic — earliest Cambrian faunal province loci. *Geology* **10**, 290–292.

McNutt, M. (1980) Implications of regional gravity for state of stress in the Earth's crust and upper mantle. *J. geophys. Res.* **85**, B11, 6377–6396.

McWilliams, M.O. (1981) Palaeomagnetism and Precambrian tectonic evolution of Gondwana. In Kröner, A. (ed.), *Precambrian Plate Tectonics*, Elsevier, Amsterdam, 649–687.

Mégnien, C. and Pomerol, C. (1980) Subsidence of the Paris basin from the Lias to the late Cretaceous. In Bally, A.W., Bender, P.L., McGetchin, T.R. and Walcott, R.I. (eds.), *Dynamics of Plate Interiors*, American Geophysical Union, Geological Society of America, Geodynamics Series, vol.1, 91–92.

Menard, H.W. (1984) Evolution of ridges by asymmetrical spreading. *Geology* **12**, 177–180.

Menard, H.W. and Chase, T.E. (1970) Fracture zones. In

Maxwell, A.E. (ed.), *The Sea*, vol.4, pt.1, Wiley Interscience, New York, 421–443.

Menard, H.W. and Smith, S.M. (1966) Hypsometry of ocean basin provinces. *J. geophys. R.* **71**, 4305–4325.

Mercier, J.-C.C. (1980) Magnitude of the continental lithospheric stresses inferred from rheomorphic petrology. *J. geophys. Res.* **85**, B11, 6293–6303.

Mercier, J.L. (1981) Extensional-compressional tectonics associated with the Aegean arc: comparisons with the Andean Cordillera of south Peru-north Bolivia. *Phil. Trans. R. Soc. London* **A300**, 337–355.

Merle, O. and Brun, J.P. (1984) The curved translation path of the Parpaillon nappe (French Alps). *J. struct. Geol.* **6**, 711–719.

Milsom, J. and Audley-Charles, M.G. (1986) Post-collision isostatic readjustment in the southern Banda arc. In Coward, M.P. and Ries, A. *Collision Tectonics, Spec. Publ. geol. Soc. London* **19**, 353–364.

Minster, J.B. and Jordan, T.H. (1978) Present-day plate motions. *J. geophys. Res.* **83**, 5331–5354.

Minster, J.B., Jordan, T.H., Molnar, P. and Haines, E. (1974) Numerical modelling of instantaneous plate tectonics. *Geophys. J. R. astron. Soc.* **36**, 541–576.

Mishra, D.C. (1982) Crustal structure and dynamics under Himalaya and Pamir ranges. *Earth planet. Sci. Lett.* **57**, 415–420.

Mitchell, A.H.G. (1978) The Grampian orogeny in Scotland: arc-continent collision and polarity reversal. *J. Geol.* **86**, 643–646.

Mitchell, A.H.G. (1981) Phanerozoic plate boundaries in mainland SE Asia, the Himalayas and Tibet. *J. geol. Soc. London* **138**, 109–122.

Mitchell, A.H.G. (1984) Post-Permian events in the Zangbo 'suture' zone, Tibet. *J. geol. Soc. London* **141**, 129–136.

Miyashiro, A. (1973) The Troodos ophiolite complex was probably formed in an island arc. *Earth planet. Sci. Lett.* **19**, 218–224.

Miyashiro, A., Aki, K. and Sengor, A.M.C. (1982) *Orogeny* John Wiley, Chichester.

Mogi, K. (1973) Relationship between shallow and deep seismicity in the western Pacific region. *Tectonophysics* **17**,1–22.

Mohr, P. (1982) Musings on continental rifts. In Pálmason, G. (ed.), *Continental and Oceanic Rifts*, American Geophysical Union, Geological Society of America, Geodynamics Series, vol.8, 293–309.

Molnar, P. and Atwater, T. (1978) Interarc spreading and Cordilleran tectonics as alternates related to the age of subducted oceanic lithosphere. *Earth planet. Sci. Lett.* **41**, 330–340.

Molnar, P. and Chen, W–P. (1978) Evidence of large Cainozoic crustal shortening of Asia. *Nature (London)* **273**, 218–220.

Molnar, P. and Oliver, R.L. (1969) Lateral variations of attenuation in the upper mantle and discontinuities in the lithosphere. *J. geophys. Res.* **73**, 1959–1982.

Molnar, P. and Tapponnier, P. (1975) Cenozoic tectonics of Asia: effects of a continental collision. *Science* **189**, 419–426.

Molnar, P. and Tapponnier, P. (1978) Active tectonics of Tibet. *J. geophys. Res.* **83**, B11, 5361–5375.

Monger, J.W.H., Souther, J.G. and Gabrielse, H. (1972) Evolution of the Canadian Cordillera: a plate tectonic model. *Am. J. Sci.* **272**, 577–602.

Moore, J.C., Cowan, D.S. and Karig, D.E. (1985) Structural styles and deformation fabrics of accretionary complexes: Penrose Conference report. *Geology* **13**, 77–79.

Morgan, W.J. (1968) Rises, trenches, great faults, and crustal blocks. *J. geophys. Res.* **73**, 1959–1982.

Morgan, W.J. (1971) Convection plumes in the lower mantle. *Nature (London)* **230**, 42–43.

Morgan, W.J. (1972) Deep mantle convection plumes and plate motions. *Bull. Am. Assoc. petrol. Geol.* **56**, 203–213.

Morgan, P. and Baker, B.H. (eds.) (1983) Processes of Continental rifting. *Tectonophysics* **94**.

Moruzi, G.A. (1968) Applications of rock mechanics in mine planning and ground control. *Can. Mining J.* **89**, F12–15.

Moore, J.C., Biju-Duval, B. and others (1982) Offscraping and underthrusting of sediment at the deformation front of the Barbados ridge: Deep Sea Drilling Project leg 78A. *Bull geol. Soc. Am.* **93**, 1065–1077.

Moseley, F. (1977) Caledonian plate tectonics and the place of the English Lake District. *Bull. geol. Soc. Am.* **88**, 764–768.

Mulleried, Fr. (1921) Klufte, Harnische und Tektonik der Dinkelberge und des Basler Tafeljuras. *Verh. naturhist.-med. Ver. Heidelberg* **15**, 1–46.

Myers, J.S. (1987) The East Greenland Nagssugtoqidian mobile belt compared with the Lewisian complex. In Park, R.G. and Tarney, J. (eds.), *Evolution of the Lewisian Complex and Comparable Precambrian High-Grade Terrains, Spec. Publ. geol. Soc. London* **27**, 235–246.

Nakamura, K. and Uyeda, S. (1980) Stress gradient in arc-back arc regions and plate subduction. *J geophys. Res.* **85** (B11) 6419–6428.

Neathery, T.L. and Thomas, W.A. (1983) Geodynamics transect of the Appalachian orogen in Alabama. In Rast, N. and Delaney, F.M. (eds.), *Profiles of Orogenic Belts*, American Geophysical Union, Geological Society of America, Geodynamics Series, vol.10, 301–307.

Neugebauer, H.J. (1983) Mechanical aspects of continental rifting *Tectonophysics* **94**, 91–108.

Neumann, E.-R. and Ramberg, I.B. (eds.) (1978) *Tectonics and Geophysics of Continental Rifts*. D. Reidel, Dordrecht.

Newmark, R.L., Zoback, M.D. and Anderson, R.N. (1984) Orientation of *insitu* stresses in the oceanic crust. *Nature (London)* **311**, 424–428.

Ni, J. and York, J.E. (1978) Late Cenozoic tectonics of the Tibetan plateau. *J. geophys. Res.* **83**, 5377–5384.

Nikonov, A.A. (1980) Manifestations of glacio-isostatic processes in northern countries during the Holocene and at present. In Mörner, N.-A. (ed.), *Earth Rheology, Isostasy and Eustasy*, John Wiley, Chichester, 341–354.

Nisbet, E.G. & Fowler, C.M.R. (1983) Model for Archaean plate tectonics. *Geology* **11**, 376–379.

Norvick, M.S. (1979) The tectonic history of the Banda arcs, eastern Indonesia: a review. *J. geol. Soc. London* **136**, 519–527.

Nur, A. and Ben-Avraham, Z. (1983) Volcanic gaps due to oblique consumption of aseismic ridges. *Tectonophysics* **99**, 355–362.

Obert, L. (1962) *In situ* determination of stress in rock.

Mining Eng. (London) **14**, 51–58.

Okada, H. and Smith, A.J. (1980) The Welsh 'geosyncline' of the Silurian was a fore-arc basin. *Nature (London)* **288**, 352–354.

Olesen, N.O., Korstgård, J.A. and Sørensen, K. (1979) A summary of lithology and structure within the Agtø map sheet (67 V.1 Nord), Nagssugtoqidian mobile belt, West Greenland. *Rapp. Grønlands geol. Unders.* **89**, 19–22.

Oliver, J., Isacks, B., Barazangi, M. and Mitronovas, W. (1973) Dynamics of the downgoing lithosphere. *Tectonophysics* **19**, 133–147.

Oxburgh, E.R. (1972) Flake tectonics and continental collision. *Nature (London)* **239**, 202–215.

Pakiser, L.C. and Zietz, I. (1965) Transcontinental crustal and upper-mantle structure. *Rev. Geophys.* **3**, 505–520.

Pálmason, G. (ed.) (1982) *Continental and Oceanic Rifts*, American Geophysical Union, Geological Society of America, Geodynamics Series, vol.8.

Paquin, C., Froidevaux, C., Bloyet, J., Ricard, Y. and Angelidis, C. (1982) Tectonic stresses on the mainland of Greece: in-situ measurements by overcoring. *Tectonophysics* **86**, 17–26.

Park, R.G. (1981*a*) Shear-zone deformation and bulk strain in granite-greenstone terrain of the western Superior province, Canada. *Precambrian Res.* **14**, 31–47.

Park, R.G. (1981*b*) Origin of horizontal structure in high-grade Archaean terrains. *Spec. Publ. geol. Soc. Aust.* **7**, 483–490.

Park, R.G. (1982) Archaean tectonics. *Geol. Rundsch.* **71**, 22–37.

Park, R.G. and Ermanovics, I.F. (1978) Tectonic evolution of two greenstone belts from the Superior province in Manitoba. *Can. J. Earth Sci.* **15**, 1808–1816.

Park, R.G. and Tarney, J. (1987) The Lewisian complex: a typical Precambrian high-grade terrain. In Park, R.G. and Tarney, J. (eds.), *Evolution of the Lewisian Complex and Comparable Precambrian High-Grade Terrains, Spec. Publ. geol. Soc. London* **27**, 13–25.

Park, R.G., Åhäll, K–I., Crane, A.C. and Daly, J.S. (1987) The structure and kinematic evolution of the Lysekil-Marstrand area, Östfold-Marstrand belt, SW Sweden. *Sver. geol. Unders.* **(in press).**

Parsons, B., Cochran, J., LeDouran, S., McKenzie, D.P. and Roufosse, M. (1983) Geoid and depth anomalies in the Atlantic ocean. *EoS* **64**, 676.

Patchett, P.J., Bylund, G. and Upton, B.G.J. (1978) Palaeomagnetism and the Grenville orogeny: new Rb-Sr ages from dolerites in Canada and Greenland. *Earth planet. Sci. Lett.* **40**, 349–364.

Peach, B.N., Horne, J., Gunn, W., Clough, C.T., Hinxman, L.W. and Teall, J.J.H. (1907) The geological structure of the north-west Highlands of Scotland. *Mem. geol. Surv. GB.*

Perrier, G. and Vialon, P. (1980) Les connaissances géophysiques sur le SE de la France: implications géodynamiques. *Géol. Alpine* **56**, 13–20.

Phillips, W.E.A., Stillman, C.J. and murphy, T. (1976) A Caledonian plate tectonic model. *J. geol. Soc. London* **132**, 579–609.

Piasecki, M.A.J. and Van Breemen, O. (1983) Field and isotopic evidence for a *c.* 750Ma tectonothermal event in Moine rocks in the Central Highland region of the Scottish Caledonides. *Trans. R. Soc. Edin. (Earth Sci.)*

73, 119–134.

Piper, J.D.A. (1982) The Precambrian palaeomagnetic record: the case for the Proterozoic supercontinent. *Earth planet. Sci. Lett.* **59**, 61–89.

Piper, J.D.A. (1985) Continental movements and breakup in late Precambrian-Cambrian times: prelude to Caledonian orogenesis. In GEE, D.G. and Sturt, B.A. (eds.), *The Caledonide Orogen — Scandinavia and Related Areas*, John Wiley, Chichester, 19–34.

Pitman, W.C. and Hayes, D.E. (1968) Sea-floor spreading in the Gulf of Alaska. *J. geophys. Res.* **73**, 6571–6580.

Plant, J.A., Watson, J. and Green, P.M. (1984) Moine-Dalradian relationships and their palaeotectonic significance. *Proc. R. Soc. London.* **A395**, 185–202.

Platt, J.P. and Lister, G.S. (1985) Structural history of high-pressure metamorphic rocks in the southern Vanoise massif, French Alps, and their relation to Alpine tectonic events. *J. struct. Geol.* **7**, 19–35.

Platt, J.P., Leggett, J.K., Young, J., Raza, H. and Alam, S. (1985) Large-scale sediment underplating in the Makran accretionary prism, southwest Pakistan. *Geology* **13**, 507–511.

Post, R.L. (1977) High temperature creep of Mt Burnet dunite. *Tectonophysics* **38**, 279–296.

Price, R.A. (1981) The Cordilleran thrust and fold belt in the southern Canadian Rocky Mountains. In McClay, K.R. and Price, N.J. (eds.), *Thrust and Nappe Tectonics, Spec. Publ. geol. Soc. London* **9**, 427–448.

Quennell, A.M. (1959) Tectonics of the Dead Sea rift. *Int. geol. Congr.* Mexico, Asosiación de Servicios geologicos Africanos, 385–405.

Ramsay, D.M., Sturt, B.A., Zwann, K.B. and Roberts, D. (1985) Caledonides of northern Norway. In Gee, D.G. and Sturt, B.A. (eds.), *The Caledonide Orogen – Scandinavia and Related Areas*, John Wiley, Chichester, 164–184.

Ramsay, J.G. (1963) Stratigraphy, structure and metamorphism in the western Alps. *Proc. Geol. Assoc.* **74**, 357–392.

Ramsay, J.G. (1981) Tectonics of the Helvetic Alps. In McClay, K.R. and Price, N.J. (eds.), *Thrust and Nappe Tectonics, Spec. Publ. geol. Soc. London* **9**, 293–309.

Ramsay, J.G., Casey, M. and Kligfield, R. (1983) Role of shear in development of the Helvetic fold-thrust belt of Switzerland. *Geology* **11**, 439–442.

Rathbone, P.A., Coward, M.P. and Harris, A.L. (1983) Cover and basement: a contrast in style and fabrics. *Mem. geol. Soc. Am.* **158**, 213–223.

Rattey, P.R. and Sanderson, D.J. (1982) Patterns of folding within nappes and thrust sheets: examples from the Variscan of southwest England. *Tectonophysics* **88**, 247–267.

Read, H.H. (1961) Aspects of Caledonian magmatism in Britain. *Liverpool Manchester geol. J.* **2**, 653–683.

Read W.R. (1988) Controls on Silesian sedimentation in the Midland Valley of Scotland. In Belsy, B.M. and Kelling, G. (eds.), *Sedimentation in a Synorogenic Basin Complex: the Upper Carboniferous of NW Europe*, **Blackie, Glasgow and London (in press).**

Reading, H.G. (1980) Characteristics and recognition of strike-slip fault systems. *Spec. Publ. Int. Assoc. Sedimentol.* **4**, 7–26.

Richardson, R.M., Solomon, S.C. and Sleep, N.H. (1976) Intraplate stress as an indicator of plate tectonic driving force. *J. geophys. Res.* **81**, 1847–1856.

Richardson, S.W. and Oxburgh, E.R. (1978) Heat flow, radiogenic heat production and crustal temperatures in England and Wales. *J. geol. Soc. London* **135**, 323–338.

Roach, R.A. and Duffell, S. (1974) Structural analysis of the Mount Wright map-area, southernmost Labrador trough, Quebec, Canada. *Bull. geol. Soc. Am.* **85**, 947–962.

Roberts, D. and Gee, D.G. (1985) An introduction to the structure of the Scandinavian Caledonides. In Gee, D.G. and Sturt, B.A. (eds.), *The Caledonide Orogen: Scandinavia and Related Areas*, John Wiley, New York, 56–68.

Roberts, J.L. and Treagus, J.E. (1977) Polyphase generation of nappe structures in the Dalradian rocks of the southwest Highlands of Scotland. *Scott. J. Geol.* **13**, 237–254.

Roeder, D.H. (1973) Subduction and orogeny. *J. geophys. Res.* **78**, (23) 5005–5024.

Ross, J.V., Avé Lallemant, H.G. and Carter, N.L. (1980) Stress dependence of recrystallized-grain and subgrain size in olivine. *Tectonophysics* **70**, 39–61.

Ruff, L. and Kanamori, H. (1983) Seismic coupling and uncoupling at subduction zones. *Tectonophysics* **99**, 99–117.

Runcorn, S.K. (1962) Palaeomagnetic evidence for continental drift and its geophysical cause. In Runcorn, S.K. (ed.), *Continental Drift*, Academic Press, New York and London, 1–40.

Ryan, W.B.F., Stanley, D.J., Hersey, J.B., Fahlquist, D.A. and Allan, T.D. (1970) The tectonics and geology of the Mediterranean Sea. In Maxwell, A. (ed.) *The Sea, Vol.4, II: The Tectonics and Geology of the Mediterranean Sea*, Wiley-Interscience, New York, 387–492.

Ruff, L. and Kanamori, H. (1983) Seismic coupling and uncoupling at subduction zones. *Tectonophysics* **99**, 99–117.

Runcorn, S.K. (1962) Palaeomagnetic evidence for continental drift and its geophysical cause. In Runcorn, S.K. (ed.), *Continental Drift*, Academic Press, New York and London, 1–40.

Saemundsson, K. (1974) Evolution of the axial rifting zone in northern Iceland. *Bull. geol. Soc. Am.* **85**, 495–504.

Samuelsson, L. and Åhäll, K.–I. (1985) Proterozoic development of Bohuslän, south-western Sweden. In Tobi, A.C. and Touret, J.L.R. (eds.), *The Deep Proterozoic Crust in the North Atlantic Provinces*, D. Reidel, Dordrecht, 345–357.

Sanderson, D.J. and Marchini, W.R.D. (1984) Transpression. *J. struct. Geol.* **6**, 449–458.

Sawyer, D.S., Swift, B.A., Sclater, J.G. and Toksoz, M.N. (1982) Extensional model for the subsidence of the northern United States Atlantic continental margin. *Geology* **10**, 134–140.

Sbar, M.L. and Sykes, L.R. (1973) Contemporary compressive stress and seismicity in eastern North America: an example of intra-plate tectonics. *Bull. geol. Soc. Am.* **84**, 1861–1882.

Scholl, D.W., von Huene, R., Vallier, T.L. and Howell, D.G. (1980) Sedimentary masses and concepts about tectonic processes at underthrust ocean margins. *Geology* **8**, 564–568.

Scholz, C.H. and Page, R. (1970) Buckling in island arcs (abstr.) *EoS* **51**, 429.

Schubert, G., Yuen, D.A., Froidevaux, C., Fleitout, C. and Sourian, M. (1978) Mantle circulation with partial shallow return flow: effects on stresses in oceanic plates and topography of the sea floor. *J. geophys. Res.* **83**, 745–758.

Schwerdtner, W.M., Stone, D., Osadetz, K., Morgan, J. and Stott, G.M. (1979) *Can. J. Earth Sci.* **16**, 1956–1977.

Sclater, J.G. (1972) New perspectives in terrestrial heat flow. *Tectonophysics* **13**, 257–291.

Sclater, J.G. and Christie, P.A.F. (1980) Continental stretching: an explanation of the post-mid-Cretaceous subsidence of the central North Sea basin. *J. geophys. Res.* **85**, 3711–3739.

Sclater, J.G. and Francheteau, J. (1970) The implications of terrestrial heat flow observations on current tectonic and geochemical models of the crust and upper mantle of the Earth. *Geophys. J. R. astron. Soc.* **20**, 509–542.

Sclater, J.G., Royden, L., Horvath, F., Burchfiel, C. and Stegena, L. (1980) The formation of the intra-Carpathian basins as determined from subsidence data. *Earth planet. Sci. Letts.* **51**, 139.

Searle, R.C. (1979) Side-scan sonar studies of North Atlantic fracture zones. *J. geol. Soc. London* **136**, 283–292.

Searle, R.C. (1983) Multiple, closely spaced transform faults in fast-slipping fracture zones. *Geology* **11**, 607–610.

Searle, R.C. (1986) GLORIA investigations of oceanic fracture zones: comparative study of the transform fault zone. *J. geol. Soc. London* **143**, 743–756.

Sellars, C.M. (1978) Recrystallisation of metals during hot-deformation *Phil. Trans. R. Soc. London* **A288**, 147–158.

Sellers, J.B. (1969) *Strain Relief Overcoring to Measure In-Situ Stresses*. US Army Corps of Engineers Report, Buffalo District, NY.

Sellers, J.B. (1977) The measurement of stress changes in rock using the vibrating wire stressmeter. In Kovari, K. (ed.), *Field Measurements in Rock Mechanics*, vol.1, Balkema, Rotterdam, 275–288.

Şengör, A.M.C. and Burke, K.C. (1978) Relative timing of rifting and volcanism on Earth and its tectonic implications. *Geophys. Res. Lett.* **5**, 419–421.

Shackleton, R.M. (1976) Pan African structures. *Phil. Trans. R. Soc. London* **A280**, 491–497.

Shackleton, R.M. (1977) Possible late Precambrian ophiolites in Africa and Brazil. *Ann. Rept. Univ. Leeds Res. Inst. African Geol.* **20**, 3–7.

Shackleton, R.M. (1981) Structure of southern Tibet: report on a traverse from Lhasa to Khatmandu organised by Academia Sinica. *J. struct. Geol.* **3**, 97–105.

Shackleton, R.M. (1986) Collision tectonics in Africa. In Coward, M.P. and Ries, A.C. (eds.), *Collision Tectonics Spec. Publ. geol. Soc. London* **19**, 329–349.

Shackleton, R.M. and Ries, A.C. (1984) The relation between regionally consistent stretching lineations and plate motions. *J. struct. Geol.* **6**, 111–117.

Shackleton, R.M., Ries, A.C. and Coward, M.P. (1982) An interpretation of the Variscan structure in SW England. *J. geol. Soc. London* **139**, 533–541.

Shatskiy, N.S. (1955) The origin of the Pachelma trench: comparative tectonics of ancient platforms. *Bull. Moscow Soc. Nat. (Geology Sect.)*, Paper No.5, **30**, 5–26.

Shelton, G. and Tullis, J. (1981) Experimental flow laws for crustal rocks. *Eos* **62**, 396.

Silver, E.A. and Smith, R.B. (1983) Comparison of terrane accretion in modern Southeast Asia and the Mesozoic North American Cordillera. *Geology* **11**, 198–202.

Sleep, N.H. and Sloss, L.L. (1980) The Michigan basin. In Bally, A.W., Bender, P.L., McGetchin, T.R. and Walcott, R.I. (eds.), *Dynamics of Plate Interiors*, Geological Society of America, American Geophysical Union, Geodynamics Series, vol.1, 93–98.

Sleep, N.L. (1975) Formation of oceanic crust. *J. geophys. Res.* **80**, 4027–4042.

Sleep, N.L. and Rosendahl, B.R. (1979) Topography and tectonics of mid-oceanic ridge axes. *J. geophys. Res.* **84** (B12) 6831.

Smith, A.G. and Briden, J.C. (1977) *Mesozoic and Cenozoic Paleocontinental Maps*. Cambridge University Press, Cambridge.

Smith, A.G. and Hallam, A. (1970) The fit of the southern continents. *Nature (London)* **225**, 139–144.

Symthe, D.K., Dobinson, A., McQuillan, R., Brewer, J.A., Matthews, D.H., Blundell, D.J. and Kelk, B. (1982) Deep structure of the Scottish Caledonides revealed by the MOIST reflection profile. *Nature (London)* **299**, 338–340.

Solomon, S.C., Richardson, R.M. and Bergman, E.A. (1980) Tectonic stress: models and magnitudes. *J. geophys. Res.* **85**, B11, 6086–6092.

Soper, N.J. and Barber, A.J. (1982) A model for the deep structure of the Moine thrust zone. *J. geol. Soc. London* **139**, 127–138.

Spray, J.G. (1983) Lithosphere-asthenosphere decoupling at spreading centres and initiation of obduction. *Nature (London)* **304**, 253–255.

Stacey, J.S. and Hedge, C.E. (1984) Geochronologic and isotopic evidence for early Proterozoic crust in the eastern Arabian shield. *Geology* **12**, 310–313.

Stearn, J.E.F. and Piper, J.D.A. (1984) Palaeomagnetism of the Sveconorwegian mobile belt of the Fennoscandian shield. *Precambrian Res.* **23**, 201–246.

Stewart, J.H. (1980) Regional tilt patterns of late Cenozoic basin-range fault blocks, western United States. *Bull. geol. Soc. Am.* **91**, 460–464.

Stockwell, C.H., McGlynn, J.C., Emslie, R.F., Sanford, B.V., Norris, A.W., Donaldson, W.F., Fahrig, W.F. and Currie, K.L. (1970) Geology of the Canadian shield. In Douglas, R.J.W. (ed.), *Geology and Economic Minerals of Canada*, Geological Survey of Canada, 43–150.

Sturt, B.A., Ramsay, D.M., Pringle, I.R. and Teggin, D.E. (1977) Precambrian gneisses in the Dalradian sequence of northeast Scotland. *J. geol. Soc. London* **134**, 41–44.

Sutton, J. (1963) Long term cycles in the evolution of the continents. *Nature (London)* **198**, 731–735.

Sykes, L.R. (1967) Mechanism of earthquakes and nature of faulting on the mid-oceanic ridges. *J. geophys. Res.* **72**, 2131.

Sykes, L.R. and Sbar, M.L. (1974) Focal mechanism solutions of intraplate earthquakes and stresses in the lithosphere. In *Geodynamics of Iceland and the North Atlantic Area*, D. Reidel, Dordrecht, 207–224.

Takeuchi, H. and Uyeda, S. (1965) A possibility of present-day regional metamorphism. *Tectonophysics* **2**, 59–68.

Talwani, M., Le Pichon, X. and Ewing, M. (1965) Crustal structure of the mid-ocean ridges, part 2. *J. geophys. Res.* **70**, 341–352.

Talwani, M., Sutton, G.H. and Worzel, J.L. (1959) A crustal section across the Puerto Rico trench. *J. geophys. Res.* **64**, 1545–1555.

Tanner, P.W.G. (1970) The Sgurr Beag slide: a major tectonic break within the Moinian of the western Highlands of Scotland. *J. geol. Soc. London* **126**, 435–463.

Tapponnier, P. and Molnar, P. (1976) Slip line field theory and large scale continental tectonics. *Nature (London)* **264**, 319–324.

Tapponnier, P. and Molnar, P. (1977) Active faulting and tectonics in China. *J. geophys. Res.* **82**(20), 2905–2930.

Tapponnier, P., Peltzer, G. and Armijo, R. (1986) On the mechanics of collision between India and Asia. In Coward, M.P. and Ries, A. (eds.), *Collision Tectonics*, *Spec. Publ. geol. Soc. London* **19**, 115–157.

Tapponnier, P., Peltzer, G., Le Dain, A.Y., Armijo, R. and Cobbold, P. (1982) Propagating extrusion tectonics in Asia: new insights from simple experiments with plasticine. *Geology* **10**, 611–616.

Tarney, J. and Weaver, B.L. (1987) Geochemistry of the Scourian complex: petrogenesis and tectonic models. In Park, R.G. and Tarney, J. (eds.), *Evolution of the Lewisian Complex and Related Precambrian High-Grade Terrains*, *Spec. Publ. geol. Soc. London* **27**, 45–56.

Tarney, J. and Windley, B.F. (1977) Chemistry, thermal gradients and evolution of the lower continental crust. *J. geol. Soc. London* **134**, 153–172.

Taylor, P.N., Moorbath, S., Goodwin, R. and Petrykowski, A.C. (1980) Crustal contamination as an indicator of the extent of early Archaean continental crust: Pb isotopic evidence from the late Archaean gneisses of West Greenland. *Geochim. Cosmochim. Acta* **44**, 1437–1453.

Thomas, P.R. (1979) New evidence for a Central Highland root zone. In Harris, A.L., Holland, C.H. and Leake, B.E. (eds.), *The Caledonides of the British Isles: reviewed*, *Spec. Publ. geol. Soc. London* **8**, 205–211.

Thomas, W.A. (1983) Continental margins, orogenic belts, and intracratonic structures. *Geology* **11**, 270–272.

Tobisch, O.T., Fleuty, M.J., Merh, S.S., Mukhopadhyay, D. and Ramsay, J.G. (1970) Deformational and metamorphic history of Moinian and Lewisian rocks between Strathconon and Glenn Affric. *Scott. J. Geol.* **6**, 243–265.

Trumpy, R. (1973) The timing of orogenic events in the central Alps. In de Jong, K.A. and Scholten, R. (eds.), *Gravity and Tectonics*, John Wiley, New York, 229–252.

Tryggvason, E. (1982) Recent ground deformation in continental and oceanic rift zones. In Palmason, G. (ed.), *Continental and Oceanic Rifts*, American Geophysical Union, Geological Society of America, Geodynamics Series, vol.8, 17–29.

Turcotte, D.L. (1974*a*) Membrane tectonics. *Geophys. J. R. aston. Soc.* **36**, 33–42.

Turcotte, D.L. (1974*b*) Are transform faults thermal contraction cracks. *J. geophys. Res.* **79**, 2573–2577.

Turcotte, D.L. and Emerman, S.H. (1983) Mechanisms of active and passive rifting. *Tectonophysics* **94**, 39–50.

Turcotte, D.L. and Oxburgh, E.R. (1976) Stress accumulation in the lithosphere. *Tectonophysics* **35**, 183–199.

Uyeda, S. (1978) *The New View of the Earth*. Freeman, San Francisco.

Vacquier, V.S., Uyeda, S., Yasui, M., Sclater, J.G., Corry, C. and Watanabe, T. (1966) Heat flow measurements in the northwestern Pacific. *Bull. Earthquake Res. Inst. Tokyo Univ.* **44**, 1519–1535.

Van Breemen, O., Aftalion, M. and Johnson, M.R.W. (1979) Age of the Loch Borolan complex, Assynt and late movements on the Moine thrust. *J. geol. Soc. London* **136**, 489–496.

van der Voo, R. (1969) Palaeomagnetic evidence for the rotation of the Iberian peninsula. *Tectonophysics* **7**, 5–56.

van der Voo, R. and Scotese, C. (1981) Palaeomagnetic evidence for a large (c. 2,000km) sinistral offset along the Great Glen fault during Carboniferous time. *Geology* **9**, 583–589.

Vening Meinesz, F.A. (1950) Les 'graben' africains, resultat de compression ou de tension dans la croûte terrestre. *Koninkl. Belg. Kol. Inst. Bull* **21**, 539–552.

Versteeve, A. (1975) Isotope geochronology in the high-grade metamorphic Precambrian of southwestern Norway. *Norg. geol. Unders.* **318**, 1–50.

Vidal, P., Auvray, B., Charlot, R. and Cogné, J. (1981) Precambrian relicts in the Armorican massif: their age and role in the evolution of the western and central European Cadomian-Hercynian belt. *Precambrian Res.* **14**, 1–20.

Vine, F.J. and Hess, H.H. (1970) Sea-floor spreading. In Maxwell, A.E., Bullard, E.C., Goldberg, E. and Worzel, J.L. (eds.), *The Sea*, vol.4, Wiley Interscience, New York.

Vita-Finzi, C. (1986) *Recent Earth Movements*. Academic Press, London and New York.

Vitorello, I. and Pollack, H.N. (1980) On the variation of continental heat flow with age and the thermal evolution of continents. *J. geophys. Res.* **85**, B2, 983–995.

Vogt, P.R. (1973) Subduction and aseismic ridges. *Nature (London)* **241**, 189–191.

Von Herzen, R.P. and Lee, W.H.K. (1969) Heat flow in oceanic regions. In Hart, P.J. (ed.), *The Earth's Crust and Upper Mantle*, American Geophysical Union, Geophysical Monograph, **13**, 88–95.

Walcott, R.I. (1970) Flexural rigidity, thickness, and viscosity of the lithosphere. *J. geophys. Res.* **75**, 3941–3954.

Walcott, R.I. (1972) Gravity, flexure, and the growth of sedimentary basins at a continental edge. *Bull. geol. Soc. Am.* **83**, 1845–1848.

Walcott, R.I. (1980) Rheological models and observational data of glacio-isostatic rebound. In Moerner, N.-A. (ed.), *Earth Rheology, Isostasy and Eustasy*, John Wiley, Chichester, 3–10.

Walsh, J.B. (1965) The effect of cracks in rocks on Poisson's ratio. *J. geophys. Res.* **71**, 5249–5257.

Warsi, W.E.K., Hilde, T.W.C. and Searle, R.C. (1983) Convergence structures of the Peru trench between 10°S and 14°S. *Tectonophysics* **99**, 313–329.

Watkins, J.S., Moore, J.C. *et al.* (1981) *Initial reports of the Deep Sea Drilling Project*. US Government Printing Office, Washington, DC.

Watson, J. and Dunning, F.W. (1979) Basement-cover relations in the British Caledonides. In Harris, A.L., Holland, C.H. and Leake, B.E. (eds.), *The Caledonides of the British Isles: Reviewed, Spec. Publ. geol. Soc. London* **8**, 67–91.

Watterson, J. (1978) Proterozoic intraplate deformation in the light of South-east Asian neotectonics. *Nature (London)* **273**, 636–640.

Watts, A.B. and Talwani, M. (1974) Gravity anomalies seaward of deep-ocean trenches and their tectonic implications. *Geophys. J. R. astron. Soc.* **36**, 57–90.

Weber, K. (1984) Variscan events: early Palaeozoic continental rift metamorphism and late Palaeozoic crustal shortening. In Hutton, D.H.W. and Sanderson, D.J. (eds.), *Variscan tectonics of the North Atlantic Region, Spec. Publ. geol. Soc. London* **14**, 3–22.

Wegener, A. (1929) *Die Entstehung der Kontinente und Ozeane* (4th edn.), Vieweg und Sohn, Braunschweig.

Weir, J.A. (1974) The sedimentology and diagenesis of the Silurian rocks on the coast west of Gatehouse, Kirkudbrightshire. *Scott. J. Geol.* **10**, 165–186.

Weissel, J.K. (1981) Magnetic lineations in marginal basins of the western Pacific. *Phil. Trans. R. Soc. London* **A300**, 223–245.

Wernicke, B. (1981) Low-angle normal faults in the Basin and Range province: nappe tectonics in an extending orogen. *Nature (London)* **291**, 645–648.

Wernicke, B. (1985) Uniform-sense normal simple shear of the continental lithosphere. *Can. J. Earth Sci.* **22**, 108–125.

Wernicke, B. and Burchfiel, B.C. (1982) Modes of extensional tectonics. *J. struct. Geol.* **4**, 105–115.

Wernicke, B., Spencer, J.E., Burchfiel, B.C. and Guth, P.L. (1982) Magnitude of crustal extension in the southern Great Basin. *Geology* **10**, 499–502.

Westbrook, G.K. (1982) The Barbados ridge complex: tectonics of a mature forearc system. In Leggett, J.K. (ed.), *Trench-Forearc Geology: Sedimentation and Tectonics on Modern and Ancient Active Plate Margins, Spec. Publ. geol. Soc. London* **10**, 275–290.

White, S.H. (1976) The effects of strain on the microstructures, fabrics, and deformation mechanisms in quartzites. *Phil. Trans. R. Soc. London* **A283**, 69–86.

White, R.S. (1982) Deformation of the Makran accretionary sediment prism in the Gulf of Oman (northwest Indian Ocean). In Leggett, J.K. (ed.), *Trench-Forearc Geology: sedimentation and tectonics on modern and ancient active plate margins. Spec. Publ. geol. Soc. London* **10**, 357–372.

Wiebols, G.A. Jaeger, J.C. and Cook, N.G.W. (1968) Rock property tests in a stiff testing machine. *Tenth Rock Mechanics Symposium*, Rice University, Houston.

Williams, G.D. and Chapman, T.J. (1986) The Bristol-Mendip foreland thrust belt. *J. geol. Soc. London* **143**, 63–74.

Wilson, J.T. (1963) Evidence from islands on the spreading of ocean floors. *Nature (London)* **197**, 536–538.

Wilson, J.T. (1965) A new class of faults and their bearing on continental drift. *Nature (London)* **207**, 343–347.

Wilson, J.T. (1966) Did the Atlantic close and then reopen. *Nature (London)* **211**, 676.

Wilson, J.T. (1968) Static or mobile Earth: the current scientific revolution. *J. Am. phil. Soc.* **112**, 309–320.

Wilson, J.T. (1973) Mantle plumes and plate motions. *Tectonophysics* **19**, 149–164.

Winchester, J.A. (1973) Pattern of regional metamorphism suggests a sinistral displacement of 160km along the Great Glen fault. *Nature (Physical Sciences) (London)* **246**, 81–84.

Winchester, J.A. (1985) Major low-angle fault displacement measured by matching amphibolite chemistry: an example from Scotland. *Geology* **13**, 604–606.

Windley, B.F. (1973) Archaean anorthosites: a review with the Fiskenaesset complex, West Greenland, as a model for interpretation. *Spec. Publ. geol. Surv. South Afr.* **3**, 312–332.

Windley, B.F. (1977) *The Evolving Continents.* John Wiley, Chichester.

Windley, B.F. (1981) Precambrian rocks in the light of the plate tectonic concept. In Kröner, A. (ed.), *Precambrian Plate Tectonics*, Elsevier, Amsterdam, 1–20.

Windley, B.F. (1987) Comparative tectonics of the western Grenville and western Himalayas. In Moore, J.M., Baer, A.J. and Davidson, A. (eds.), *New Perspectives on the Grenville Problem, Spec. Pap. geol. Assoc. Can.* (in press).

Windley, B.F. and Smith, J.V. (1976) Archaean high-grade complexes and modern continental margins. *Nature (London)* **260**, 671–675.

Wood, R. and Barton, P. (1983) Crustal thinning and subsidence in the North Sea. *Nature (London)* **302**, 134–136.

Woodcock, N.H. (1984) The Pontesford lineament, Welsh borderland. *J. geol. Soc. London* **141**, 1001–1014.

Wyllie, P.J. (1971) *The Dynamic Earth: Textbook in Geosciences.* John Wiley, New York.

Wynne-Edwards, H. (1972) The Grenville province. In Price, R.A. and Douglas, R.J.W. (eds.), *Variations in Tectonic Styles in Canada, Spec. Pap. geol. Assoc. Can.* **11**, 263–334.

Zak, I. and Freund, R. (1966) Recent strike-slip movements along the Dead Sea rift. *Israel J. Earth Sci.* **15**, 33–37.

Zeuner, F.E. (1958) *Dating the Past* (4th edn.) Methuen, London.

Ziegler, P.A. (1975) Geologic evolution of North Sea and its tectonic framework. *Bull. Am. Assoc. petrol. Geol.* **59**, 1073–1097.

Ziegler, P.A. (1982) Faulting and graben formation in western and central Europe. *Phil. Trans. R. Soc. London* **A305**, 113–143.

Ziegler, P.A. (1985) Late Caledonian framework of western and central Europe. In Gee, D.G. and Sturt, B.A. (eds.), *The Caledonide Orogen — Scandinavia and Related Areas*, John Wiley, New York, 3–18.

Zijderveld, J.D.A., de Jong, J.A. and van der Voo, R. (1970a) Rotation of Sardinia: paleomagnetic evidence from Permian rocks. *Nature (London)* **226**, 933–934.

Zijderveld, J.D.A., Hazeu, G.J.A., Nardin, M. and van der Voo, R. (1970b) Shear in the Tethys and the palaeomagnetism in the southern Alps, including new results. *Tectonophysics* **10**, 639–661.

Zoback, M.L. and Zoback, M. (1980) State of stress in the coterminous United States. *J. geophys. Res.* **85** (B11) 6113–6156.

Zoback, M.L., Anderson, R.E. and Thompson, G.A. (1981) Cainozoic evolution of the state of stress and style of tectonism of the Basin and Range province of the western United States. *Phil. Trans. R. Soc. London* **A300**, 407–434.

Zoback, M.D., Tsukahara, H. and Hickman, S. (1980) Stress measurements at depth in the vicinity of the San Andreas fault: implications for the magnitude of shear stress at depth. *J. geophys. Res.* **85**, 6157–6173.

Zwart, H.J. (1967) The duality of orogenic belts. *Geol. Mijnbouw* **46**, 283–309.

Index